Bayesian
Artificial
Intelligence

SECOND EDITION

Chapman & Hall/CRC
Computer Science and Data Analysis Series

The interface between the computer and statistical sciences is increasing, as each discipline seeks to harness the power and resources of the other. This series aims to foster the integration between the computer sciences and statistical, numerical, and probabilistic methods by publishing a broad range of reference works, textbooks, and handbooks.

SERIES EDITORS
David Blei, Princeton University
David Madigan, Rutgers University
Marina Meila, University of Washington
Fionn Murtagh, Royal Holloway, University of London

Proposals for the series should be sent directly to one of the series editors above, or submitted to:

Chapman & Hall/CRC
4th Floor, Albert House
1-4 Singer Street
London EC2A 4BQ
UK

Published Titles

Computer Science and Data Analysis Series

Bayesian Artificial Intelligence

SECOND EDITION

Kevin B. Korb
Ann E. Nicholson

CRC Press
Taylor & Francis Group
Boca Raton   London   New York

CRC Press is an imprint of the
Taylor & Francis Group, an **informa** business

A CHAPMAN & HALL BOOK

CRC Press
Taylor & Francis Group
6000 Broken Sound Parkway NW, Suite 300
Boca Raton, FL 33487-2742

© 2011 by Taylor and Francis Group, LLC
CRC Press is an imprint of Taylor & Francis Group, an Informa business

Library of Congress Cataloging-in-Publication Data

Korb, Kevin B.
    Bayesian artificial intelligence / Kevin B. Korb, Ann E. Nicholson. -- 2nd ed.
       p. cm. -- (Chapman & Hall/CRC computer science and data analysis series)
    Includes bibliographical references and index.
    ISBN 978-1-4398-1591-5 (hardback)
    1. Bayesian statistical decision theory--Data processing. 2. Machine learning. 3.
Neural networks (Computer science) I. Nicholson, Ann E. II. Title.

    QA279.5.K67 2010
    519.5'42--dc22
                                              2010043669

Visit the Taylor & Francis Web site at
http://www.taylorandfrancis.com

and the CRC Press Web site at
http://www.crcpress.com

To Judea Pearl and Chris Wallace

Contents

III   KNOWLEDGE ENGINEERING   293

10 Knowledge Engineering with Bayesian Networks   297

List of Figures

List of Tables

Preface

Bayesian Artificial Intelligence, in our understanding, is the incorporation of Bayesian inferential methods in the development of a software architecture for an Artificial Intelligence (AI). We believe that important ingredients of such an architecture will be Bayesian networks and the Bayesian learning of Bayesian networks (Bayesian causal discovery) from observation and experiment. In this book we present the elements of Bayesian network technology, automated causal discovery, learning probabilities from data, and examples and ideas about how to employ these technologies in developing probabilistic expert systems, which we call Knowledge Engineering with Bayesian Networks.

This is a very practical project, because data mining with Bayesian networks (applied causal discovery) and the deployment of Bayesian networks in industry and government are two of the most promising areas in applied AI today. But it is also a very theoretical project, because the achievement of a Bayesian AI would be a major theoretical achievement.

With our title there are a number of subjects we could naturally include, but have not. Thus, another necessary aspect of an effective Bayesian AI will be the learning of concepts, and hierarchies of concepts. Bayesian methods for concept formation exist (e.g., Chris Wallace's Snob, Wallace and Boulton, 1968), but we do not treat them here. We could also have discussed function discovery, polynomial curve fitting, time series modeling, etc. We have chosen to hew close to the theme of using and discovering Bayesian networks both because this is our own main research area and because, important as the other Bayesian learning methods are, we believe the Bayesian network technology is central to the overall project. We have added for this edition a treatment of the use of Bayesian networks for prediction (classification).

Our text differs from others available on Bayesian networks in a number of ways. We aim at a practical and accessible introduction to the main concepts in the technology, while paying attention to foundational issues. Most texts in this area require somewhat more mathematical sophistication than ours; we presuppose only a basic understanding of algebra and calculus. Also, we give roughly equal weight to the causal discovery of networks and to the Bayesian inference procedures using a network once found. Most texts either ignore causal discovery or treat it lightly. Richard Neapolitan's book, *Learning Bayesian Networks* (2003), is an exception, but it is more technically demanding than ours. Another distinguishing feature of our text is that we advocate a causal interpretation of Bayesian networks, and we discuss the use of Bayesian networks for causal modeling. We also illustrate various applications of the technology at length, drawing upon our own applied research. We hope that these illustrations will be of some interest and indicate some of the possibilities for the technology. Our text is aimed at advanced undergraduates in computer science who have some background in artificial intelligence and at those who wish to engage in applied or pure research in Bayesian network technology.

A few remarks about notation before we begin. The notation special to Bayesian networks (or to our treatment of them) will be introduced as we proceed; first introductions of notation (including general mathematical notation) and acronyms will be recorded, with page numbers, in Appendix A. When first introducing new terminology we shall employ boldface to point it out; thus, for example, the first appearance (after this) of "Bayesianism" will be in boldface.

Here we describe the simplest aspects of the notation we adopt. First, variables (nodes in the network) will be named, with the names being capitalized and usually italicized (e.g., Y, *Alarm*, *Cancer*). Sets of variables will be set in boldface (e.g., $\mathbf{X}, \mathbf{Y}$). The values that variables take will not be capitalized, but will be italicized; thus, to assert that the alarm is on, we might write *Alarm = on*. Values abbreviated to single letters, such as *True (T)* and *False (F)*, however, will be capitalized. Where no confusion is likely to arise, variables and values may be abbreviated.

The book Web site is `http://www.csse.monash.edu.au/bai` and contains a variety of aids for study, including example Bayesian networks and data sets. Instructors can email us for sample solutions to many of the problems in the text.

New in the second edition

We have found and corrected numerous mistakes in the first edition, most of the credit for which belongs to readers and students (see below). We claim full credit for all the errors remaining or newly introduced. Other new introductions include: §4.7, object-oriented Bayesian networks, Chapter 7, Bayesian Network Classifiers, which describes naive Bayes models and other classifiers; §9.7, which addresses two foundational problems with causal discovery as well as Markov blanket discovery; §9.8, which treats methods of evaluating causal discovery programs; substantial new material in Chapter 10, including discussions of many common modeling errors; and new applications and case studies in Chapters 5 and 11. The uses of causal interventions to understand and reason with causal Bayesian networks receive fuller coverage in a number of places. The Evaluation Chapter of the first edition has been divided, enhanced and scattered: evaluation of prediction goes into Chapter 7, evaluation of causal discovery into Chapter 9 and sensitivity analysis and related issues into Chapter 10. The result is a somewhat thicker book and one we hope is more useful for students and practitioners.

Acknowledgments

There are many whom we wish to acknowledge. For assistance reviewing portions of the text or reporting mistakes we thank: David Albrecht, Lloyd Allison, Helen Armstrong, Tali Boneh, Darren Boulton, Mark Burgman, Jefferson Coelho, Steven Gardner, Li Guoliang, Lucas Hope, Finn Jensen, Emily Korb, Anders Madsen, Ion Muslea, Richard Neapolitan, Rodney O'Donnell, Edin Sarajlic, Vitaly Sender, Kaye Stacey, Vicki Steinle, Charles Twardy, Wouter Vinck, Chris Wallace, David Wooff and the CRC Press reviewer. Uffe Kjærulff (Hugin), Brent Boerlage (Netica), Marek Druzdzel (GeNIe), Anders Madsen (Hugin) and Lionel Jouffe (BayesiaLab) helped us with their software packages, while Kevin Murphy assisted with the software package summary in Appendix B. Our research partners in various projects include: Nathalie Jitnah, Scott Thomson, Jason Carlton, Darren Boulton, Brendon Taylor, Kym McGain (Bayesian poker); Ryan McGowan, Daniel Willis, Ian Brown (ambulation monitoring); Kaye Stacey, Tali Boneh, Liz Sonenberg, Vicki Steinle, Tim Wilkin (intelligent tutoring); Tali Boneh, Liz Sonenberg (Matilda); Lucas Hope (VE); Ingrid Zukerman, Ricky McConachy (NAG); Russell Kennett, Chris Ryan, Tali Boneh, John Bally, Gary Weymouth (meteorology); Charles Twardy, Danny Liew, Lucas Hope, Bin Han (cardiovascular risk assessment, TakeHeart II); Owen Woodberry, Carmel Pollino, Barry Hart (Goulburn Catchment ERA); Chris Wallace, Julian Neil, Lucas Hope, Helen Armstrong, Charles Twardy, Rodney O'Donnell, Lloyd Allison, Steven Mascaro, Julia Flores, Ying Ying Wen, Honghua Dai (causal discovery); Erik Nyberg, Lucas Hope, Charles Twardy, Toby Handfield, Graham Oppy, Chris Hitchcock, Phil Dawid, Karl Axnick (causation and intervention). Various colleagues have been influential in our intellectual development leading us to this endeavor; we wish in particular to acknowledge: David Albrecht, Mike Brady, Tom Dean, Colin Howson, Finn Jensen, Leslie Kaelbling, Jak Kirman, Uffe Kjærulff, Noretta Koertge, Richard Neapolitan, Stuart Russell, Wesley Salmon, Neil Thomason, Ingrid Zukerman. We thank Alan Dorin for creating our cover image. Our dedication reflects our indebtedness to two of the great teachers, Judea Pearl and Chris Wallace. Finally, on a personal level Ann would like to thank her parents, Paul, Robbie and Clare, and Kevin would like to thank Emily and Su.

About the Authors

Kevin B. Korb, Ph.D., earned his doctorate in the philosophy of science at Indiana University, Bloomington (1992), working on the philosophical foundations for the automation of Bayesian reasoning. Since then he has lectured at Monash University, Melbourne, Australia, in computer science, combining his interests in philosophy of science and artificial intelligence in work on understanding and automating inductive inference, the use of MML in learning causal theories, artificial evolution of cognitive and social behavior and modeling Bayesian and human reasoning in the automation of argumentation.

Ann E. Nicholson, D.Phil., did her undergraduate computer science studies at the University of Melbourne, Australia, and her doctorate in the robotics research group at Oxford University, UK (1992), working on dynamic Bayesian networks for discrete monitoring. She then spent two years at Brown University, Providence, Rhode Island, as a post-doctoral research fellow before taking up a lecturing position at Monash University, Melbourne, Australia, in computer science. Her general research area is AI methods for reasoning under uncertainty, while her current research focus is on knowledge engineering with Bayesian networks and applications of Bayesian networks.

Part I

PROBABILISTIC REASONING

1

Bayesian Reasoning

1.1 Reasoning under uncertainty

Artificial intelligence (AI), should it ever exist, will be an intelligence developed by humans, implemented as an artifact. The level of intelligence demanded by Alan Turing's famous test (1950) — the ability to fool ordinary (unfoolish) humans about whether the other end of a dialogue is being carried on by a human or by a computer — is some indication of what AI researchers are aiming for. Such an AI would surely transform our technology and economy. We would be able to automate a great deal of human drudgery and paperwork. Since computers are universal, programs can be effortlessly copied from one system to another (to the consternation of those worried about intellectual property rights!), and the labor savings of AI support for bureaucratic applications of rules, medical diagnosis, research assistance, manufacturing control, etc. promises to be enormous. If a serious AI is ever developed.

There is little doubt that an AI will need to be able to reason logically. An inability to discover, for example, that a system's conclusions have reached inconsistency is more likely to be debilitating than the discovery of an inconsistency itself. For a long time there has also been widespread recognition that practical AI systems shall have to cope with **uncertainty** — that is, they shall have to deal with incomplete evidence leading to beliefs that fall short of knowledge, with fallible conclusions and the need to recover from error, called **non-monotonic reasoning**. Nevertheless, the AI community has been slow to recognize that any serious, general-purpose AI will need to be able to reason probabilistically, what we call here **Bayesian reasoning**.

There are at least three distinct forms of uncertainty which an intelligent system operating in anything like our world shall need to cope with:

1. **Ignorance.** The limits of our knowledge lead us to be uncertain about many things. Does our poker opponent have a flush or is she bluffing?

2. **Physical randomness or indeterminism.** Even if we know everything that we might care to investigate about a coin and how we impart spin to it when we toss it, there will remain an inescapable degree of uncertainty about whether it will land heads or tails when we toss it. A die-hard determinist might claim otherwise, that some unimagined amount of detailed investigation might some-day reveal which way the coin will fall; but such a view is for the foreseeable future a mere act of scientist faith. We are all practical indeterminists.

3. **Vagueness.** Many of the predicates we employ appear to be vague. It is often

unclear whether to classify a dog as a spaniel or not, a human as brave or not, a thought as knowledge or opinion.

Bayesianism is the philosophy that asserts that in order to understand human opinion as it ought to be, constrained by ignorance and uncertainty, the probability calculus is the single most important tool for representing appropriate strengths of belief. In this text we shall present Bayesian computational tools for reasoning with and about strengths of belief as probabilities; we shall also present a Bayesian view of physical randomness. In particular we shall consider a probabilistic account of causality and its implications for an intelligent agent's reasoning about its physical environment. We will not address the third source of uncertainty above, vagueness, which is fundamentally a problem about semantics and one which has no good analysis so far as we are aware.

1.2   Uncertainty in AI

The successes of formal logic have been considerable over the past century and have been received by many as an indication that logic should be the primary vehicle for **knowledge representation** and reasoning within AI. **Logicism** in AI, as this has been called, dominated AI research in the 1960s and 1970s, only losing its grip in the 1980s when artificial neural networks came of age. Nevertheless, even during the heyday of logicism, any number of practical problems were encountered where logic would not suffice, because uncertain reasoning was a key feature of the problem. In the 1960s, medical diagnosis problems became one of the first attempted application areas of AI programming. But there is no symptom or prognosis in medicine which is strictly logically implied by the existence of any particular disease or syndrome; so the researchers involved quickly developed a set of "probabilistic" relations. Because probability calculations are hard — in fact, NP hard in the number of variables (Cooper, 1990) (i.e., computationally intractable; see §1.11) — they resorted to implementing what has subsequently been called "naive Bayes" (or, "Idiot Bayes"), that is, probabilistic updating rules which assume that symptoms are independent of each other given diseases.[1]

The independence constraints required for these systems were so extreme that the systems were received with no wide interest. On the other hand, a very popular set of expert systems in the 1970s and 1980s were based upon Buchanan and Shortliffe's MYCIN, or the uncertainty representation within MYCIN which they called **certainty factors** (Buchanan and Shortliffe, 1984). Certainty factors (CFs) were obtained by first eliciting from experts a "degree of increased belief" which some evidence e should imply for a hypothesis h, $MB(h, e) \in [0, 1]$, and also a corresponding

---

[1] We will look at naive Bayes models for prediction in Chapter 7.

"degree of increased disbelief," $MD(h,e) \in [0,1]$. These were then combined:

$$CF(h,e) = MB(h,e) - MD(h,e) \in [-1,1]$$

This division of changes in "certainty" into changes in belief and disbelief reflects the curious notion that belief and disbelief are not necessarily related to one another (cf. Buchanan and Shortliffe, 1984, section 11.4). A popular AI text, for example, sympathetically reports that "it is often the case that an expert might have confidence 0.7 (say) that some relationship is true and have no feeling about it being not true" (Luger and Stubblefield, 1993, p. 329). The same point can be put more simply: experts are often inconsistent. Our goal in Bayesian modeling is, at least largely, to find the most accurate representation of a real system about which we may be receiving inconsistent expert advice, rather than finding ways of modeling the inconsistency itself.

Regardless of how we may react to this interpretation of certainty factors, no operational semantics for CFs were provided by Buchanan and Shortliffe. This meant that no real guidance could be given to experts whose opinions were being solicited. Most likely, they simply assumed that they were being asked for conditional probabilities of h given e and of $\neg h$ given e. And, indeed, there finally was a probabilistic semantics given for certainty factors: David Heckerman (1986) proved that a consistent probabilistic interpretation of certainty factors[2] would once again require strong independence assumptions: in particular that, when combining multiple pieces of evidence, the different pieces of evidence must always be independent of each other. Whereas this appears to be a desirable simplification of **rule-based** systems, allowing rules to be "modular," with the combined impact of diverse evidence being a compositional function of their separate impacts it is easy to demonstrate that the required independencies are frequently unavailable. The price of rule-based simplicity is irrelevance.

Bayesian networks provide a natural representation of probabilities which allow for (and take advantage of, as we shall see in Chapter 2) any independencies that may hold, while not being limited to problems satisfying strong independence requirements. The combination of substantial increases in computer power with the Bayesian network's ability to use any existing independencies to computational advantage make the approximations and restrictive assumptions of earlier uncertainty formalisms pointless. So we now turn to the main game: understanding and representing uncertainty with probabilities.

1.3 Probability calculus

The probability calculus allows us to represent the independencies which other systems require, but also allows us to represent any dependencies which we may need.

---

[2]In particular, a mapping of certainty factors into likelihood ratios.

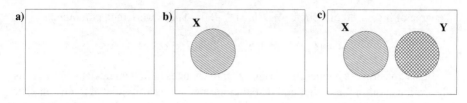

FIGURE 1.1: (a) The event space U; (b) $P(X)$; (c) $P(X \cup Y)$.

The probability calculus was specifically invented in the 17th century by Fermat and Pascal in order to deal with the problems of physical uncertainty introduced by gambling. But it did not take long before it was noticed that the concept of probability could be used in dealing also with the uncertainties introduced by ignorance, leading Bishop Butler to declare in the 18th century that "probability is the very guide to life." So now we introduce this formal language of probability, in a very simple way using Venn diagrams.

Let U be the universe of possible events; that is, if we are uncertain about which of a number of possibilities is true, we shall let U represent all of them collectively (see Figure 1.1(a)). Then the maximum probability must apply to the true event lying within U. By convention we set the maximum probability to 1, giving us Kolmogorov's first axiom for probability theory (Kolmogorov, 1933):

Axiom 1.1 $P(U) = 1$

This probability mass, summing or integrating to 1, is distributed over U, perhaps evenly or perhaps unevenly. For simplicity we shall assume that it is spread evenly, so that the probability of any region is strictly proportional to its area. For any such region X its area cannot be negative, even if X is empty; hence we have the second axiom (Figure 1.1(b)):

Axiom 1.2 *For all* $X \subseteq U, P(X) \geq 0$

We need to be able to compute the probability of combined events, X and Y. This is trivial if the two events are mutually exclusive, giving us the third and last axiom (Figure 1.1(c)), known as **additivity**:

Axiom 1.3 *For all* $X, Y \subseteq U$, *if* $X \cap Y = \emptyset$, *then* $P(X \cup Y) = P(X) + P(Y)$

Any function over a field of subsets of U satisfying the above axioms will be a probability function.[3]

A simple theorem extends addition to events which overlap (i.e., sets which intersect):

Theorem 1.1 *For all* $X, Y \subseteq U, P(X \cup Y) = P(X) + P(Y) - P(X \cap Y)$.

---

[3]A set-theoretic field is a set of sets containing U and $\emptyset$ and is closed under union, intersection and complementation.

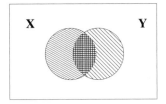

FIGURE 1.2: Conditional probability: $P(X|Y) = P(X \cap Y)/P(Y)$.

This can be intuitively grasped from Figure 1.2: the area of $X \cup Y$ is less than area of X plus the area of Y because when adding the area of intersection $X \cap Y$ has been counted twice; hence, we simply remove the excess to find $P(X \cup Y)$ for any two events X and Y.

The concept of **conditional probability** is crucial for the useful application of the probability calculus. It is usually introduced by definition:

Definition 1.1 Conditional probability

$$P(X|Y) = \frac{P(X \cap Y)}{P(Y)}$$

That is, given that the event Y has occurred, or will occur, the probability that X will also occur is $P(X|Y)$. Clearly, if Y is an event with zero probability, then this conditional probability is undefined. This is not an issue for probability distributions which are **positive**, since, by definition, they are non-zero over every event. A simple way to think about probabilities conditional upon Y is to imagine that the universe of events U has shrunk to Y. The conditional probability of X on Y is just the measure of what is left of X relative to what is left of Y; in Figure 1.2 this is just the ratio of the darker area (representing $X \cap Y$) to the area of Y. This way of understanding conditional probability is justified by the fact that the conditional function $P(\cdot|Y)$ is itself a probability function[4] — that is, it provably satisfies the three axioms of probability.

Another final probability concept we need to introduce is that of **independence** (or, marginal independence). Two events X and Y are probabilistically independent (in notation, $X \perp\!\!\!\perp Y$) whenever conditioning on one leaves the probability of the other unchanged:

Definition 1.2 Independence $X \perp\!\!\!\perp Y \equiv P(X|Y) = P(X)$

This is provably symmetrical: $X \perp\!\!\!\perp Y \equiv Y \perp\!\!\!\perp X$. The simplest examples of independence come from gambling. For example, two rolls of dice are normally independent. Getting a one with the first roll will neither raise nor lower the probability of getting a one the second time. If two events are **dependent**, then one coming true will *alter* the probability of the other. Thus, the probability of getting a diamond flush in poker

---

[4]$P(\cdot|Y)$ is just the function equal to $P(X|Y)$ for all $X \subset U$.

(five diamonds in five cards drawn) is *not* simply $(1/4)^5 = 1/1024$: the probability that the first card drawn being a diamond is $1/4$, but the probability of subsequent cards being diamonds is influenced by the fact that there are then fewer diamonds left in the deck.

Conditional independence generalizes this concept to X and Y being independent given some additional event Z:

Definition 1.3 Conditional independence $X \perp\!\!\!\perp Y | Z \equiv P(X|Y,Z) = P(X|Z)$

This is a true generalization because, of course, Z can be the empty set $\emptyset$, when it reduces to marginal independence. Conditional independence holds when the event Z tells us everything that Y does about X and possibly more; once you know Z, learning Y is uninformative. For example, suppose we have two diagnostic tests for cancer X, an inexpensive but less accurate one, Y, and a more expensive and more accurate one, Z. If Z is more accurate partly because it effectively incorporates all of the diagnostic information available from Y, then knowing the outcome of Z will render an additional test of Y irrelevant — Y will be "screened off" from X by Z.

1.3.1 Conditional probability theorems

We introduce without proof two theorems on conditional probability which will be of frequent use:

Theorem 1.2 Total Probability *Assume the set of events $\{A_i\}$ is a partition of U; i.e., $\bigcup_i A_i = U$ and for any distinct i and j $A_i \cap A_j = \emptyset$. Then*

$$P(U) = \sum_i P(A_i)$$

We can equally well partition the probability of any particular event B instead of the whole event space. In other words, under the above conditions (and if $\forall i A_i \neq \emptyset$),

$$P(B) = \sum_i P(B \cap A_i)$$

We shall refer to either formulation under the title "Total Probability".

Theorem 1.3 The Chain Rule *Given three events A, B, C in a chain of influence (i.e., A and C independent given B),*

$$P(C|A) = P(C|B)P(B|A) + P(C|\neg B)P(\neg B|A)$$

assuming the conditional probabilities are defined. This allows us to divide the probabilistic influence of C on A across the different states of a third variable. (Here, the third variable is binary, but the theorem is easily generalized to variables of arbitrary arity.)

1.3.2 Variables

Although we have introduced probabilities over events, in most of our discussion we shall be concerned with probabilities over **random variables**. A random variable is a variable which reports the outcome of some measurement process. It can be related to events, to be sure. For example, instead of talking about which event in a partition $\{A_i\}$ turns out to be the case, we can equivalently talk about which state x_i the random variable X takes, which we write $X = x_i$. The set of states a variable X can take form its state space, written Ω_X, and its size (or arity) is $|\Omega_X|$.

The discussion thus far has been implicitly of discrete variables, those with a finite state space. However, we need also to introduce the concept of probability distributions over **continuous variables**, that is, variables which range over real numbers, like *Temperature*. For the most part in this text we shall be using probability distributions over discrete variables (events), for two reasons. First, the Bayesian network technology is primarily oriented towards handling discrete state variables, for example the inference algorithms of Chapter 3. Second, for most purposes continuous variables can be **discretized**. For example, temperatures can be divided into ranges of ± 5 degrees for many purposes; and if that is too crude, then they can be divided into ranges of ± 1 degree, etc.

Despite our ability to evade probabilities over continuous variables much of the time, we shall occasionally need to discuss them. We introduce these probabilities by first starting with a **density function** $f(X)$ defined over the continuous variable X. Intuitively, the density assigns a weight or measure to each possible value of X and can be approximated by a finely partitioned histogram reporting samples from X. Although the density is not itself a probability function, it can be used to generate one so long as $f(\cdot)$ satisfies the conditions:

$$f(x) \geq 0 \tag{1.1}$$

$$\int_{-\infty}^{+\infty} f(x)dx = 1 \tag{1.2}$$

In words: each point value is positive or zero and all values integrate to 1. In that case we can define the **cumulative probability distribution** $F(\cdot)$ by

$$F(x) = P(X \leq x) = \int_{x' \leq x} f(x')dx' \tag{1.3}$$

This function assigns probabilities to ranges from each possible value of x down to negative infinity. Note that we can analogously define probabilities over any continuous interval of values for X, so long as the interval is not degenerate (equal to a point). In effect, we obtain a probability distribution by discretizing the continuous variable — i.e., by looking at the mass of the density function over intervals.

1.4   Interpretations of probability

There have been two main contending views about how to understand probability. One asserts that probabilities are fundamentally dispositional properties of non-deterministic physical systems, the classical such systems being gambling devices, such as dice. This view is particularly associated with **frequentism**, advocated in the 19th century by John Venn (1866), identifying probabilities with long-run frequencies of events. The obvious complaint that short-run frequencies clearly do not match probabilities (e.g., if we toss a coin only once, we would hardly conclude that its probability of heads is either one or zero) does not actually get anywhere, since no claim is made identifying short-run frequencies with probabilities. A different complaint does bite, however, namely that the distinction between short-run and long-run is vague, leaving the commitments of this frequentist interpretation unclear. Richard von Mises in the early 20th century fixed this problem by formalizing the frequency interpretation (von Mises, 1919), identifying probabilities with frequency limits in infinite sequences satisfying certain assumptions about randomness. Some version of this frequency interpretation is commonly endorsed by statisticians.

A more satisfactory theoretical account of physical probability arises from Karl Popper's observation (1959) that the frequency interpretation, precise though it was, fails to accommodate our intuition that probabilities of singular events exist and are meaningful. If, in fact, we do toss a coin once and once only, and if this toss should *not* participate in some infinitude (or even large number) of appropriately similar tosses, it would not for that reason fail to have some probability of landing heads. Popper identified physical probabilities with the **propensities** (dispositions) of physical systems ("chance setups") to produce particular outcomes, whether or not those dispositions were manifested repeatedly. An alternative that amounts to much the same thing is to identify probabilities with counterfactual frequencies generated by hypothetically infinite repetitions of an experiment (van Fraassen, 1989).

Whether physical probability is relativized to infinite random sequences, infinite counterfactual sequences or chance setups, these accounts all have in common that the assertion of a probability is relativized to *some* definite physical process or the outcomes it generates.

The traditional alternative to the concept of physical probability is to think of probabilities as reporting our subjective degrees of belief. This view was expressed by Thomas Bayes (1958) (Figure 1.3) and Pierre Simon de Laplace (1951) two hundred years ago. This is a more general account of probability in that we have subjective belief in a huge variety of propositions, many of which are not at all clearly tied to a physical process capable even in principle of generating an infinite sequence of outcomes. For example, most of us have a pretty strong belief in the Copernican hypothesis that the earth orbits the sun, but this is based on evidence not obviously the same as the outcome of a sampling process. We are not in any position to generate solar systems repeatedly and observe the frequency with which their planets revolve around the sun, for example. Bayesians nevertheless are prepared to talk about the probability of the truth of the Copernican thesis and can give an account of the relation between that probability and the evidence for and against it. Since these probabilities are typically subjective, not clearly tied to physical models, most frequentists (hence, most statisticians) deny their meaningfulness. It is not insignificant that this leaves their (usual) belief in Copernicanism unexplained.

The first thing to make clear about this dispute between physicalists and Bayesians is that Bayesianism can be viewed as *generalizing* physicalist accounts of probability. That is, it is perfectly compatible with the Bayesian view of probability as measuring degrees of subjective belief to adopt what David Lewis (1980) dubbed the **Principal Principle** whenever you learn that the physical probability of an outcome is r, set your subjective probability for that outcome to r. This is really just common sense: you may think that the probability of a friend shaving his head is 0.01, but if you learn that he will do so if and only if a fair coin yet to be flipped lands heads, you'll revise your opinion accordingly.

So, the Bayesian and physical interpretations of probability are compatible, with the Bayesian interpretation *extending* the application of probability beyond what is directly justifiable in physical terms. That is the view we adopt here. But what justifies this extension?

FIGURE 1.3: Reverend Thomas Bayes (1702–1761).

1.5   Bayesian philosophy

1.5.1   Bayes' theorem

The origin of Bayesian philosophy lies in an interpretation of **Bayes' Theorem:**

Theorem 1.4 Bayes' Theorem

$$P(h|e) = \frac{P(e|h)P(h)}{P(e)}$$

This is a non-controversial (and simple) theorem of the probability calculus. Under its usual Bayesian interpretation, it asserts that the probability of a hypothesis h conditioned upon some evidence e is equal to its **likelihood** $P(e|h)$ times its probability prior to any evidence $P(h)$, **normalized** by dividing by $P(e)$ (so that the conditional probabilities of all hypotheses sum to 1). Proof is trivial.

The further claim that this is a right and proper way of adjusting our beliefs in our hypotheses given new evidence is called **conditionalization**, and it is controversial.

Definition 1.4 Conditionalization *After applying Bayes' theorem to obtain $P(h|e)$ adopt that as your posterior degree of belief in h — or, $Bel(h) = P(h|e)$.*

Conditionalization, in other words, advocates belief updating via probabilities conditional upon the available evidence. It identifies **posterior probability** (the probability function after incorporating the evidence, which we are writing $Bel(\cdot)$) with **conditional probability** (the prior probability function conditional upon the evidence, which is $P(\cdot|e)$). Put thus, conditionalization may also seem non-controvertible. But there are certainly situations where conditionalization very clearly does not work. The two most basic such situations simply violate what are frequently explicitly stated as assumptions of conditionalization: (1) There must exist joint priors over the hypothesis and evidence spaces. Without a joint prior, Bayes' theorem cannot be used, so conditionalization is a non-starter. (2) The evidence conditioned upon, e, is all and only the evidence learned. This is called the **total evidence condition**. It is a significant restriction, since in many settings it cannot be guaranteed.

The first assumption is also significant. Many take it as the single biggest objection to Bayesianism to raise the question "Where do the numbers come from?" For example, the famous anti-Bayesian Clark Glymour (1980) doesn't complain about Bayesian reasoning involving gambling devices, when the outcomes are engineered to start out equiprobable, but doubts that numbers can be found for more interesting cases. To this kind of objection Bayesians react in a variety of ways. In fact, the different varieties of response pretty much identify the different schools of Bayesianism. Objectivists, such as Rudolf Carnap (1962) and Ed Jaynes (1968), attempt to define prior probabilities based upon the structure of language. Extreme subjectivists, such as de Finetti (1964), assert that it makes no difference what source your priors have:

given that de Finetti's representation theorem shows that non-extreme priors converge in the limit (under reasonable constraints), it just doesn't matter what priors you adopt.

The practical application of Bayesian reasoning does not appear to depend upon settling this kind of philosophical problem. A great deal of useful application can be done simply by refusing to adopt a dogmatic position and accepting common-sense prior probabilities. For example, if there are ten possible suspects in a murder mystery, a fair starting point for any one of them is a 1 in 10 chance of guilt; or, again, if burglaries occur in your neighborhood of 10,000 homes about once a day, then the probability of your having been burglarized within the last 24 hours might reasonably be given a prior probability of 1/10000.

Colin Howson points out that conditionalization is a valid rule of inference if and only if $Bel(e|h) = P(e|h)$, that is, if and only if your prior and posterior probability functions share the relevant conditional probabilities (cf. Howson, 2001). This is certainly a pertinent observation, since encountering some possible evidence may well inform us more about defects in our own conditional probability structure than about the hypothesis at issue. Since Bayes' theorem has $P(h|e)$ being proportional to $P(e|h)$, if the evidence leads us to revise $P(e|h)$, we will be in no position to conditionalize.

How to generate prior probabilities or new conditional probability structure is not dictated by Bayesian principles. Bayesian principles advise how to update probabilities once such a conditional probability structure has been adopted, given appropriate priors. Expecting Bayesian principles to answer all questions about reasoning is expecting too much. Nevertheless, we shall show that Bayesian principles implemented in computer programs can deliver a great deal more than the nay-sayers have ever delivered.

Definition 1.5 Jeffrey conditionalization *Suppose your observational evidence does not correspond specifically to proposition e, but can be represented as a posterior shift in belief about e. In other words, posterior belief in e is not full but partial, having shifted from $P(e)$ to $Bel(e)$. Then, instead of Bayesian conditionalization, apply Jeffrey's update rule for* **probability kinematics***: $Bel(h) = P(h|e)Bel(e) + P(h|\neg e)Bel(\neg e)$ (Jeffrey, 1983).*

Jeffrey's own example is one where your hypothesis is about the color of a cloth, the evidence proposition *e* describes the precise quality of your visual experience under good light, but you are afforded a view of the cloth only under candlelight, in such a way that you cannot exactly articulate what you have observed. Nevertheless, you have learned *something*, and this is reflected in a shift in belief about the quality of your visual experience. Jeffrey conditionalization is very intuitive, but again is not strictly valid. As a practical matter, the need for such partial updating is common in Bayesian modeling.

1.5.2   Betting and odds

Odds are the ratio between the cost of a bet in favor of a proposition and the reward should the bet be won. Thus, assuming a stake of $1 (and otherwise simply rescaling the terms of the bet), a bet at 1:19 odds costs $1 and returns $20 should the proposition come true (with the reward being $20 minus the cost of the bet).[5] The odds may be set at any ratio and may, or may not, have something to do with one's probabilities. Bookies typically set odds for and against events at a slight discrepancy with their best estimate of the probabilities, for their profit lies in the difference between the odds for and against.

While odds and probabilities may deviate, probabilities and **fair odds** $O(\cdot)$ are strictly interchangeable concepts. The fair odds in favor of h are defined simply as the ratio of the probability that h is true to the probability that it is not:

Definition 1.6 Fair odds
$$O(h) = \frac{P(h)}{1 - P(h)}$$

Given this, it is an elementary matter of algebraic manipulation to find $P(h)$ in terms of odds:
$$P(h) = \frac{O(h)}{1 + O(h)} \tag{1.4}$$

Thus, if a coin is fair, the probability of heads is 1/2, so the odds in favor of heads are 1:1 (usually described as "50:50"). Or, if the odds of getting "snake eyes" (two 1's) on the roll of two dice are 1:35, then the probability of this is:

$$\frac{1/35}{1 + 1/35} = \frac{1/35}{36/35} = 1/36$$

as will always be the case with fair dice. Or, finally, suppose that the probability an agent ascribes to the Copernican hypothesis (CH) is zero; then the odds that agent is giving to Copernicus having been wrong $(\neg CH)$ are *infinite*:

$$O(\neg CH) = \frac{1}{0} = \infty$$

At these odds, incidentally, it is trivial that the agent can never reach a degree of belief in CH above zero on any finite amount of evidence, if relying upon conditionalization for updating belief.

With the concept of fair odds in hand, we can reformulate Bayes' theorem in terms of (fair) odds, which is often useful:

Theorem 1.5 Odds-Likelihood Bayes' Theorem

$$O(h|e) = \frac{P(e|h)}{P(e|\neg h)} O(h)$$

---

[5]It is common in sports betting to invert the odds, quoting the odds *against* a team winning, for example. This makes no difference; the ratio is simply reversed.

This is readily proven to be equivalent to Theorem 1.4. In English it asserts that the odds on h conditional upon the evidence e are equal to the prior odds on h times the **likelihood ratio** $P(e|h) : P(e|\neg h)$. Clearly, the fair odds in favor of h will rise if and only if the likelihood ratio is greater than one.

1.5.3 Expected utility

Generally, agents are able to assign utility (or, value) to the situations in which they find themselves. We know what we like, we know what we dislike, and we also know when we are experiencing neither of these. Given a general ability to order situations, and bets with definite probabilities of yielding particular situations, Frank Ramsey (1931) demonstrated that we can identify particular utilities with each possible situation, yielding a **utility function**.

If we have a utility function $U(O_i|A)$ over every possible outcome of a particular action A we are contemplating, and if we have a probability for each such outcome $P(O_i|A)$, then we can compute the probability-weighted average utility for that action — otherwise known as the **expected utility** of the action:

Definition 1.7 Expected utility

$$EU(A) = \sum_i U(O_i|A) \times P(O_i|A)$$

It is commonly taken as axiomatic by Bayesians that agents ought to *maximize their expected utility*. That is, when contemplating a number of alternative actions, agents ought to decide to take that action which has the maximum expected utility. If you are contemplating eating strawberry ice cream or else eating chocolate ice cream, presumably you will choose that flavor which you prefer, other things being equal. Indeed, if you chose the flavor you liked *less*, we should be inclined to think that other things are *not* equal — for example, you are under some kind of external compulsion — or perhaps that you are not being honest about your preferences. Utilities have behavioral consequences *essentially*: any agent who consistently ignores the putative utility of an action or situation arguably does not have that utility.

Regardless of such foundational issues, we now have the conceptual tools necessary to understand what is fair about fair betting. **Fair bets** are fair because their expected utility is zero. Suppose we are contemplating taking the fair bet B on proposition h for which we assign probability $P(h)$. Then the expected utility of the bet is:

$$EU(B) = U(h|B)P(h|B) + U(\neg h|B)P(\neg h|B)$$

Typically, betting on a proposition has no effect on the probability that it is true (although this is not necessarily the case!), so $P(h|B) = P(h)$. Hence,

$$EU(B) = U(h|B)P(h) + U(\neg h|B)(1 - P(h))$$

Assuming a stake of 1 unit for simplicity, then by definition $U(h|B) = 1 - P(h)$ (i.e., this is the utility of h being true given the bet for h) while $U(\neg h|B) = -P(h)$, so,

$$EU(B) = (1 - P(h))P(h) - P(h)(1 - P(h)) = 0$$

Given that the bet has zero expected utility, the agent should be no more inclined to take the bet in favor of h than to take the opposite bet against h.

1.5.4   Dutch books

The original Dutch book argument of Ramsey (1931) (see also de Finetti, 1964) claims to show that subjective degrees of belief, if they are to be rational, *must* obey the probability calculus. It has the form of a *reductio ad absurdum* argument:

1. A rational agent should be willing to take either side of any combination of fair bets.
2. A rational agent should never be willing to take a combination of bets which guarantees a loss.
3. Suppose a rational agent's degrees of belief violate one or more of the axioms of probability.
4. Then it is provable that some combination of fair bets will lead to a guaranteed loss.
5. Therefore, the agent is both willing and not willing to take this combination of bets.

Now, the inferences to (4) and (5) in this argument are not in dispute (see §1.11 for a simple demonstration of (4) for one case). A *reductio* argument needs to be resolved by finding a prior assumption to blame, and concluding that it is false. Ramsey, and most Bayesians to date, supposed that the most plausible way of relieving the contradiction of (5) is by refusing to suppose that a rational agent's degrees of belief may violate the axioms of probability. This result can then be generalized beyond settings of explicit betting by taking "bets with nature" as a metaphor for decision-making generally. For example, walking across the street is in some sense a bet about our chances of reaching the other side.

Some anti-Bayesians have preferred to deny (1), insisting for example that it would be uneconomic to invest in bets with zero expected value (e.g., Chihara and Kennedy, 1979). But the ascription of the radical incoherence in (5) simply to the willingness of, say, bored aristocrats to place bets that will net them nothing clearly will not do: the effect of incoherence is entirely out of proportion with the proposed cause of effeteness.

Alan Hájek (2008) has pointed out a more plausible objection to (2). In the scenarios presented in Dutch books there is always some combination of bets which guarantees a net loss whatever the outcomes on the individual bets. But equally there is always some combination of bets which guarantees a net gain — a "Good Book." So, one agent's half-empty glass is another's half-full glass! Rather than dismiss the Dutch-bookable agent as irrational, we might commend it for being open to a guaranteed win! So, Hájek's point seems to be that there is a fundamental symmetry in Dutch book arguments which leaves open the question whether violating probability axioms is rational or not. Certainly, when metaphorically extending betting to a "struggle" with Nature, it becomes rather implausible that She is really out to Dutch book us!

Hájek's own solution to the problem posed by his argument is to point out that whenever an agent violates the probability axioms there will be some variation of its system of beliefs which is guaranteed to win money whenever the original system is guaranteed to win, and which is also capable of winning in some situations when the original system is not. So the variant system of belief in some sense dominates the original: it is everywhere at least as good as the original and in some places better. In order to guarantee that your system of beliefs cannot be dominated, you must be probabilistically coherent (see §1.11). This, we believe, successfully rehabilitates the Dutch book in a new form.

Rather than rehabilitate, a more obviously Bayesian response is to consider the probability of a bookie hanging around who has the smarts to pump our agent of its money and, again, of a simpleton hanging around who will sign up the agent for guaranteed winnings. In other words, for rational choice surely what matters is the relative expected utility of the choice. Suppose, for example, that we are offered a set of bets which has a guaranteed loss of $10. Should we take it? The Dutch book assumes that accepting the bet is irrational. But, if the one and only alternative available is another bet with an expected loss of $1,000, then it no longer seems so irrational. An implicit assumption of the Dutch book has always been that betting is voluntary and when all offered bets are turned down the expected utility is zero. The further implicit assumption pointed out by Hájek's argument is that there is always a shifty bookie hanging around ready to take advantage of us. No doubt that is not always the case, and instead there is only some probability of it. Yet referring the whole matter of justifying the use of Bayesian probability to expected utility smacks of circularity, since expectation is understood in terms of Bayesian probability.

Aside from invoking the rehabilitated Dutch book, there is a more pragmatic approach to justifying Bayesianism, by looking at its importance for dealing with cases of practical problem solving. We take Bayesian principles to be normative, and especially to be a proper guide, under some range of circumstances, to evaluating hypotheses in the light of evidence. The form of justification that we think is ultimately most compelling is the "method of reflective equilibrium," generally attributed to Goodman (1973) and Rawls (1971), but adumbrated by Aristotle in his *Nicomachian Ethics*. In a nutshell, it asserts that the normative principles to accept are those which best accommodate our basic, unshakable intuitions about what is good and bad (e.g., paradigmatic judgments of correct inference in simple domains, such as gambling) and which best integrate with relevant theory and practice. We now present some cases which Bayesian principle handles readily, and better than any alternative normative theory.

1.5.5   Bayesian reasoning examples

1.5.5.1   Breast cancer

Suppose the women attending a particular clinic show a long-term chance of 1 in 100 of having breast cancer. Suppose also that the initial screening test used at the clinic has a false positive rate of 0.2 (that is, 20% of women without cancer will test positive for cancer) and that it has a false negative rate of 0.1 (that is, 10% of women with cancer will test negative). The laws of probability dictate from this last fact that the probability of a positive test given cancer is 90%. Now suppose that you are such a woman who has just tested positive. What is the probability that you have cancer?

This problem is one of a class of probability problems which has become notorious in the cognitive psychology literature (cf. Tversky and Kahneman, 1974). It seems that very few people confronted with such problems bother to pull out pen and paper and compute the right answer via Bayes' theorem; even fewer can get the right answer without pen and paper. It appears that for many the probability of a positive test (which is observed) given cancer (i.e., 90%) dominates things, so they figure that they have quite a high chance of having cancer. But substituting into Theorem 1.4 gives us:

$$P(Cancer|Pos) = \frac{P(Pos|Cancer)P(Cancer)}{P(Pos)}$$

Note that the probability of *Pos* given *Cancer* — which is the likelihood 0.9 — is only *one* term on the right hand side; the other crucial term is the prior probability of cancer. Cognitive psychologists studying such reasoning have dubbed the dominance of likelihoods in such scenarios "base-rate neglect," since the base rate (prior probability) is being suppressed (Kahneman and Tversky, 1973). Filling in the formula and computing the conditional probability of *Cancer* given *Pos* gives us quite a different story:

$$
\begin{aligned}
P(Cancer|Pos) &= \frac{P(Pos|Cancer)P(Cancer)}{P(Pos)} \\
&= \frac{P(Pos|Cancer)P(Cancer)}{P(Pos|Cancer)P(Cancer) + P(Pos|\neg Cancer)P(\neg Cancer)} \\
&= \frac{0.9 \times 0.01}{0.9 \times 0.01 + 0.2 \times 0.99} \\
&= \frac{0.009}{0.009 + 0.198} \\
&\approx 0.043
\end{aligned}
$$

Now the discrepancy between 4% and 80 or 90% is no small matter, particularly if the consequence of an error involves either unnecessary surgery or (in the reverse case) leaving a cancer untreated. But decisions similar to these are constantly being made based upon "intuitive feel" — i.e., without the benefit of paper and pen, let alone Bayesian networks (which are simpler to use than paper and pen!).

1.5.5.2 People v. Collins

The legal system is replete with misapplications of probability and with incorrect claims of the irrelevance of probabilistic reasoning as well.

In 1964 an interracial couple was convicted of robbery in Los Angeles, largely on the grounds that they matched a highly improbable profile, a profile which fit witness reports (Sullivan, Sullivan). In particular, the two robbers were reported to be

- A man with a mustache

- Who was black and had a beard
- And a woman with a ponytail
- Who was blonde
- The couple was interracial
- And were driving a yellow car

The prosecution suggested that these characteristics had the following probabilities of being observed at random in the LA area:

1. A man with a mustache 1/4
2. Who was black and had a beard 1/10
3. And a woman with a ponytail 1/10
4. Who was blonde 1/3
5. The couple was interracial 1/1000
6. And were driving a yellow car 1/10

The prosecution called an instructor of mathematics from a state university who apparently testified that the "product rule" applies to this case: where mutually independent events are being considered jointly, the joint probability is the product of the individual probabilities.[6] This last claim is, in fact, correct (see Problem 2 below); what is false is the idea that the product rule is relevant to this case. If we label the individual items of evidence e_i $(i = 1, \ldots, 6)$, the joint evidence e, and the hypothesis that the couple was guilty h, then what is claimed is

$$P(e|\neg h) = \prod_i P(e_i|\neg h) = 1/12000000$$

The prosecution, having made this inference, went on to assert that the probability the couple were innocent was no more than 1/12000000. The jury convicted.

As we have already suggested, the product rule does *not* apply in this case. Why not? Well, because the individual pieces of evidence are obviously *not* independent. If, for example, we know of the occupants of a car that one is black and the other has blonde hair, what then is the probability that the occupants are an interracial couple? Clearly not 1/1000! If we know of a man that he has a mustache, is the probability of having a beard unchanged? These claims are preposterous, and it is simply shameful that a judge, prosecutor and defence attorney could not recognize how preposterous they are — let alone the mathematics "expert" who testified to them. Since e_2 implies e_1, while e_2, e_3, e_4 jointly imply e_5 (to a fair approximation), a far better estimate for $P(e|\neg h)$ is $P(e_2|\neg h)P(e_3|\neg h)P(e_4|\neg h)P(e_6|\neg h) = 1/3000$.

To be sure, if we accepted that the probability of innocence were a mere 1/3000 we might well accept the verdict. But there is a more fundamental error in the prosecution reasoning than neglecting the conditional dependencies in the evidence. If,

---

[6]Coincidentally, this is just the kind of independence required for certainty factors to apply.

unlike the judge, prosecution and jury, we take a peek at Bayes' theorem, we discover that the probability of guilt $P(h|e)$ is *not* equal to $1 - P(e|\neg h)$; instead

$$P(h|e) = \frac{P(e|h)P(h)}{P(e|h)P(h) + P(e|\neg h)P(\neg h)}$$

Now if the couple in question *were* guilty, what are the chances the evidence accumulated would have been observed? That's a rather hard question to answer, but feeling generous towards the prosecution, let us simplify and say 1. That is, let us accept that $P(e|h) = 1$. Plugging in our assumptions we have thus far:

$$P(h|e) = \frac{P(h)}{P(h) + P(\neg h)/3000}$$

We are missing the crucial prior probability of a random couple being guilty of the robbery. Note that we cannot here use the prior probability of, for example, an interracial couple being guilty, since the fact that they are interracial is a piece of the evidence. The most plausible approach to generating a prior of the needed type is to count the number of couples in the LA area and give them an equal prior probability. In other words, if N is the number of possible couples in the LA area, $P(h) = 1/N$. So, what is N? The population at the time was about 6.5 million people (Demographia, Demographia). If we conservatively take half of them as being eligible to be counted (e.g., being adult humans), this gives us 1,625,000 eligible males and as many females. If we simplify by supposing that they are all in heterosexual partnerships, that will introduce a slight bias in favor of innocence; if we also simplify by ignoring the possibility of people traveling in cars with friends, this will introduce a larger bias in favor of guilt. The two together give us 1,625,000 available couples, suggesting a prior probability of guilt of 1/1625000. Plugging this in we get:

$$P(h|e) = \frac{1/1625000}{1/1625000 + (1 - 1/1625000)/3000} \approx 0.002$$

In other words, even ignoring the huge number of trips with friends rather than partners, we obtain a 99.8% chance of innocence and so a very large probability of a nasty error in judgment. The good news is that the conviction (of the man only!) was subsequently overturned, partly on the basis that the independence assumptions are false. The bad news is that the appellate court finding also suggested that probabilistic reasoning is just irrelevant to the task of establishing guilt, which is a nonsense. One right conclusion about this case is that, assuming the likelihood has been *properly* worked out, a sensible prior probability must also be taken into account. In some cases judges have specifically ruled out all consideration of prior probabilities, while allowing testimony about likelihoods! Probabilistic reasoning which simply ignores half of Bayes' theorem is dangerous indeed!

Note that we do not claim that 99.8% is the best probability of innocence that can be arrived at for the case of People v. Collins. What we *do* claim is that, for the particular facts represented as having a particular probabilistic interpretation, this is far closer to a reasonable probability than that offered by the prosecution, namely

1/12000000. We also claim that the forms of reasoning we have here illustrated are *crucial* for interpreting evidence in general: namely, whether the offered items of evidence are conditionally independent and what the prior probability of guilt may be.

1.6 The goal of Bayesian AI

The most commonly stated goal for artificial intelligence is that of producing an artifact which performs difficult intellectual tasks at or beyond a human level of performance. Of course, machine chess programs have satisfied this criterion for some time now. Although some AI researchers have claimed that therefore an AI has been produced — that denying this is an unfair shifting of the goal line — it is absurd to think that we ought to be satisfied with programs which are strictly special-purpose and which achieve their performance using techniques that deliver nothing when applied to most areas of human intellectual endeavor.

Turing's test for intelligence appears to be closer to satisfactory: fooling ordinary humans with verbal behavior not restricted to any domain would surely demonstrate some important *general* reasoning ability. Many have pointed out that the conditions for Turing's test, strictly verbal behavior without any afferent or efferent nervous activity, yield at best some kind of disembodied, ungrounded intelligence. John Searle's Chinese Room argument (Searle, 1980) for example, can be interpreted as making such a case; for this kind of interpretation of Searle see Harnad (1989) and Korb (1991). A more convincing criterion for human-like intelligence is to require of an artificial intelligence that it be capable of powering a robot-in-the-world in such a way that the robot's performance cannot be distinguished from human performance in terms of behavior (disregarding, for example, whether the skin can be so distinguished). The program that can achieve this would surely satisfy any sensible AI researcher, or critic, that an AI had been achieved.

We are not, however, actually motivated by the idea of behaviorally cloning humans. If all we wish to do is reproduce humans, we would be better advised to employ the tried and true methods we have always had available. Our motive is to understand *how* such performance can be achieved. We are interested in knowing how humans perform the many interesting and difficult cognitive tasks encompassed by AI — such as, natural language understanding and generation, planning, learning, decision making — but we are also interested in knowing how they might be performed otherwise, and in knowing how they might be performed optimally. By building artifacts which model our best understanding of how humans do these things (which can be called **descriptive artificial intelligence**) and also building artifacts which model our best understanding of what is optimal in these activities (**normative artificial intelligence**), we can further our understanding of the nature of intelligence and also produce some very useful tools for science, government and industry.

As we have indicated through example, medical, legal, scientific, political and most other varieties of human reasoning either consider the relevant probabilistic factors and accommodate them or run the risk of introducing egregious and damaging errors. The goal of a Bayesian artificial intelligence is to produce a thinking agent which does as well or better than humans in such tasks, which can adapt to stochastic and changing environments, recognize its own limited knowledge and cope sensibly with these varied sources of uncertainty.

1.7   Achieving Bayesian AI

Given that we have this goal, how can we achieve it? The first step is to develop algorithms for doing Bayesian conditionalization properly and, insofar as possible, efficiently. This step has already been achieved, and the relevant algorithms are described in Chapters 2 and 3. The next step is to incorporate methods for computing expected utilities and develop methods for maximizing utility in decision making. We describe algorithms for this in Chapter 4. We would like to test these ideas in application: we describe some Bayesian network applications in Chapter 5.

These methods for probability computation are fairly well developed and their improvement remains an active area of research in AI today. The biggest obstacles to Bayesian AI having a broad and deep impact outside of the research community are the difficulties in developing applications, difficulties with eliciting knowledge from experts, and integrating and validating the results. One issue is that there is no clear methodology for developing, testing and deploying Bayesian network technology in industry and government — there is no recognized discipline of "software engineering" for Bayesian networks. We make a preliminary effort at describing one — Knowledge Engineering with Bayesian Networks (KEBN) in Part III, including its illustration in case studies of Bayesian network development in Chapter 11.

Another important response to the difficulty of building Bayesian networks by hand is the development of methods for their automated learning — the machine learning of Bayesian networks (aka "data mining"). In Part II we introduce and develop the main methods for learning Bayesian networks with reference to the theory of causality underlying them. These techniques logically come before the knowledge engineering methodology, since that draws upon and integrates machine learning with expert elicitation.

1.8   Are Bayesian networks Bayesian?

Many AI researchers like to point out that Bayesian networks are not inherently Bayesian at all; some have even claimed that the label is a misnomer. At the 2002 Australasian Data Mining Workshop, for example, Geoff Webb made the former claim. Under questioning it turned out he had two points in mind: (1) Bayesian networks are frequently "data mined" (i.e., learned by some computer program) via non-Bayesian methods. (2) Bayesian networks at bottom represent probabilities; but probabilities can be interpreted in any number of ways, including as some form of frequency; hence, the networks are not intrinsically either Bayesian or non-Bayesian, they simply represent values needing further interpretation.

These two points are entirely correct. We shall ourselves present non-Bayesian methods for automating the learning of Bayesian networks from statistical data. We shall also present Bayesian methods for the same, together with some evidence of their superiority. The interpretation of the probabilities represented by Bayesian networks is open so long as the philosophy of probability is considered an open question. Indeed, much of the work presented here ultimately depends upon the probabilities being understood as *physical probabilities*, and in particular as propensities or probabilities determined by propensities. Nevertheless, we happily invoke the Principal Principle: where we are convinced that the probabilities at issue reflect the true propensities in a physical system we are certainly going to use them in assessing our own degrees of belief.

The advantages of the Bayesian network representations are largely in simplifying conditionalization, planning decisions under uncertainty and explaining the outcome of stochastic processes. These purposes all come within the purview of a clearly Bayesian interpretation of what the probabilities mean, and so, we claim, the Bayesian network technology which we here introduce is aptly named: it provides the technical foundation for a truly Bayesian artificial intelligence.

1.9 Summary

How best to reason about uncertain situations has always been of concern. From the 17th century we have had available the basic formalism of probability calculus, which is far and away the most promising formalism for coping with uncertainty. Probability theory has been used widely, but not deeply, since then. That is, the elementary ideas have been applied to a great variety of problems — e.g., actuarial calculations for life insurance, coping with noise in measurement, business decision making, testing scientific theories, gambling — but the problems have typically been of highly constrained size, because of the computational infeasibility of conditionalization when dealing with large problems. Even in dealing with simplified problems, humans have had difficulty handling the probability computations. The development of Bayesian network technology automates the process and so promises to free us from such difficulties. At the same time, improvements in computer capacity, together with the ability of Bayesian networks to take computational advantage of any available independencies between variables, promise to both widen and deepen the domain of probabilistic reasoning.

1.10   Bibliographic notes

An excellent source of information about different attempts to formalize reasoning about uncertainty — including certainty factors, non-monotonic logics, Dempster-Shafer calculus, as well as probability — is the anthology *Readings in Uncertain Reasoning* edited by Shafer and Pearl (1990). Three polemics against non-Bayesian approaches to uncertainty are those by Drew McDermott (1987), Peter Cheeseman (1988) and Kevin Korb (1995). For understanding Bayesian philosophy, Ramsey's original paper "Truth and Probability" is beautifully written, original and compelling (1931); for a more comprehensive and recent presentation of Bayesianism see Howson and Urbach's *Scientific Reasoning* (2007). For Bayesian decision analysis see Richard Jeffrey's *The Logic of Decision* (1983). DeGroot and Schervish (2002) provide an accessible introduction to both the probability calculus and statistics.

Karl Popper's original presentation of the propensity interpretation of probability is (Popper, 1959). This view is related to the elaboration of a probabilistic account of causality in recent decades. Wesley Salmon (1984) provides an overview of probabilistic causality.

Naive Bayes models, despite their simplicity, have done surprisingly well as predictive classifiers for data mining problems; see Chapter 7.

1.11   Technical notes

A Dutch book

Here is a simple Dutch book. Suppose someone assigns $P(A) = -0.1$, violating probability Axiom 2. Then $O(A) = -0.1/(1 - (-0.1)) = -0.1/1.1$. The reward for a bet on A with a \$1 stake is $\$(1 - P(U)) = \$1.1$ if A comes true and $\$ - P(U) = \$0.1$ if A is false. That's everywhere positive and so is a "Good Book." The Dutch book simply requires this agent to take the fair bet *against A*, which has the payoffs $-\$1.1$ if A is true and $-\$0.1$ otherwise.

The rehabilitated Dutch book

Following Hájek, we can show that incoherence (violating the probability axioms) leads to being "dominated" by someone who is coherent — that is, the coherent bettor can take advantage of offered bets that the incoherent bettor cannot and otherwise will do as well.

Suppose Ms. Incoherent assigns $P_I(U) < 1$ (where U is the universal event that *must* occur), for example. Then Ms. Incoherent will take any bet for U at odds of $P_I(U)/(1 - P_I(U))$ or greater. But Ms. Coherent has assigned $P_C(U) = 1$, of course, and so can take any bet for U at any odds offered greater than zero. So for the odds within the range $[0, \frac{P_I(U)}{1-P_I(U)}]$ Ms. Coherent is guaranteed a profit whereas Ms. Incoherent is sitting on her hands.

NP hardness

A problem is Non-deterministic Polynomial-time (NP) if it is solvable in polynomial time on a non-deterministic Turing machine. A problem is Non-deterministic Polynomial time hard (NP hard) if every problem that is NP can be translated into this NP hard problem in polynomial time. If there is a polynomial time solution to any NP hard problem, then because of polynomial time translatability for all other NP problems, there must be a polynomial time solution to all NP problems. No one knows of a polynomial time solution to any NP hard problem; the best known solutions are exponentially explosive. Thus, "NP hard" problems are generally regarded as computationally intractable. (The classic introduction to computational complexity is Garey and Johnson (1979).)

1.12   Problems

Probability Theory

Problem 1

Prove that the conditional probability function $P(\cdot|e)$, if well defined, is a probability function (i.e., satisfies the three axioms of Kolmogorov).

Problem 2

Given that two pieces of evidence e_1 and e_2 are conditionally independent given the hypothesis — i.e., $P(e_1|e_2,h) = P(e_1|h)$ — prove the "product rule": $P(e_1,e_2|h) = P(e_1|h) \times P(e_2|h)$.

Problem 3

Prove the theorems of §1.3.1, namely the Total Probability theorem and the Chain Rule.

Problem 4

There are five containers of milk on a shelf; unbeknownst to you, two of them have passed their use-by date. You grab two at random. What's the probability that neither have passed their use-by date? Suppose someone else has got in just ahead of you, taking one container, after examining the dates. What's the probability that the two you take at random after that are ahead of their use-by dates?

Problem 5

The probability of a child being a boy (or a girl) is 0.5 (let us suppose). Consider all the families with exactly two children. What is the probability that such a family has two girls given that it has at least one girl?

Problem 6

The frequency of male births at the Royal Women's Hospital is about 51 in 100. On a particular day, the last eight births have been female. The probability that the next birth will be male is:

1. About 51%
2. Clearly greater than 51%
3. Clearly less than 51%
4. Almost certain
5. Nearly zero

Bayes' Theorem

Problem 7

After winning a race, an Olympic runner is tested for the presence of steroids. The test comes up positive, and the athlete is accused of doping. Suppose it is known that 5% of all victorious Olympic runners do use performance-enhancing drugs. For this particular test, the probability of a positive finding given that drugs are used is 95%. The probability of a false positive is 2%. What is the (posterior) probability that the athlete did in fact use steroids, given the positive outcome of the test?

Problem 8

You consider the probability that a coin is double-headed to be 0.01 (call this option h'); if it isn't double-headed, then it's a fair coin (call this option h). For whatever reason, you can only test the coin by flipping it and examining the coin (i.e., you can't simply examine both sides of the coin). In the worst case, how many tosses do you need before having a posterior probability for either h or h' that is greater than 0.99, i.e., what's the maximum number of tosses until that happens?

Problem 9

(Adapted from Fischoff and Bar-Hillel (1984).) Two cab companies, the Blue and the Green, operate in a given city. Eighty-five percent of the cabs in the city are Blue; the remaining 15% are Green. A cab was involved in a hit-and-run accident at night. A witness identified the cab as a Green cab. The court tested the witness' ability to distinguish between Blue and Green cabs under night-time visibility conditions. It found that the witness was able to identify each color correctly about 80% of the time, but confused it with the other color about 20% of the time.

What are the chances that the errant cab was indeed Green, as the witness claimed?

Odds and Expected Value

Problem 10

Construct a Dutch book against someone who violates the Axiom of Additivity. That is, suppose a Mr. Fuzzy declares about the weather tomorrow that $P(Sunny) = 0.5, P(Inclement) = 0.5$, and $P(Sunny\ or\ inclement) = 0.5$. Mr. Fuzzy and you agree about what will count as sunny and as inclement weather and you both agree that they are incompatible states. How can you construct a Dutch book against Fuzzy, using only fair bets?

Problem 11

A bookie offers you a ticket for $5.00 which pays $6.00 if Manchester United beats Arsenal and nothing otherwise. What are the odds being offered? To what probability of Manchester United winning does that correspond?

Problem 12

You are offered a Keno ticket in a casino which will pay you $1 million if you win! It only costs you $1 to buy the ticket. You choose 4 numbers out of a 9x9 grid of distinct numbers. You win if all of your 4 numbers come up in a random draw of four from the 81 numbers. What is the expected dollar value of this gamble?

Applications

Problem 13

(Note: this is the case of Sally Clark, convicted in the UK in 1999, found innocent on appeal in 2003, and tragically died in 2007 of alcohol poisoning. See Innocent, 2002.) A mother was arrested after her second baby died a few months old, apparently of sudden infant death syndrome (SIDS), exactly as her first child had died a year earlier. According to prosecution testimony, about 2 in 17200 babies die of SIDS. So, according to their argument, there is only a probability of $(2/17200)^2 \approx 1/72000000$ that two such deaths would happen in the same family by chance alone. In other words, according to the prosecution, the woman was guilty beyond a reasonable doubt. The jury returned a guilty verdict, even though there was no significant evidence of guilt presented beyond this argument. Which of the following is the truth of the matter? Why?

1. Given the facts presented, the probability that the woman is guilty is greater than 99%, so the jury decided correctly.
2. The argument presented by the prosecution is irrelevant to the mother's guilt or innocence.
3. The prosecution argument is relevant but inconclusive.

4. The prosecution argument only establishes a probability of guilt of about 16%.

5. Given the facts presented, guilt and innocence are equally likely.

Problem 14

A DNA match between the defendant and a crime scene blood sample has a probability of 1/100000 if the defendant is innocent. There is no other significant evidence.

1. What is the probability of guilt?
2. Suppose we agree that the prior probability of guilt under the (unspecified) circumstances is 10%. What then is the probability of guilt?
3. The suspect has been picked up through a universal screening program applied to all Australians seeking a Medicare card. So far, 10 million people have been screened. What then is the probability of guilt?

2

Introducing Bayesian Networks

2.1   Introduction

Having presented both theoretical and practical reasons for artificial intelligence to use probabilistic reasoning, we now introduce the key computer technology for dealing with probabilities in AI, namely **Bayesian networks.** Bayesian networks (BNs) are graphical models for reasoning under uncertainty, where the nodes represent variables (discrete or continuous) and arcs represent direct connections between them. These direct connections are often causal connections. In addition, BNs model the quantitative strength of the connections between variables, allowing probabilistic beliefs about them to be updated automatically as new information becomes available.

In this chapter we will describe how Bayesian networks are put together (the **syntax**) and how to interpret the information encoded in a network (the **semantics**). We will look at how to model a problem with a Bayesian network and the types of reasoning that can be performed.

2.2   Bayesian network basics

A **Bayesian network** is a graphical structure that allows us to represent and reason about an uncertain domain. The nodes in a Bayesian network represent a set of random variables, $\mathbf{X} = X_1, ..X_i, ...X_n$, from the domain. A set of directed **arcs** (or links) connects pairs of nodes, $X_i \rightarrow X_j$, representing the direct dependencies between variables. Assuming discrete variables, the strength of the relationship between variables is quantified by conditional probability distributions associated with each node. The only constraint on the arcs allowed in a BN is that there must not be any directed cycles: you cannot return to a node simply by following directed arcs. Such networks are called directed acyclic graphs, or simply **dags**.

There are a number of steps that a **knowledge engineer**[1] must undertake when building a Bayesian network. At this stage we will present these steps as a sequence; however it is important to note that in the real-world the process is not so simple. In Chapter 10 we provide a fuller description of BN knowledge engineering.

---

[1]Knowledge engineer in the jargon of AI means a practitioner applying AI technology.

Throughout the remainder of this section we will use the following simple medical diagnosis problem.

Example problem: Lung cancer. *A patient has been suffering from shortness of breath (called* dyspnoea*) and visits the doctor, worried that he has lung cancer. The doctor knows that other diseases, such as tuberculosis and bronchitis, are possible causes, as well as lung cancer. She also knows that other relevant information includes whether or not the patient is a smoker (increasing the chances of cancer and bronchitis) and what sort of air pollution he has been exposed to. A positive X-ray would indicate either TB or lung cancer.*[2]

2.2.1 Nodes and values

First, the knowledge engineer must identify the variables of interest. This involves answering the question: what are the nodes to represent and what values can they take, or what state can they be in? For now we will consider only nodes that take discrete values. The values should be both **mutually exclusive** and **exhaustive**, which means that the variable must take on exactly one of these values at a time. Common types of discrete nodes include:

- Boolean nodes, which represent propositions, taking the binary values true (T) and false (F). In a medical diagnosis domain, the node *Cancer* would represent the proposition that a patient has cancer.
- Ordered values. For example, a node *Pollution* might represent a patient's pollution exposure and take the values {*low, medium, high*}.
- Integral values. For example, a node called *Age* might represent a patient's age and have possible values from 1 to 120.

Even at this early stage, modeling choices are being made. For example, an alternative to representing a patient's exact age might be to clump patients into different age groups, such as {*baby, child, adolescent, young, middleaged, old*}. The trick is to choose values that represent the domain efficiently, but with enough detail to perform the reasoning required. More on this later!

TABLE 2.1

Preliminary choices of nodes and values for the lung cancer example.

Node name	Type	Values
Pollution	Binary	{*low, high*}
Smoker	Boolean	{*T, F*}
Cancer	Boolean	{*T, F*}
Dyspnoea	Boolean	{*T, F*}
X-ray	Binary	{*pos, neg*}

---

[2]This is a modified version of the so-called "Asia" problem Lauritzen and Spiegelhalter, 1988, given in §2.5.3.

For our example, we will begin with the restricted set of nodes and values shown in Table 2.1. These choices already limit what can be represented in the network. For instance, there is no representation of other diseases, such as TB or bronchitis, so the system will not be able to provide the probability of the patient having them. Another limitation is a lack of differentiation, for example between a heavy or a light smoker, and again the model assumes at least some exposure to pollution. Note that all these nodes have only two values, which keeps the model simple, but in general there is no limit to the number of discrete values.

2.2.2 Structure

The structure, or topology, of the network should capture qualitative relationships between variables. In particular, two nodes should be connected directly if one affects or causes the other, with the arc indicating the direction of the effect. So, in our medical diagnosis example, we might ask what factors affect a patient's chance of having cancer? If the answer is "Pollution and smoking," then we should add arcs from *Pollution* and *Smoker* to *Cancer*. Similarly, having cancer will affect the patient's breathing and the chances of having a positive X-ray result. So we add arcs from *Cancer* to *Dyspnoea* and *XRay*. The resultant structure is shown in Figure 2.1. It is important to note that this is just one possible structure for the problem; we look at alternative network structures in §2.4.3.

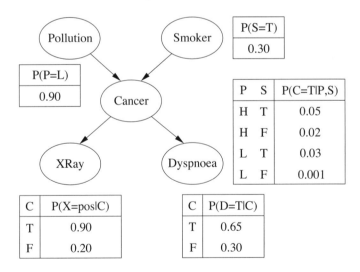

FIGURE 2.1: A BN for the lung cancer problem.

Structure terminology and layout

In talking about network structure it is useful to employ a family metaphor: a node is a **parent** of a **child**, if there is an arc from the former to the latter. Extending the metaphor, if there is a directed chain of nodes, one node is an **ancestor** of another if it appears earlier in the chain, whereas a node is a **descendant** of another node if it comes later in the chain. In our example, the *Cancer* node has two parents, *Pollution* and *Smoker*, while *Smoker* is an ancestor of both *X-ray* and *Dyspnoea*. Similarly, *X-ray* is a child of *Cancer* and descendant of *Smoker* and *Pollution*. The set of parent nodes of a node X is given by *Parents(X)*.

Another useful concept is that of the **Markov blanket** of a node, which consists of the node's parents, its children, and its children's parents. Other terminology commonly used comes from the "tree" analogy (even though Bayesian networks in general are graphs rather than trees): any node without parents is called a **root** node, while any node without children is called a **leaf** node. Any other node (non-leaf and non-root) is called an **intermediate node**. Given a causal understanding of the BN structure, this means that root nodes represent original causes, while leaf nodes represent final effects. In our cancer example, the causes *Pollution* and *Smoker* are root nodes, while the effects *X-ray* and *Dyspnoea* are leaf nodes.

By convention, for easier visual examination of BN structure, networks are usually laid out so that the arcs generally point from top to bottom. This means that the BN "tree" is usually depicted upside down, with roots at the top and leaves at the bottom![3]

2.2.3   Conditional probabilities

Once the topology of the BN is specified, the next step is to quantify the relationships between connected nodes – this is done by specifying a conditional probability distribution for each node. As we are only considering discrete variables at this stage, this takes the form of a conditional probability *table* (CPT).

First, for each node we need to look at all the possible combinations of values of those parent nodes. Each such combination is called an **instantiation** of the parent set. For each distinct instantiation of parent node values, we need to specify the probability that the child will take each of its values.

For example, consider the *Cancer* node of Figure 2.1. Its parents are *Pollution* and *Smoking* and take the possible joint values $\{< H,T >, < H,F >, < L,T >, < L,F >\}$. The conditional probability table specifies in order the probability of cancer for each of these cases to be: $< 0.05, 0.02, 0.03, 0.001 >$. Since these *are* probabilities, and must sum to one over all possible states of the *Cancer* variable, the probability of no cancer is already implicitly given as one minus the above probabilities in each case; i.e., the probability of no cancer in the four possible parent instantiations is $< 0.95, 0.98, 0.97, 0.999 >$.

Root nodes also have an associated CPT, although it is degenerate, containing only one row representing its prior probabilities. In our example, the prior for a patient being a smoker is given as 0.3, indicating that 30% of the population that the

---

[3]Oddly, this is the antipodean standard in computer science; we'll let you decide what that may mean about computer scientists!

doctor sees are smokers, while 90% of the population are exposed to only low levels of pollution.

Clearly, if a node has many parents or if the parents can take a large number of values, the CPT can get very large! The size of the CPT is, in fact, exponential in the number of parents. Thus, for Boolean networks a variable with n parents requires a CPT with 2^{n+1} probabilities.

2.2.4 The Markov property

In general, modeling with Bayesian networks requires the assumption of the **Markov property**: there are no direct dependencies in the system being modeled which are not already explicitly shown via arcs. In our *Cancer* case, for example, there is no way for smoking to influence dyspnoea except by way of causing cancer (or not) — there is no hidden "backdoor" from smoking to dyspnoea. Bayesian networks which have the Markov property are also called **Independence-maps** (or, **I-maps** for short), since every independence suggested by the lack of an arc is real in the system.

Whereas the independencies suggested by a lack of arcs are generally required to exist in the system being modeled, it is not generally required that the arcs in a BN correspond to real dependencies in the system. The CPTs may be parameterized in such a way as to nullify any dependence. Thus, for example, every fully-connected Bayesian network can represent, perhaps in a wasteful fashion, any joint probability distribution over the variables being modeled. Of course, we shall prefer **minimal models** and, in particular, **minimal I-maps**, which are I-maps such that the deletion of any arc violates I-mapness by implying a non-existent independence in the system.

If, in fact, every arc in a BN happens to correspond to a direct dependence in the system, then the BN is said to be a **Dependence-map** (or, **D-map** for short). A BN which is both an I-map and a D-map is said to be a **perfect map**.

2.3 Reasoning with Bayesian networks

Now that we know how a domain and its uncertainty may be represented in a Bayesian network, we will look at how to use the Bayesian network to reason about the domain. In particular, when we observe the value of some variable, we would like to **condition** upon the new information. The process of conditioning (also called **probability propagation** or **inference** or **belief updating**) is performed via a "flow of information" through the network. Note that this information flow is *not* limited to the directions of the arcs. In our probabilistic system, this becomes the task of computing the posterior probability distribution for a set of **query** nodes, given values for some **evidence** (or **observation**) nodes.

2.3.1   Types of reasoning

Bayesian networks provide full representations of probability distributions over their variables. That implies that they can be conditioned upon any subset of their variables, supporting any direction of reasoning.

For example, one can perform **diagnostic reasoning**, i.e., reasoning from symptoms to cause, such as when a doctor observes *Dyspnoea* and then updates his belief about *Cancer* and whether the patient is a *Smoker*. Note that this reasoning occurs in the *opposite* direction to the network arcs.

Or again, one can perform **predictive reasoning**, reasoning from new information about causes to new beliefs about effects, following the directions of the network arcs. For example, the patient may tell his physician that he is a smoker; even before any symptoms have been assessed, the physician knows this will increase the chances of the patient having cancer. It will also change the physician's expectations that the patient will exhibit other symptoms, such as shortness of breath or having a positive X-ray result.

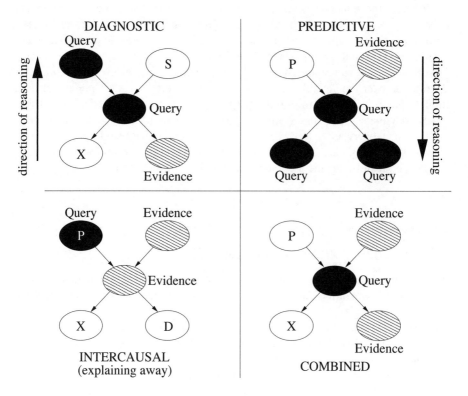

FIGURE 2.2: Types of reasoning.

A further form of reasoning involves reasoning about the mutual causes of a common effect; this has been called **intercausal reasoning**. A particular type called

explaining away is of some interest. Suppose that there are exactly two possible causes of a particular effect, represented by a **v-structure** in the BN. This situation occurs in our model of Figure 2.1 with the causes *Smoker* and *Pollution* which have a common effect, *Cancer* (of course, reality is more complex than our example!). Initially, according to the model, these two causes are independent of each other; that is, a patient smoking (or not) does not change the probability of the patient being subject to pollution. Suppose, however, that we learn that Mr. Smith has cancer. This will raise our probability for both possible causes of cancer, increasing the chances both that he is a smoker and that he has been exposed to pollution. Suppose then that we discover that he is a smoker. This new information explains the observed cancer, which in turn *lowers* the probability that he has been exposed to high levels of pollution. So, even though the two causes are initially independent, with knowledge of the effect the presence of one explanatory cause renders an alternative cause less likely. In other words, the alternative cause has been *explained away*.

Since any nodes may be query nodes and any may be evidence nodes, sometimes the reasoning does not fit neatly into one of the types described above. Indeed, we can combine the above types of reasoning in any way. Figure 2.2 shows the different varieties of reasoning using the Cancer BN. Note that the last combination shows the simultaneous use of diagnostic and predictive reasoning.

2.3.2   Types of evidence

So Bayesian networks can be used for calculating new beliefs when new information – which we have been calling **evidence** – is available. In our examples to date, we have considered evidence as a definite finding that a node X has a particular value, x, which we write as $X = x$. This is sometimes referred to as **specific evidence**. For example, suppose we discover the patient is a smoker, then *Smoker=T*, which is specific evidence.

However, sometimes evidence is available that is not so definite. The evidence might be that a node Y has the value y_1 *or* y_2 (implying that all other values are impossible). Or the evidence might be that Y is *not* in state y_1 (but may take any of its other values); this is sometimes called a **negative evidence**.

In fact, the new information might simply be any new probability distribution over Y. Suppose, for example, that the radiologist who has taken and analyzed the X-ray in our cancer example is uncertain. He thinks that the X-ray looks positive, but is only 80% sure. Such information can be incorporated equivalently to Jeffrey conditionalization of §1.5.1, in which case it would correspond to adopting a new posterior distribution for the node in question. In Bayesian networks this is also known as **virtual evidence**. Since it is handled via likelihood information, it is also known as **likelihood evidence**. We defer further discussion of virtual evidence until Chapter 3, where we can explain it through the effect on belief updating.

2.3.3   Reasoning with numbers

Now that we have described qualitatively the types of reasoning that are possible using BNs, and types of evidence, let's look at the actual numbers. Even before we obtain any evidence, we can compute a prior belief for the value of each node; this is the node's prior probability distribution. We will use the notation Bel(X) for the posterior probability distribution over a variable X, to distinguish it from the prior and conditional probability distributions (i.e., $P(X)$, $P(X|Y)$).

The exact numbers for the updated beliefs for each of the reasoning cases described above are given in Table 2.2. The first set are for the priors and conditional probabilities originally specified in Figure 2.1. The second set of numbers shows what happens if the smoking rate in the population increases from 30% to 50%, as represented by a change in the prior for the *Smoker* node. Note that, since the two cases differ only in the prior probability of smoking ($P(S = T) = 0.3$ versus $P(S = T) = 0.5$), when the evidence itself is about the patient being a smoker, then the prior becomes irrelevant and both networks give the same numbers.

TABLE 2.2
Updated beliefs given new information with smoking rate 0.3 (top set) and 0.5 (bottom set).

Node P(S)=0.3	No Evidence	Reasoning Case				
		Diagnostic D=T	Predictive S=T	Intercausal C=T	C=T S=T	Combined D=T S=T
Bel(P=high)	0.100	0.102	0.100	0.249	0.156	0.102
Bel(S=T)	0.300	0.307	1	0.825	1	1
Bel(C=T)	0.011	0.025	0.032	1	1	0.067
Bel(X=pos)	0.208	0.217	0.222	0.900	0.900	0.247
Bel(D=T)	0.304	1	0.311	0.650	0.650	1
P(S)=0.5						
Bel(P=high)	0.100	0.102	0.100	0.201	0.156	0.102
Bel(S=T)	0.500	0.508	1	0.917	1	1
Bel(C=T)	0.174	0.037	0.032	1	1	0.067
Bel(X=pos)	0.212	0.226	0.311	0.900	0.900	0.247
Bel(D=T)	0.306	1	0.222	0.650	0.650	1

Belief updating can be done using a number of exact and approximate inference algorithms. We give details of these algorithms in Chapter 3, with particular emphasis on how choosing different algorithms can affect the efficiency of both the knowledge engineering process and the automated reasoning in the deployed system. However, most existing BN software packages use essentially the same algorithm and it is quite possible to build and use BNs without knowing the details of the belief updating algorithms.

2.4   Understanding Bayesian networks

We now consider how to interpret the information encoded in a BN — the probabilistic **semantics** of Bayesian networks.

2.4.1   Representing the joint probability distribution

Most commonly, BNs are considered to be representations of joint probability distributions. There is a fundamental assumption that there is a useful underlying structure to the problem being modeled that can be captured with a BN, i.e., that not every node is connected to every other node. If such domain structure exists, a BN gives a more compact representation than simply describing the probability of every joint instantiation of all variables. **Sparse** Bayesian networks (those with relatively few arcs, which means few parents for each node) represent probability distributions in a computationally tractable way.

Consider a BN containing the n nodes, X_1 to X_n, taken in that order. A particular value in the joint distribution is represented by $P(X_1 = x_1, X_2 = x_2, \ldots, X_n = x_n)$, or more compactly, $P(x_1, x_2, \ldots, x_n)$. The **chain rule** of probability theory allows us to factorize joint probabilities so:

$$
\begin{aligned}
P(x_1, x_2, \ldots, x_n) &= P(x_1) \times P(x_2|x_1) \ldots, \times P(x_n|x_1, \ldots, x_{n-1}) \\
&= \prod_i P(x_i|x_1, \ldots, x_{i-1})
\end{aligned}
\tag{2.1}
$$

Recalling from §2.2.4 that the structure of a BN implies that the value of a particular node is conditional *only* on the values of its parent nodes, this reduces to

$$
P(x_1, x_2, \ldots, x_n) = \prod_i P(x_i|Parents(X_i))
$$

provided $Parents(X_i) \subseteq \{X_1, \ldots, X_{i-1}\}$. For example, by examining Figure 2.1, we can simplify its joint probability expressions. E.g.,

$$
\begin{aligned}
P(X = pos &\wedge D = T \wedge C = T \wedge P = low \wedge S = F) \\
&= P(X = pos|D = T, C = T, P = low, S = F) \\
&\quad \times P(D = T|C = T, P = low, S = F) \\
&\quad \times P(C = T|P = low, S = F)P(P = low|S = F)P(S = F) \\
&= P(X = pos|C = T)P(D = T|C = T)P(C = T|P = low, S = F) \\
&\quad \times P(P = low)P(S = F)
\end{aligned}
$$

2.4.2   Pearl's network construction algorithm

The condition that $Parents(X_i) \subseteq \{X_1, \ldots, X_{i-1}\}$ allows us to construct a network from a given ordering of nodes using Pearl's network construction algorithm (1988, section 3.3). Furthermore, the resultant network will be a unique minimal I-map, assuming the probability distribution is positive. The construction algorithm (Algorithm 2.1) simply processes each node in order, adding it to the existing network and adding arcs from a minimal set of parents such that the parent set renders the current node conditionally independent of every other node preceding it.

Algorithm 2.1 Pearl's Network Construction Algorithm

1. *Choose the set of relevant variables $\{X_i\}$ that describe the domain.*
2. *Choose an ordering for the variables, $< X_1, \ldots, X_n >$.*
3. *While there are variables left:*
   (a) *Add the next variable X_i to the network.*
   (b) *Add arcs to the X_i node from some minimal set of nodes already in the net, $Parents(X_i)$, such that the following conditional independence property is satisfied:*

   $$P(X_i | X'_1, \ldots, X'_m) = P(X_i | Parents(X_i))$$

   where $X'_1, \ldots, X'_m$ are all the variables preceding X_i.
   (c) *Define the CPT for X_i.*

2.4.3 Compactness and node ordering

Using this construction algorithm, it is clear that a different node order may result in a different network structure, with both nevertheless representing the same joint probability distribution.

In our example, several different orderings will give the original network structure: *Pollution* and *Smoker* must be added first, but in either order, then *Cancer*, and then *Dyspnoea* and *X-ray*, again in either order.

On the other hand, if we add the symptoms first, we will get a markedly different network. Consider the order $< D, X, C, P, S >$. D is now the new root node. When adding X, we must consider "Is *X-ray* independent of *Dyspnoea*?" Since they have a common cause in *Cancer*, they will be dependent: learning the presence of one symptom, for example, raises the probability of the other being present. Hence, we have to add an arc from D to X. When adding *Cancer*, we note that *Cancer* is directly dependent upon both *Dyspnoea* and *X-ray*, so we must add arcs from both. For *Pollution*, an arc is required from C to P to carry the direct dependency. When the final node, *Smoker*, is added, not only is an arc required from C to S, but another from P to S. In our story S and P are independent, but in the new network, without this final arc, P and S are made dependent by having a common cause, so that effect must be counterbalanced by an additional arc. The result is two additional arcs and three new probability values associated with them, as shown in Figure 2.3(a). Given the order $< D, X, P, S, C >$, we get Figure 2.3(b), which is fully connected and requires as many CPT entries as a brute force specification of the full joint distribution! In such cases, the use of Bayesian networks offers no representational, or computational, advantage.

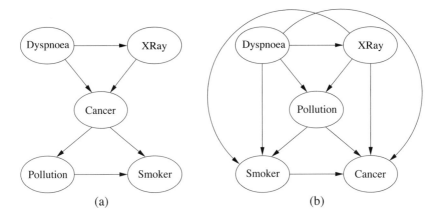

FIGURE 2.3: Alternative structures obtained using Pearl's network construction algorithm with orderings: (a) $< D,X,C,P,S >$; (b) $< D,X,P,S,C >$.

It is desirable to build the most compact BN possible, for three reasons. First, the more compact the model, the more tractable it is. It will have fewer probability values requiring specification; it will occupy less computer memory; probability updates will be more computationally efficient. Second, overly dense networks fail to represent independencies explicitly. And third, overly dense networks fail to represent the *causal* dependencies in the domain. We will discuss these last two points just below.

We can see from the examples that the compactness of the BN depends on getting the node ordering "right." The optimal order is to add the root causes first, then the variable(s) they influence directly, and continue until leaves are reached.[4] To understand *why*, we need to consider the relation between probabilistic and causal dependence.

2.4.4 Conditional independence

Bayesian networks which satisfy the Markov property (and so are I-maps) explicitly express conditional independencies in probability distributions. The relation between conditional independence and Bayesian network structure is important for understanding how BNs work.

2.4.4.1 Causal chains

Consider a causal chain of three nodes, where A causes B which in turn causes C, as shown in Figure 2.4(a). In our medical diagnosis example, one such causal chain is "smoking causes cancer which causes dyspnoea." Causal chains give rise to condi-

---

[4]Of course, one may not know the causal order of variables. In that case the automated discovery methods discussed in Part II may be helpful.

tional independence, such as for Figure 2.4(a):

$$P(C|A \wedge B) = P(C|B)$$

This means that the probability of C, given B, is exactly the same as the probability of C, given both B *and* A. Knowing that A has occurred doesn't make any difference to our beliefs about C *if we already know that B has occurred*. We also write this conditional independence as: $A \perp\!\!\!\perp C|B$.

In Figure 2.1(a), the probability that someone has dyspnoea depends directly only on whether they have cancer. If we don't know whether some woman has cancer, but we do find out she is a smoker, that would increase our belief both that she has cancer and that she suffers from shortness of breath. However, if we already *knew* she had cancer, then her smoking wouldn't make any difference to the probability of dyspnoea. That is, dyspnoea is conditionally independent of being a smoker *given* the patient has cancer.

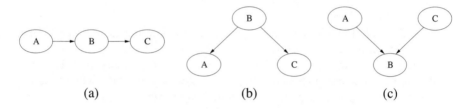

FIGURE 2.4: (a) Causal chain; (b) common cause; (c) common effect.

2.4.4.2 Common causes

Two variables A and C having a common cause B is represented in Figure 2.4(b). In our example, cancer is a common cause of the two symptoms, a positive X-ray result and dyspnoea. Common causes (or common ancestors) give rise to the same conditional independence structure as chains:

$$P(C|A \wedge B) = P(C|B) \equiv A \perp\!\!\!\perp C|B$$

If there is no evidence or information about cancer, then learning that one symptom is present will increase the chances of cancer which in turn will increase the probability of the other symptom. However, if we already know about cancer, then an additional positive X-ray won't tell us anything new about the chances of dyspnoea.

2.4.4.3 Common effects

A common effect is represented by a network v-structure, as in Figure 2.4(c). This represents the situation where a node (the effect) has two causes. Common effects (or their descendants) produce the exact opposite conditional independence structure to that of chains and common causes. That is, the parents are marginally independent ($A \perp\!\!\!\perp C$), but become dependent given information about the common effect (i.e., they are **conditionally dependent**):

$$P(A|C \wedge B) \neq P(A|B) \equiv A \not\!\perp\!\!\!\perp C|B$$

Thus, if we observe the effect (e.g., cancer), and then, say, we find out that one of the causes is absent (e.g., the patient does not smoke), this *raises* the probability of the other cause (e.g., that he lives in a polluted area) — which is just the inverse of explaining away.

Compactness again

So we can now see *why* building networks with an order violating causal order can, and generally will, lead to additional complexity in the form of extra arcs. Consider just the subnetwork { *Pollution, Smoker, Cancer* } of Figure 2.1. If we build the sub-network in that order we get the simple v-structure *Pollution* → *Smoker* ← *Cancer*. However, if we build it in the order < *Cancer, Pollution, Smoker* >, we will first get *Cancer* → *Pollution*, because they are dependent. When we add *Smoker*, it will be dependent upon *Cancer*, because in reality there is a direct dependency there. But we shall also have to add a spurious arc to *Pollution*, because otherwise *Cancer* will act as a common cause, inducing a spurious dependency between *Smoker* and *Pollution*; the extra arc is necessary to reestablish marginal independence between the two.

2.4.5   d-separation

We have seen how Bayesian networks represent conditional independencies and how these independencies affect belief change during updating. The conditional independence in $A \perp\!\!\!\perp C|B$ means that knowing the value of B **blocks** information about C being relevant to A, and vice versa. Or, in the case of Figure 2.4(c), *lack* of information about B blocks the relevance of C to A, whereas learning about B **activates** the relation between C and A.

These concepts apply not only between pairs of nodes, but also between sets of nodes. More generally, given the Markov property, it is possible to determine whether a set of nodes **X** is independent of another set **Y**, given a set of evidence nodes **E**. To do this, we introduce the notion of **d-separation** (from **direction-dependent separation**).

Definition 2.1 Path (Undirected Path) *A* **path** *between two sets of nodes* **X** *and* **Y** *is any sequence of nodes between a member of* **X** *and a member of* **Y** *such that every adjacent pair of nodes is connected by an arc (regardless of direction) and no node appears in the sequence twice.*

Definition 2.2 Blocked path *A path is* **blocked**, *given a set of nodes* **E**, *if there is a node Z on the path for which at least one of three conditions holds:*

1. *Z is in* **E** *and Z has one arc on the path leading in and one arc out (chain).*

2. *Z is in* **E** *and Z has both path arcs leading out (common cause).*

3. *Neither Z nor any descendant of Z is in* **E**, *and both path arcs lead in to Z (common effect).*

Definition 2.3 d-separation *A set of nodes* **E** **d-separates** *two other sets of nodes* **X** *and* **Y** (**X** ⊥**Y** | **E**) *if every path from a node in* **X** *to a node in* **Y** *is* **blocked** *given* **E**.

If **X** and **Y** are **d-separated** by **E**, then **X** and **Y** are **conditionally independent** given **E** (given the Markov property). Examples of these three blocking situations are shown in Figure 2.5. Note that we have simplified by using single nodes rather than sets of nodes; also note that the evidence nodes **E** are shaded.

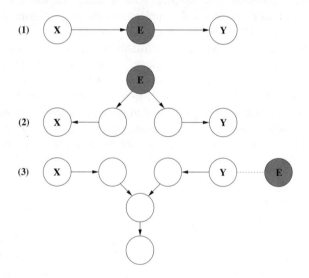

FIGURE 2.5: Examples of the three types of situations in which the path from X to Y can be blocked, given evidence E. In each case, X and Y are **d-separated** by E.

Consider d-separation in our cancer diagnosis example of Figure 2.1. Suppose an observation of the Cancer node is our evidence. Then:

1. P is d-separated from X and D. Likewise, S is d-separated from X and D (blocking condition 1).

2. While X is d-separated from D (condition 2).

3. However, if C had not been observed (and also not X or D), then S would have been d-separated from P (condition 3).

Definition 2.4 d-connection *Sets* **X** *and* **Y** *are* **d-connected** *given set* **E** (**X** ⊥̸**Y** | **E**) *if there is a path from a node in* **X** *to a node in* **Y** *which is not* **blocked** *given* **E**.

2.5 More examples

In this section we present further simple examples of BN modeling from the literature. We encourage the reader to work through these examples using BN software (see Appendix B).

2.5.1 Earthquake

Example statement: *You have a new burglar alarm installed. It reliably detects burglary, but also responds to minor earthquakes. Two neighbors, John and Mary, promise to call the police when they hear the alarm. John always calls when he hears the alarm, but sometimes confuses the alarm with the phone ringing and calls then also. On the other hand, Mary likes loud music and sometimes doesn't hear the alarm. Given evidence about who has and hasn't called, you'd like to estimate the probability of a burglary (from Pearl (1988)).*

A BN representation of this example is shown in Figure 2.6. All the nodes in this BN are Boolean, representing the true/false alternatives for the corresponding propositions. This BN models the assumptions that John and Mary do not perceive a burglary directly and they do not feel minor earthquakes. There is no explicit representation of loud music preventing Mary from hearing the alarm, nor of John's confusion of alarms and telephones; this information is summarized in the probabilities in the arcs from *Alarm* to *JohnCalls* and *MaryCalls*.

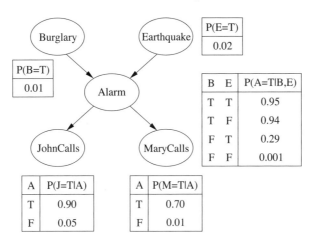

FIGURE 2.6: Pearl's Earthquake BN.

2.5.2 Metastatic cancer

Example statement: *Metastatic cancer is a possible cause of brain tumors and is also an explanation for increased total serum calcium. In turn, either of these could explain a patient falling into a coma. Severe headache is also associated with brain tumors. (This example has a long history in the literature, namely Cooper, 1984, Pearl, 1988, Spiegelhalter, 1986.)*

A BN representation of this metastatic cancer example is shown in Figure 2.7. All the nodes are Booleans. Note that this is a *graph*, not a tree, in that there is more than one path between the two nodes *M* and *C* (via *S* and *B*).

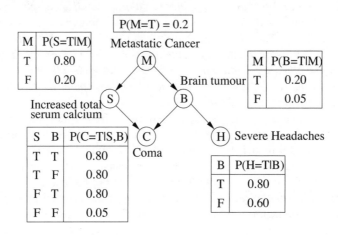

FIGURE 2.7: Metastatic cancer BN.

2.5.3 Asia

Example Statement: *Suppose that we wanted to expand our original medical diagnosis example to represent explicitly some other possible causes of shortness of breath, namely tuberculosis and bronchitis. Suppose also that whether the patient has recently visited Asia is also relevant, since TB is more prevalent there.*

Two alternative BN structures for the so-called Asia example are shown in Figure 2.8. In both networks all the nodes are Boolean. The left-hand network is based on the Asia network of Lauritzen and Spiegelhalter (1988). Note the slightly odd intermediate node *TBorC*, indicating that the patient has either tuberculosis or bronchitis. This node is not strictly necessary; however it reduces the number of arcs elsewhere, by summarizing the similarities between TB and lung cancer in terms of their relationship to positive X-ray results and dyspnoea. Without this node, as can be seen on the right, there are two parents for *X-ray* and three for *Dyspnoea*, with the same probabilities repeated in different parts of the CPT. The use of such an intermediate node is an example of "divorcing," a model structuring method described in §10.3.6.

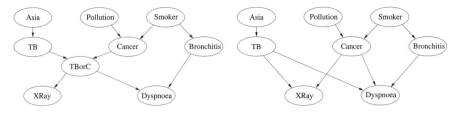

FIGURE 2.8: Alternative BNs for the "Asia" example.

2.6 Summary

Bayes' theorem allows us to update the probabilities of variables whose state has not been observed given some set of new observations. Bayesian networks automate this process, allowing reasoning to proceed in any direction across the network of variables. They do this by combining qualitative information about direct dependencies (perhaps causal relations) in arcs and quantitative information about the strengths of those dependencies in conditional probability distributions. Computational speed gains in updating accrue when the network is sparse, allowing d-separation to take advantage of conditional independencies in the domain (so long as the Markov property holds). Given a known set of conditional independencies, Pearl's network construction algorithm guarantees the development of a minimal network, without redundant arcs. In the next chapter, we turn to specifics about the algorithms used to update Bayesian networks.

2.7 Bibliographic notes

The text that marked the new era of Bayesian methods in artificial intelligence is Judea Pearl's *Probabilistic Reasoning in Intelligent Systems* (1988). This text played no small part in attracting the authors to the field, amongst many others. Richard Neapolitan's *Probabilistic Reasoning in Expert Systems* (1990) complements Pearl's book nicely, and it lays out the algorithms underlying the technology particularly well. Two more current introductions are Jensen and Nielsen's *Bayesian Networks and Decision Graphs* (2007), Kjærulff and Madsen's *Bayesian Networks and Influence Diagrams: A Guide to Construction and Analysis* (2008); both their level and treatment is similar to ours; however, they do not go as far with the machine learning and knowledge engineering issues we treat later. More technical discussions can be found in Cowell et al.'s *Probabilistic Networks and Expert Systems* (1999), Richard Neapolitan's *Learning Bayesian Networks* (2003) and Koller and Friedman's *Probabilistic Graphical Models: Principles and Techniques* (2009).

A Quick Guide to Using BayesiaLab

Installation: Web Site www.bayesia.com. Download BayesiaLab zip file which is available for all platforms that support the Sun Java Runtime Environment (JRE) (Windows, Mac OS X, and Linux). This gives you a BayesiaLab.zip. Extract the contents of the zip file, to your computer's file system. The Sun Java Runtime environment is required to run BayesiaLab. The Sun JRE can be downloaded from the Sun Java Web site: java.com To run BayesiaLab, navigate to the installation directory on the command line, and run java -Xms128M -Xmx512M -jar BayesiaLab.jar

Network Files: BNs are stored in .xbl files, with icon ⬚. BayesiaLab comes with a Graphs folder of example networks. To open an existing network, select ⬚ or select Network→Open menu option.

Evidence: Evidence can only be added and removed in "Validation Mode". To enter this mode either click on the ☰ icon or click View→Validation Mode in the main menu.

To add evidence:

1. Double-click on the node for which you want to add evidence.
2. A "monitor" for the node will appear in the list in the right-hand portion of the BayesiaLab window. In the node's monitor, double-click on the variable, for which you would like to add evidence.

To remove evidence:

- In the node's monitor, double-click on the variable, for which you would like to remove evidence; or

- Click on ✎ to remove all evidence (called "observations" in BayesiaLab).

Editing/Creating a BN: BNs can only be created or edited in "Modeling Mode".

To enter this mode either click on the ⬚ icon or click View→Modeling Mode in the main menu. Note that BayesiaLab beliefs are given out of 100, not as direct probabilities (i.e. not numbers between 0 and 1).

- Add a node by selecting ⬤ and then left-clicking, onto the canvas where you want to place the node.

- Add an arc by selecting ⬚, then dragging the arc from the parent node to the child node.

- Double click on node, then click on the Probability Distribution tab to bring up the CPT. Entries can be added or changed by clicking on the particular cells.

Saving a BN: Select ⬚ or the Network→Save menu option.

FIGURE 2.9: A quick guide to using BayesiaLab.

A Quick Guide to Using GeNIe

Installation: Web Site www.genie.sis.pitt.edu. Download GeNIe which is available for Windows. This gives you a genie2_setup.exe, an installer executable. Double-clicking the executable, will start the installation wizard.

Network Files: BNs are stored in .xdsl files, with icon ![icon]. GeNIe comes with an Examples folder of example networks. To open an existing network, select ![folder icon] or select File→Open Network menu option, or double-click on the file.

Compilation: Once a GeNIe BN has been opened, before you can see the initial beliefs, you must first compile it:

- Click on ![lightning icon] ; or
- Select Network→Update Beliefs menu option.

Once the network is compiled, you can view the state of each node by hovering over the node's tick icon (![tick icon]), with your mouse.

Evidence: To add evidence:

- Left click on the node, and select Node→Set Evidence in GeNIe's menu system; or
- Right click on the node, and select Set Evidence in the right-click menu

To remove evidence:

- Right click on the node and select Clear Evidence; or
- Select Network→Clear All Evidence menu-option.

There is an option (Network→Update Immediately) to automatically recompile and update beliefs when new evidence is set.

Editing/Creating a BN: Double-clicking on a node will bring up a window showing node features.

- Add a node by selecting ![circle icon] and then "drag-and-drop" with the mouse, onto the canvas, or right-clicking on the canvas and then selecting Insert Here→Chance from the menu.

- Add an arc by selecting ![arrow icon] , then left-click first on the parent node, then the child node.
- Double click on node, then click on the Definition tab to bring up the CPT. Entries can be added or changed by clicking on the particular cells.

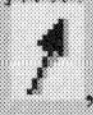

Saving a BN: Select ![save icon] or the File→Save menu option.

FIGURE 2.10: A quick guide to using GeNIe.

A Quick Guide to Using Hugin

Installation: Web Site www.hugin.com. Download Hugin Lite, which is available for MS Windows (95 / 98 / NT4 / 2000 / XP), Solaris Sparc, Solaris X86 and Linux. This gives you HuginLite63.exe, a self-extracting zip archive. Double-clicking will start the extraction process.

Network Files: BNs are stored in .net files, with icon 🔲. Hugin comes with a samples folder of example networks. To open an existing network, select 📂, or select File→Open menu option, or double-click on the file.

Compilation: Once a Hugin BN has been opened, before you can see the initial beliefs or add evidence, you must first compile it (which they call "switch to run mode"): click on 📂, or select Network→Run(in edit mode), or Recompile (in run mode) menu option.

This causes another window to appear on the left side of the display (called the Node Pane List), showing the network name, and all the node names. You can display/hide the states and beliefs in several ways. You can select a particular node by clicking on the '+' by the node name, or all nodes with View→Expand Node List, or using icon 📂. Unselecting is done similarly with '-', or View→Collapse Node List, or using icon 📂.

Selecting a node means all its states will be displayed, together with a bar and numbers showing the beliefs. Note that Hugin beliefs are given as percentages out of 100, not as direct probabilities (i.e., not numbers between 0 and 1).

Editing/Creating a BN: You can only change a BN when you are in "edit" mode, which you can enter by selecting the edit mode icon 🔲, or selecting Network→Edit. Double-clicking on a node will bring up a window showing node features, or use icon 📂.

- Add a node by selecting either 📂 (for discrete node) or 📂 (for continuous node), Edit→Discrete Chance Tool or Edit→Continuous Chance Tool. In each case, you then "drag-and-drop" with the mouse.
- Add an arc by selecting either 📂, or Edit→Link Tool, then left-click first on the parent node, then the child node.
- Click on the 🔲, icon to split the window horizontally between a Tables Pane (above), showing the CPT of the currently selected node, and the network structure (below).

Saving a BN: Select 📂, or the File→Save menu option. Note that the Hugin Lite demonstration version limits you to networks with up to 50 nodes and learn from maximum 500 cases; for larger networks, you need to buy a license.

Junction trees: To change the triangulation method select Network→Network Properties→Compilation, then turn on "Specify Triangulation Method." To view, select the Show Junction Tree option.

FIGURE 2.11: A quick guide to using Hugin.

A Quick Guide to Using Netica

Installation: Web Site `www.norsys.com`. Download Netica, which is available for MS Windows (95 / 98 / NT4 / 2000 / XP / Vista), and MacIntosh OSX. This gives you `Netica_Win.exe`, a self-extracting zip archive. Double-clicking will start the extraction process.

Network Files: BNs are stored in `.dne` files, with icon ■. Netica comes with a folder of example networks, plus a folder of tutorial examples. To open an existing network:

- Select 🗁
- Select `File→Open` menu option; or
- Double-click on the BN `.dne` file.

Compilation: Once a Netica BN has been opened, before you can see the initial beliefs or add evidence, you must first compile it:

- Click on 🗁; or
- Select `Network→Compile` menu option.

Once the network is compiled, numbers and bars will appear for each node state. Note that Netica beliefs are given out of 100, not as direct probabilities (i.e., not numbers between 0 and 1).

Evidence: To add evidence:

- Left-click on the node state name; or
- Right-click on node and select particular state name.

To remove evidence:

- Right-click on node and select `unknown`; or
- Select 🗷 ; or
- Select `Network→Remove findings` menu option.

There is an option (`Network→Automatic Update`) to automatically re-compile and update beliefs when new evidence is set.

Editing/Creating a BN: Double-clicking on a node will bring up a window showing node features.

- Add a node by selecting either 🔲; or `Modify→Add nature node`, then "drag-and-drop" with the mouse.
- Add an arc by selecting either 🔲; or `Modify→Add link`, then left-click first on the parent node, then the child node.
- Double-click on node, then click on the `Table` button to bring up the CPT. Entries can be added or changed by clicking on the particular cells.

Saving a BN: Select 🖬 or the `File→Save` menu option. Note that the Netica Demonstration version only allows you to save networks with up to 15 nodes. For larger networks, you need to buy a license.

FIGURE 2.12: A quick guide to using Netica.

2.8 Problems

Modeling

These modeling exercises should be done using a BN software package (see our **Quick Guides to Using Netica** in Figure 2.12, **Hugin** in Figure 2.11, **GeNIe** in Figure 2.10, or **BayesiaLab** in Figure 2.9, and also Appendix B).

Also note that various information, including Bayesian network examples in Netica's .dne format, can be found at the book Web site:

        http://www.csse.monash.edu.au/bai

Problem 1

Construct a network in which explaining away operates, for example, incorporating multiple diseases sharing a symptom. Operate and demonstrate the effect of explaining away. *Must* one cause explain away the other? Or, can the network be parameterized so that this doesn't happen?

Problem 2

"Fred's LISP dilemma." *Fred is debugging a LISP program. He just typed an expression to the LISP interpreter and now it will not respond to any further typing. He can't see the visual prompt that usually indicates the interpreter is waiting for further input. As far as Fred knows, there are* only *two situations that could cause the LISP interpreter to stop running: (1) there are problems with the computer hardware; (2) there is a bug in Fred's code. Fred is also running an editor in which he is writing and editing his LISP code; if the hardware is functioning properly, then the text editor should still be running. And if the editor is running, the editor's cursor should be flashing. Additional information is that the hardware is pretty reliable, and is OK about 99% of the time, whereas Fred's LISP code is often buggy, say 40% of the time.*[5]

1. Construct a Belief Network to represent and draw inferences about Fred's dilemma.

   First decide what your domain variables are; these will be your network nodes. Hint: 5 or 6 Boolean variables should be sufficient. Then decide what the causal relationships are between the domain variables and add directed arcs in the network from cause to effect. Finanly, you have to add the conditional probabilities for nodes that have parents, and the prior probabilities for nodes without parents. Use the information about the hardware reliability and how often Fred's code is buggy. Other probabilities haven't been given to you explicitly; choose values that seem reasonable and explain why in your documentation.

---

[5]Based on an example used in Dean, T., Allen, J. and Aloimonos, Y. *Artificial Intelligence Theory and Practice* (Chapter 8), Benjamin/Cumming Publishers, Redwood City, CA. 1995.

2. Show the belief of each variable before adding any evidence, i.e., about the LISP visual prompt not being displayed.

3. Add the evidence about the LISP visual prompt not being displayed. After doing belief updating on the network, what is Fred's belief that he has a bug in his code?

4. Suppose that Fred checks the screen and the editor's cursor is still flashing. What effect does this have on his belief that the LISP interpreter is misbehaving because of a bug in his code? Explain the change in terms of diagnostic and predictive reasoning.

Problem 3

"A Lecturer's Life." *Dr. Ann Nicholson spends 60% of her work time in her office. The rest of her work time is spent elsewhere. When Ann is in her office, half the time her light is off (when she is trying to hide from students and get research done). When she is not in her office, she leaves her light on only 5% of the time. 80% of the time she is in her office, Ann is logged onto the computer. Because she sometimes logs onto the computer from home, 10% of the time she is not in her office, she is still logged onto the computer.*

1. Construct a Bayesian network to represent the "Lecturer's Life" scenario just described.

2. Suppose a student checks Dr. Nicholson's login status and sees that she is logged on. What effect does this have on the student's belief that Dr. Nicholson's light is on?

Problem 4

"Jason the Juggler." *Jason, the robot juggler, drops balls quite often when its battery is low. In previous trials, it has been determined that when its battery is low it will drop the ball 9 times out of 10. On the other hand when its battery is not low, the chance that it drops a ball is much lower, about 1 in 100. The battery was recharged recently, so there is only a 5% chance that the battery is low. Another robot, Olga the observer, reports on whether or not Jason has dropped the ball. Unfortunately Olga's vision system is somewhat unreliable. Based on information from Olga, the task is to represent and draw inferences about whether the battery is low depending on how well Jason is juggling.*[6]

1. Construct a Bayesian network to represent the problem.

2. Which probability tables show where the information on how Jason's success is related to the battery level, and Olga's observational accuracy, are encoded in the network?

---

[6]Variation of Exercise 19.6 in Nilsson, N.J. *Artificial Intelligence: A New Synthesis*, Copyright (1998). With permission from Elsevier.

3. Suppose that Olga reports that Jason has dropped the ball. What effect does this have on your belief that the battery is low? What type of reasoning is being done?

Problem 5

Come up with your own problem involving reasoning with evidence and uncertainty. Write down a text description of the problem, then model it using a Bayesian network. Make the problem sufficiently complex that your network has at least 8 nodes and is multiply-connected (i.e., not a tree or a polytree).

1. Show the beliefs for each node in the network before any evidence is added.
2. Which nodes are d-separated with no evidence added?
3. Which nodes in your network would be considered evidence (or observation) nodes? Which might be considered the query nodes? (Obviously this depends on the domain and how you might use the network.)
4. Show how the beliefs change in a form of diagnostic reasoning when evidence about at least one of the domain variables is added. Which nodes are d-separated with this evidence added?
5. Show how the beliefs change in a form of predictive reasoning when evidence about at least one of the domain variables is added. Which nodes are d-separated with this evidence added?
6. Show how the beliefs change through "explaining away" when particular combinations of evidence are added.
7. Show how the beliefs change when you change the priors for a root node (rather than adding evidence).

Conditional Independence

Problem 6

Consider the following Bayesian network for another version of the medical diagnosis example, where *B=Bronchitis*, *S=Smoker*, *C=Cough*, *X=Positive X-ray* and *L=Lung cancer* and all nodes are Booleans.

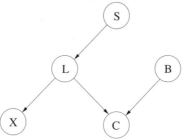

List the pairs of nodes that are conditionally independent in the following situations:

1. There is no evidence for any of the nodes.

2. The cancer node is set to true (and there is no other evidence).

3. The smoker node is set to true (and there is no other evidence).

4. The cough node is set to true (and there is no other evidence).

Variable Ordering

Problem 7

Consider the Bayesian network given for the previous problem.

1. What variable ordering(s) could have been used to produce the above network using the network construction algorithm (Algorithm 2.1)?

2. Given different variable orderings, what network structure would result from this algorithm? Use only pen and paper for now! Compare the number of parameters required by the CPTs for each network.

d-separation

Problem 8

Consider the following graph.

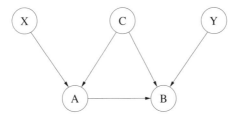

1. Find all the sets of nodes that d-separate X and Y (not including either X or Y in such sets).

2. Try to come up with a real-world scenario that might be modeled with such a network structure.

Problem 9

Design an internal representation for a Bayesian network structure; that is, a representation for the nodes and arcs of a Bayesian network (but not necessarily the parameters — prior probabilities and conditional probability tables). Implement a function which generates such a data structure from the Bayesian network described by a Netica dne input file. Use this function in the subsequent problems. (Sample dne files are available from the book Web site.)

Problem 10

Implement the network construction algorithm (Algorithm 2.1). Your program should take as input an ordered list of variables and prompt for additional input from the keyboard about the conditional independence of variables as required. It should generate a Bayesian network in the internal representation designed above. It should also print the network in some human-readable form.

Problem 11

Given as input the internal Bayesian network structure N (in the representation you have designed above), write a function which returns all undirected paths (Definition 2.1) between two sets **X** and **Y** of nodes in N.

Test your algorithm on various networks, including at least

- The d-separation network example from Problem 8, `dsepEg.dne`
- `Cancer_Neapolitan.dne`
- `ALARM.dne`

Summarize the results of these experiments.

Problem 12

Given the internal Bayesian network structure N, implement a **d-separation oracle** which, for any three sets of nodes input to it, **X**, **Y**, and **Z**, returns:

- **true** if $X \perp Y|Z$ (i.e., **Z** d-separates **X** and **Y** in N);
- **false** if $X \not\perp Y|Z$ (i.e., **X** and **Y** given **Z** are d-connected in N);
- some diagnostic (a value other than **true** or **false**) if an error in N is encountered.

Run your algorithm on a set of test networks, including at least the three network specified for Problem 11. Summarize the results of these experiments.

Problem 13

Modify your network construction algorithm from Problem 9 above to use the d-separation oracle from the last problem, instead of input from the user. Your new algorithm should produce exactly the same network as that used by the oracle whenever the variable ordering provided it is compatible with the oracle's network. Experiment with different variable orderings. Is it possible to generate a network which is simpler than the oracle's network?

3

Inference in Bayesian Networks

3.1 Introduction

The basic task for any probabilistic inference system is to compute the posterior probability distribution for a set of query nodes, given values for some evidence nodes. This task is called **belief updating** or **probabilistic inference**. Inference in Bayesian networks is very flexible, as evidence can be entered about any node while beliefs in any other nodes are updated.

In this chapter we will cover the major classes of inference algorithms — exact and approximate — that have been developed over the past 20 years. As we will see, different algorithms are suited to different network structures and performance requirements. Networks that are simple chains merely require repeated application of Bayes' Theorem. Inference in simple tree structures can be done using local computations and message passing between nodes. When pairs of nodes in the BN are connected by multiple paths the inference algorithms become more complex. For some networks, exact inference becomes computationally infeasible, in which case approximate inference algorithms must be used. In general, both exact and approximate inference are theoretically computationally complex (specifically, NP hard). In practice, the speed of inference depends on factors such as the structure of the network, including how highly connected it is, how many undirected loops there are and the locations of evidence and query nodes.

Many inference algorithms have not seen the light of day beyond the research environment that produced them. Good exact and approximate inference algorithms are implemented in BN software, so knowledge engineers do not have to. Hence, our main focus is to characterize the main algorithms' computational performance to both enhance understanding of BN modeling and help the knowledge engineer assess which algorithm is best suited to the application. It is important that the belief updating is not merely a "black-box" process, as there are knowledge engineering issues that can only be resolved through an understanding of the inference process.

We will conclude the chapter with a discussion of how to use Bayesian networks for **causal modeling**, that is for reasoning about the effect of active interventions in the causal process being represented by the network. Such reasoning is important for hypothetical or counterfactual reasoning and for planning and control applications. Unfortunately, current BN tools do not explicitly support causal modeling; however, they can be used for such reasoning and we will describe how to do so.

3.2   Exact inference in chains

3.2.1   Two node network

We begin with the very simplest case, a two node network $X \to Y$.

If there is evidence about the parent node, say $X = x$, then the posterior probability (or belief) for Y, which we denote $Bel(Y)$, can be read straight from the value in CPT, $P(Y|X = x)$.

If there is evidence about the child node, say $Y = y$, then the inference task of updating the belief for X is done using a simple application of Bayes' Theorem.

$$
\begin{aligned}
Bel(X = x) &= P(X = x|Y = y) \\
&= \frac{P(Y = y|X = x)P(X = x)}{P(Y = y)} \\
&= \alpha P(x)\lambda(x)
\end{aligned}
$$

where

$$
\alpha = \frac{1}{P(Y = y)}
$$

$P(x)$ is the prior and $\lambda(x) = P(Y = y|X = x)$ is the **likelihood**. Note that we don't need to know the prior for the evidence. Since the beliefs for all the values of X must sum to one (due to the Total Probability Theorem 1.2), we can compute α, as a **normalizing constant**.

Example: $Flu \to HighTemp$

Suppose that we have this very simple model of flu causing a high temperature, with prior $P(Flu = T) = 0.05$, and CPT values $P(HighTemp = T|Flu = T) = 0.9$, $P(HighTemp = T|Flu = F) = 0.2$. If a person has a high temperature (i.e., the evidence available is $HighTemp = T$), the computation for this diagnostic reasoning is as follows.

$$
\begin{aligned}
Bel(Flu = T) &= \alpha P(Flu = T)\lambda(Flu = T) \\
&= \alpha \times 0.05 \times 0.9 \\
&= \alpha 0.045 \\
Bel(Flu = F) &= \alpha P(Flu = F)\lambda(Flu = F) \\
&= \alpha \times 0.95 \times 0.2 \\
&= \alpha 0.19
\end{aligned}
$$

We can compute α via

$$
Bel(Flu = T) + Bel(Flu = F) = 1 = \alpha 0.045 + \alpha 0.19
$$

giving

$$
\alpha = \frac{1}{0.19 + 0.045}
$$

This allows us to finish the belief update:

$$Bel(Flu = T) = \frac{0.045}{0.19 + 0.045} = 0.1915$$

$$Bel(Flu = F) = \frac{0.19}{0.19 + 0.045} = 0.8085$$

3.2.2 Three node chain

We can apply the same method when we have three nodes in a chain, $X \rightarrow Y \rightarrow Z$.

If we have evidence about the root node, $X = x$, updating in the same direction as the arcs involves the simple application of the chain rule (Theorem 1.3), using the independencies implied in the network.

$$Bel(Z) = P(Z|X = x) = \sum_{Y=y} P(Z|Y)P(Y|X = x)$$

If we have evidence about the leaf node, $Z = z$, the diagnostic inference to obtain $Bel(X)$ is done with the application of Bayes' Theorem and the chain rule.

$$
\begin{aligned}
Bel(X = x) &= P(X = x|Z = z) \\
&= \frac{P(Z = z|X = x)P(X = x)}{P(Z = z)} \\
&= \frac{\sum_{Y=y} P(Z = z|Y = y, X = x)P(Y = y|X = x)P(X = x)}{P(Z = z)} \\
&= \frac{\sum_{Y=y} P(Z = z|Y = y)P(Y = y|X = x)P(X = x)}{P(Z = z)} \quad (Z \perp\!\!\!\perp X|Y) \\
&= \alpha P(x)\lambda(x)
\end{aligned}
$$

where

$$\lambda(x) = P(Z = z|X = x) = \sum_{Y=y} P(Z = z|Y = y)P(Y = y|X = x)$$

Example: $Flu \rightarrow HighTemp \rightarrow HighTherm$

Here our flu example is extended by having an observation node $HighTherm$ representing the reading given by a thermometer. The possible inaccuracy in the thermometer reading is represented by the following CPT entries:

- $P(HighTherm = T|HighTemp = T) = 0.95$ (5% chance of a false negative reading)
- $P(HighTherm = T|HighTemp = F) = 0.15$ (15% chance of a false positive reading).

Suppose that a high thermometer reading is taken, i.e., the evidence is $HighTherm = T$.

$$
\begin{aligned}
\lambda(Flu = T) &= P(HighTherm = T | HighTemp = T)P(HighTemp = T | Flu = T)\\
&\quad + P(HighTherm = T | HighTemp = F)P(HighTemp = F | Flu = T)\\
&= 0.95 \times 0.9 + 0.15 \times 0.1\\
&= 0.855 + 0.015 = 0.87\\
Bel(Flu = T) &= \alpha P(Flu = T)\lambda(Flu = T)\\
&= \alpha 0.05 \times 0.87 = \alpha 0.0435\\
\lambda(Flu = F) &= P(HighTherm = T | HighTemp = T)P(HighTemp = T | Flu = F)\\
&\quad + P(HighTherm = T | HighTemp = F)P(HighTemp = F | Flu = F)\\
&= 0.95 \times 0.2 + 0.15 \times 0.8\\
&= 0.19 + 0.12 = 0.31\\
Bel(Flu = F) &= \alpha P(Flu = F)\lambda(Flu = F)\\
&= \alpha 0.95 \times 0.31 = \alpha 0.2945
\end{aligned}
$$

So,

$$
\alpha = \frac{1}{0.0435 + 0.2945}
$$

hence

$$
\begin{aligned}
Bel(Flu = T) &= \frac{0.0435}{0.0435 + 0.2945} = 0.1287\\
Bel(Flu = F) &= \frac{0.2945}{0.0435 + 0.2945} = 0.8713 \; (= 1 - Bel(Flu = T))
\end{aligned}
$$

Clearly, inference in chains is straightforward, using application of Bayes' Theorem and the conditional independence assumptions represented in the chain. We will see that these are also the fundamental components of inference algorithms for more complex structures.

3.3 Exact inference in polytrees

Next we will look at how inference can be performed when the network is a simple structure called a **polytree** (or "**forest**"). Polytrees have at most one path between any pair of nodes; hence they are also referred to as **singly-connected** networks.

Assume X is the query node, and there is some set of evidence nodes **E** (not including X). The task is to update $Bel(X)$ by computing $P(X|\mathbf{E})$.

Figure 3.1 shows a diagram of a node X in a generic polytree, with all its connections to parents (the U_i), children (the Y_j), and the children's other parents (the Z_{ij}). The local belief updating for X must incorporate evidence from all other parts of the network. From this diagram we can see that evidence can be divided into:

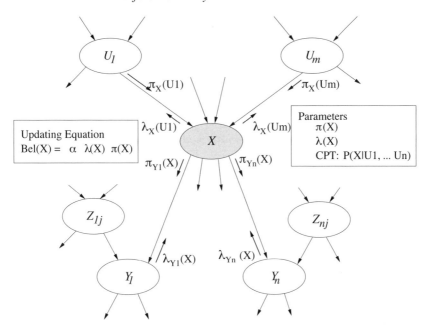

FIGURE 3.1: A generic polytree showing how local belief updating of node X is achieved through incorporation of evidence through its parents (the U_i) and children (the Y_j). Also shown are the message passing parameters and messages.

- The **predictive support** for X, from evidence nodes connected to X through its parents, $U_1, \ldots U_m$; and
- The **diagnostic support** for X, from evidence nodes connected to X through its children $Y_1, \ldots Y_n$.

3.3.1 Kim and Pearl's message passing algorithm

A "bare bones" description of Kim and Pearl's message passing algorithm appears below as Algorithm 3.1. The derivation of the major steps is beyond the scope of this text; suffice it to say, it involves the repeated application of Bayes' Theorem and use of the conditional independencies encoded in the network structure. Those details can be found elsewhere (e.g., Pearl, 1988, Neapolitan, 2003). Instead, we will present the major features of the algorithm, illustrating them by working through a simple example.

The parameters and messages used in this algorithm are also shown in Figure 3.1. The basic idea is that at each iteration of the algorithm, $Bel(X)$ is updated locally using three types of parameters — $\lambda(X)$, $\pi(X)$ and the CPTs (equation (3.1)). $\lambda(X)$ and $\pi(X)$ are computed using the π and λ messages received from X's parents and children respectively. π and λ messages are also sent out from X so that its neighbors can perform updates. Let's look at some of the computations more closely.

Algorithm 3.1 Kim and Pearl's Message Passing Algorithm

This algorithm requires the following three types of parameters to be maintained.

- *The current strength of the predictive support π contributed by each incoming link $U_i \rightarrow X$:*

$$\pi_X(U_i) = P(U_i|E_{U_i \setminus X})$$

where $E_{U_i \setminus X}$ is all evidence connected to U_i except via X.

- *The current strength of the diagnostic support λ contributed by each outgoing link $X \rightarrow Y_j$:*

$$\lambda_{Y_j}(X) = P(E_{Y_j \setminus X}|X)$$

where $E_{Y_j \setminus X}$ is all evidence connected to Y_j through its parents except via X.

- *The fixed CPT $P(X|U_i, \dots, U_n)$ (relating X to its immediate parents).*

These parameters are used to do local belief updating in the following three steps, which can be done in any order.

(Note: in this algorithm, x_i means the ith state of node X, while $u_1 \dots u_n$ is used to represent an instantiation of the parents of X, $U_1 \dots U_n$, in the situations where there is a summation of all possible instantiations.)

1. **Belief updating**.

   Belief updating for a node X is activated by messages arriving from either children or parent nodes, indicating changes in their belief parameters.

   When node X is activated, inspect $\pi_X(U_i)$ (messages from parents), $\lambda_{Y_j}(X)$ (messages from children). Apply with

   $$Bel(x_i) = \alpha \lambda(x_i)\pi(x_i) \tag{3.1}$$

   where,

   $$\lambda(x_i) = \begin{cases} 1 & \text{if evidence is } X = x_i \\ 0 & \text{if evidence is for another } x_j \\ \prod_j \lambda_{Y_j}(x_i) & \text{otherwise} \end{cases} \tag{3.2}$$

   $$\pi(x_i) = \sum_{u_1,\dots,u_n} P(x_i|u_1,\dots,u_n) \prod_i \pi_X(u_i) \tag{3.3}$$

   and α is a normalizing constant rendering $\sum_{x_i} Bel(X = x_i) = 1$.

2. **Bottom-up propagation**.

   Node X computes new λ messages to send to its parents.

   $$\lambda_X(u_i) = \sum_{x_i} \lambda(x_i) \sum_{u_k : k \neq i} P(x_i|u_1,\dots,u_n) \prod_{k \neq i} \pi_X(u_k) \tag{3.4}$$

3. **Top-down propagation**.

Node X computes new π messages to send to its children.

$$\pi_{Y_j}(x_i) = \begin{cases} 1 & \text{if evidence value } x_i \text{ is entered} \\ 0 & \text{if evidence is for another value } x_j \\ \alpha[\prod_{k \neq j} \lambda_{Y_k}(x_i)] \sum_{u_1,\ldots,u_n} P(x_i|u_1,\ldots,u_n) \prod_i \pi_X(u_i) \\ = \frac{\alpha Bel(x_i)}{\lambda_{Y_j}(x_i)} \end{cases} \quad (3.5)$$

First, equation (3.2) shows how to compute the $\lambda(x_i)$ parameter. Evidence is entered through this parameter, so it is 1 if x_i is the evidence value, 0 if the evidence is for some other value x_j, and is the product of all the λ messages received from its children if there is no evidence entered for X. The $\pi(x_i)$ parameter (3.3) is the product of the CPT and the π messages from parents.

The λ message to one parent combines (i) information that has come from children via λ messages and been summarized in the $\lambda(X)$ parameter, (ii) the values in the CPT and (iii) any π messages that have been received from any other parents.

The $\pi_{Y_j}(x_i)$ message down to child Y_j is 1 if x_i is the evidence value and 0 if the evidence is for some other value x_j. If no evidence is entered for X, then it combines (i) information from children other than Y_j, (ii) the CPT and (iii) the π messages it has received from its parents.

The algorithm requires the following initializations (i.e., before any evidence is entered).

- Set all λ values, λ messages and π messages to 1.
- Root nodes: If node W has no parents, set $\pi(W)$ to the prior, $P(W)$.

The message passing algorithm can be used to compute the beliefs for all nodes in the network, even before any evidence is available.

When specific evidence $W = w_i$ is obtained, given that node W can take values $\{w_1, w_2, \ldots, w_n\}$,

- Set $\lambda(W) = (0, 0, \ldots, 0, 1, 0, \ldots, 0)$ with the 1 at the ith position.

This π/λ notation used for the messages is that introduced by Kim and Pearl and can appear confusing at first. Note that the format for both types of messages is $\pi_{Child}(Parent)$ and $\lambda_{Child}(Parent)$. So,

- π messages are sent in the direction of the arc, from parent to child, hence the notation is $\pi_{Receiver}(Sender)$;
- λ messages are sent from child to parent, against the direction of the arc, hence the notation is $\lambda_{Sender}(Receiver)$.

Note also that π plays the role of prior and λ the likelihood in Bayes' Theorem.

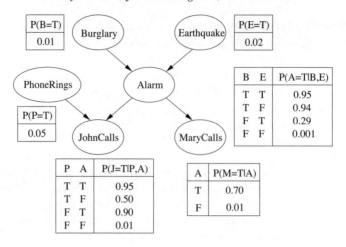

FIGURE 3.2: Extended earthquake BN.

3.3.2   Message passing example

Here, we extend the BN given in Figure 2.6 for Pearl's earthquake problem, by adding a node *PhoneRings* to represent explicitly the phone ringing that John sometimes confuses with the alarm. This extended BN is shown in Figure 3.2, with the parameters and messages used in the message passing algorithm shown in Figure 3.3.

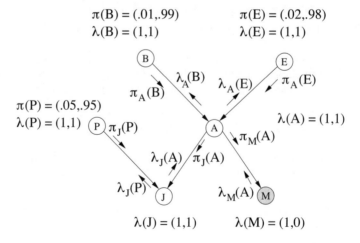

FIGURE 3.3: Extended earthquake BN showing parameter initialization and π and λ messages for Kim and Pearl's message passing algorithm.

We will work through the message passing updating for the extended earthquake problem, first without evidence, then with evidence entered for node M. The data propagation messages are shown in Figure 3.4. The updating and propagation sequencing presented here are not necessarily those that a particular algorithm would produce, but rather the most efficient sequencing to do the belief updating in the minimum number of steps.

PROPAGATION, NO EVIDENCE

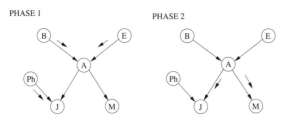

PROPAGATION, EVIDENCE for node M

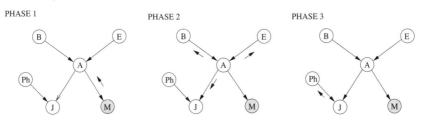

FIGURE 3.4: Message passing algorithm propagation stages for without evidence (above) and with evidence for node M (below).

First, before any evidence is entered, various parameters are initialized: $\pi(B)$, $\pi(E)$, $\pi(P)$, from the priors, and $\lambda(J)$, $\lambda(M)$ with $(1,1)$ as leaf nodes without evidence.

During this propagation before evidence, all of the diagnostic messages (the λ messages) will be the unit vector. We can see from the updating equations that we only multiply by these messages, so multiplying by the unit vector will not change any other parameters or messages. So we will not show the unit vector λ message propagation at this stage.

Using the belief updating equation (3.1), $Bel(B)$, $Bel(E)$ and $Bel(P)$ are computed, while sending new messages $\pi_A(B)$, $\pi_A(E)$ and $\pi_J(P)$ are computed using (3.5) and sent out.

Node A has received all its π messages from its parents, so it can update $Bel(A)$ and compute its own π messages to pass down to its children J and M. At this stage, we *could* update $Bel(J)$, as it has just received a π message from P; however, it has yet to receive a π message from A, so we won't do that this cycle.

After the second message propagation phase, $Bel(J)$ and $Bel(M)$ can be computed, as all π messages been received from their parents.

Next, we look at how evidence $M = T$, entered by setting $\lambda(M) = (1,0)$, can be propagated through the network. First, the message $\lambda_M(A)$ is computed and sent up to A. $\lambda(A)$ and in turn $Bel(A)$ are then recomputed, and new messages sent to A's parents via $\lambda_A(B)$ and $\lambda_A(E)$, and to its other child J via $\pi_J(A)$.

These new messages allow $Bel(B)$, $Bel(E)$ and $Bel(J)$ to be recomputed, and the final message $\lambda_J(P)$ computed and sent from J to P. The last computation is the updating of $Bel(P)$. Note that the minimum number of propagation steps for this evidence example is three, since that is the distance of the furthest node P from the evidence node M.

3.3.3   Algorithm features

All the computations in the message passing algorithm are local: the belief updating and new outgoing messages are all computed using incoming messages and the parameters. While this algorithm is efficient in a sense because of this locality property, and it lends itself to parallel, distributed implementations, we can see that there is a summation over all joint instantiations of the parent nodes, which is exponential in the number of parents. Thus, the algorithm is computationally infeasible when there are too many parents. And the longer the paths from the evidence node or nodes, the more cycles of data propagation must be performed to update all other nodes.

Note also that in the presentation of this algorithm, we have followed Pearl's presentation (1988) and normalized all the messages. This is a computational overhead that is not strictly necessary, as all the normalizing can be done when computing the marginals (i.e., when computing $Bel(x_i)$). The normalization constants, α, are the same for all the marginals, being the inverse of the probability $P(\mathbf{E})$, the computation of which is often useful for other purposes.

3.4   Inference with uncertain evidence

Thus far we have assumed that any evidence is a direct observation of the value of a variable that will result in the belief for a node being set to 1 for that value and 0 for all other values. This is the specific evidence described in §2.3.2, which is entered in the message passing algorithm as a vector with a 1 in the position of the evidence value and 0 in the other positions. Once specific evidence is entered for a node, the belief for that node is "clamped" and doesn't change no matter what further information becomes available.

However, the inference algorithms should also be able to handle evidence that has uncertainty associated with it, as in Jeffrey conditionalization (Definition 1.5). In our earthquake example, suppose that after you get a call from your neighbor Mary saying she has heard your alarm going off, a colleague who is in your office at the time says he thinks he heard earlier on the radio that there was a minor earthquake in your area, but he is only 80% sure.

We introduced this notion very briefly in §2.3.2, where we mentioned that this sort of evidence is called "virtual" evidence or "likelihood" evidence. We deferred further explanation of these terms until here, as they relate to how inference is performed.

Some of the major BN software packages (e.g., Netica, Hugin, GeNIe and BayesiaLab) provide the facility for adding likelihood evidence, as well as specific and negative evidence. In fact, we describe how to enter likelihood evidence in both Netica and Hugin in Figure 3.19. In our opinion, the explanations of this sort of evidence in both the literature and available software documentation are confusing and incomplete. It is important for people using this feature in the software to understand how likelihood evidence affects the inference and also how to work out the numbers to enter.

First, there is the issue of how to interpret uncertainty about an observation. The uncertain information could be represented by adopting it as the new distribution over the variable in question. This would mean for the earthquake node somehow setting $Bel(Earthquake = T) = 0.8$. However, we certainly do not want to "clamp" this belief, since this probabilistic judgement should be integrated with any further independent information relevant to the presence of an earthquake (e.g., the evidence about Mary calling).

3.4.1 Using a virtual node

Let's look at incorporating an uncertain observation for the simplest case, a single Boolean node X, with a uniform prior, that is $P(X=T)=P(X=F)=0.5$. We add a **virtual node** V, which takes values $\{T,F\}$,[1] as a child of X, as shown in Figure 3.5. The uncertainty in the observation of X is represented by the CPT; for an 80% sure observation this gives $P(V = T|X = T) = 0.8$ and $P(V = T|X = F) = 0.2$.[2] Now specific evidence is entered that $V = T$. We can use Bayes' Theorem to perform the inference, as follows.

$$
\begin{aligned}
Bel(X = T) &= \alpha P(V = T|X = T)P(X = T) \\
&= \alpha 0.8 \times 0.5 \\
Bel(X = F) &= \alpha P(V = T|X = T)P(X = F) \\
&= \alpha 0.2 \times 0.5
\end{aligned}
$$

Since $Bel(X = T) + Bel(X = F) = 1$, this gives us $\alpha = 2$, and hence $Bel(X = T) = 0.8$ and $Bel(X = F) = 0.2$, as desired.

It is important to note here that due to the normalization, it is not the likelihoods for $P(V = T|X = T)$ and $P(V = T|X = F)$ that determine the new belief, but rather the **ratio**:

$$
P(V = T|X = T) : P(V = T|X = F)
$$

For example, we would get the same answers for $P(V = T|X) = (0.4, 0.1)$ or $P(V = T|X) = (0.5, 0.125)$.

---

[1]Note that V takes these values *irrespective* of the values of the observed node.
[2]Note that the semantics of this CPT are not well specified. For example, what is the meaning of the CPT entries for $P(V = F|X)$?

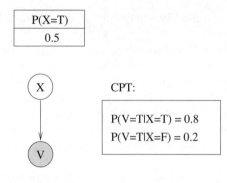

virtual node

FIGURE 3.5: Virtual evidence handled through virtual node V for single node X, with uncertainty in CPT.

So far, we have seen that with the use of a "virtual" node we can represent uncertainty in an observation by a likelihood ratio. For example, my colleague attributing 80% credibility to there having been an earthquake means that he is four times more likely to think an earthquake occurred if that was in fact the case than he is of mistakenly thinking it occurred.

But we have considered only the situation of a node without parents, with uniform priors. If the priors are *not* uniform, then simply mapping the uncertainty into a likelihood ratio as described does not give a new belief that correctly represents the specified observational uncertainty. For example, in our earthquake example, if $P(E) = (0.02, 0.98)$, then

$$
\begin{aligned}
Bel(E = T) &= \alpha P(V = T | E = T) P(E = T) \\
&= \alpha 0.8 \times 0.02 \\
Bel(E = F) &= \alpha P(V = T | E = T) P(E = F) \\
&= \alpha 0.2 \times 0.98
\end{aligned}
$$

which gives $\alpha \approx 4.72$, and $Bel(E = T) = 0.075$ and $Bel(E = F) = 0.925$ so the posterior belief in *Earthquake* is only 7.5%. However, if $P(E) = (0.9, 0.1)$, then

$$
\begin{aligned}
Bel(E = T) &= \alpha P(V = T | E = T) P(E = T) \\
&= \alpha 0.8 \times 0.9 \\
Bel(E = F) &= \alpha P(V = T | E = T) P(E = F) \\
&= \alpha 0.2 \times 0.1
\end{aligned}
$$

which gives us $Bel(E = T) = 0.973$ and $Bel(E = F) = 0.027$.

It is clear that directly mapping the observational uncertainty into a likelihood ratio only results in the identical belief for the node in question (as intended in Jeffrey conditionalization) when the priors are uniform. When the priors are non-uniform, a likelihood ratio of 4:1 will increase the belief, but not necessarily to the intended degree.

Some people feel this is an advantage of this representation, supposedly indicating that the observation is an external "opinion" from someone whose priors are unknown. However, if we wish to represent unknown priors, we have no business inferring a posterior at all!

If we *really* want the uncertain evidence to shift the beliefs to $Bel(X = T) = P(X = T|V) = 0.8$, then we can still use a virtual node and likelihood ratio approach, but we need to compute the ratio properly. Let's return to our earthquake example. Recalling Definition 1.6, $P(E|V) = 0.8$ means $O(E = T|V) = 4$. Since (by odds-likelihood Bayes' theorem 1.5)

$$O(E = T|V) = \frac{P(V = T|E = T)}{P(V = T|E = F)} O(E = T)$$

and, as given in the example,

$$O(E = T) = 0.02/0.98$$

we get the likelihood ratio

$$\frac{P(V = T|E = T)}{P(V = T|E = F)} = \frac{0.8}{0.2} \times \frac{0.98}{0.02} = 196$$

This ratio is *much* higher than the one used with uniform priors, which is necessary in order to shift the belief in *Earthquake* from its very low prior of 0.02 to 0.8. We note that some software packages do in fact allow the posterior to be fixed in this way, e.g. Netica, which calls it "calibration."

3.4.2 Virtual nodes in the message passing algorithm

When implementing a message passing inference algorithm, we don't actually need to add the V node. Instead, the virtual node is connected by a virtual link as a child to the node it is "observing." In the message passing algorithm, these links only carry information one way, from the virtual node to the observed variable. For our earthquake example, the virtual node V represents the virtual evidence on E. There are no parameters $\lambda(V)$, but instead it sends a $\lambda_V(E)$ message to E. The virtual node and the message it sends to E is shown in Figure 3.6.

3.4.3 Multiple virtual evidence

The situation may also arise where there is a set, or a sequence, of multiple uncertain observations for the same variable. For example, there may be multiple witnesses for the same event, or a sensor may be returning a sequence of readings for the same variable over time.

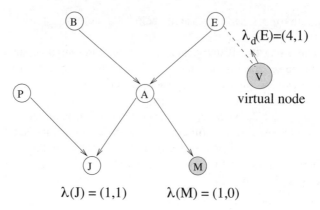

$$\lambda(J) = (1,1) \qquad \lambda(M) = (1,0)$$

FIGURE 3.6: Virtual evidence handled through virtual node V for earthquake node E, with λ message in message passing algorithm.

This is straightforward to represent using one virtual child node for each observation, as shown in Figure 3.7(a). The inference algorithms work in exactly the same way, passing up one likelihood vector for each observation. However, if the observations are not independent, then the dependencies must be represented explicitly by modeling with actual (rather than "virtual") nodes. Figure 3.7(b) shows an example of modeling dependencies, where the first witness tells all the other witnesses what she thought she saw.

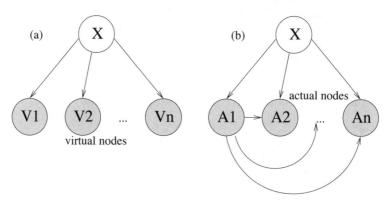

FIGURE 3.7: Multiple uncertain evidence for the same node can be handled by having one virtual node for each: (a) when the observations are independent (b) dependencies between observations should be represented explicitly.

3.5 Exact inference in multiply-connected networks

In the most general case, the BN structure is a (directed acyclic) graph, rather than simply a tree. This means that at least two nodes are connected by more than one path in the underlying undirected graph. Such a network is **multiply-connected** and occurs when some variable can influence another through more than one causal mechanism.

The metastatic cancer network shown in Figure 3.8 is an example. The two causes of *C* (*S* and *B*) share a common parent, *M*. For this BN, there is an undirected loop around nodes *M*, *B*, *C* and *S*. In such networks, the message passing algorithm for polytrees presented in the previous section does not work. Intuitively, the reason is that with more than one path between two nodes, the same piece of evidence about one will reach the other through two paths and be counted twice. There are many ways of dealing with this problem in an exact manner; we shall present the most popular of these.

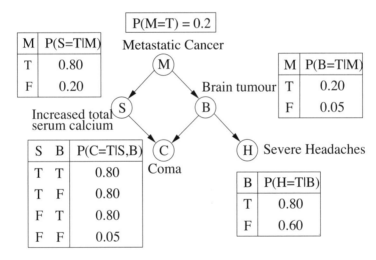

FIGURE 3.8: Example of a multiply-connected network: metastatic cancer BN (§2.5.2).

3.5.1 Clustering methods

Clustering inference algorithms transform the BN into a probabilistically equivalent polytree by merging nodes, removing the multiple paths between the two nodes along which evidence may travel. In the cancer example, this transformation can be done by creating a new node, say *Z*, that combines nodes *B* and *S*, as shown in Figure 3.9. The new node has four possible values, {*TT, TF, FT, FF*}, corresponding to all possible instantiations of *B* and *S*. The CPTs for the transformed network are also shown in

Figure 3.9. They are computed from the CPTs of the original graph as follows.

$$P(Z|M) \quad = \quad P(S,B|M) \text{ by definition of Z}$$
$$= \quad P(S|M)P(B|M) \text{ since S and B are independent given M}$$

Similarly,

$$P(H|Z) \quad = \quad P(H|S,B) = P(H|B) \text{ since H is independent of S given B}$$
$$P(C|Z) \quad = \quad P(C|S,B) \text{ by definition of Z}$$

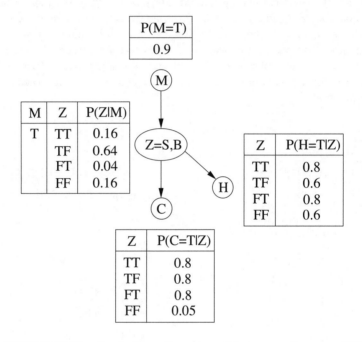

FIGURE 3.9: Result of ad hoc clustering of the metastatic cancer BN.

It is always possible to transform a multiply-connected network into a polytree. In the extreme case, all the non-leaf nodes can be merged into a single compound node, as shown in Figure 3.10(a) where all the nodes D_i are merged into a single super-node. The result is a simple tree. But other transformations are also possible, as shown in Figure 3.10(b) and (c), where two different polytrees are produced by different clusterings. It is better to have smaller clusters, since the CPT size for the cluster grows exponentially in the number of nodes merged into it. However the more highly connected the original network, the larger the clusters required.

Exact clustering algorithms perform inference in two stages.

1. Transform the network into a polytree
2. Perform belief updating on that polytree

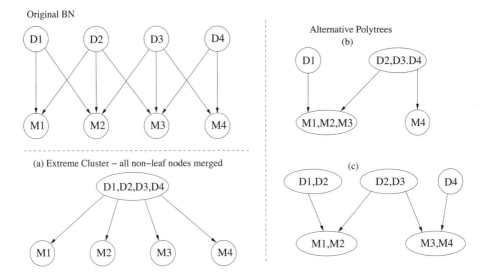

FIGURE 3.10: The original multiply-connected BN can be clustered into: (a) a tree; different polytrees (b) and (c).

The transformation in Step 1 may be slow, since a large number of new CPT values may need to be computed. It may also require too much memory if the original network is highly connected. However, the transformation only needs to be done once, unless the structure or the parameters of the original network are changed.

The belief updating in the new polytree is usually quite fast, as an efficient polytree message passing algorithm may now be applied. The caveat does need to be added, however, that since clustered nodes have increased complexity, this can slow down updating as well.

3.5.2 Junction trees

The clustering we've seen thus far has been ad hoc. The junction tree algorithm provides a methodical and efficient method of clustering, versions of which are implemented in the main BN software packages (see §B.3).

Algorithm 3.2 The Junction Tree Clustering Algorithm

1. **Moralize:** *Connect all parents and remove arrows; this produces a so-called* **moral graph**.

2. **Triangulate:** *Add arcs so that every cycle of length > 3 has a chord (i.e., so there is a subcycle composed of exactly three of its nodes); this produces a* **triangulated graph**.

3. **Create new structure:** *Identify maximal cliques in the triangulated graph to become new compound nodes, then connect to form the so-called* **junction tree**.

4. **Create separators:** *Each arc on the junction tree has an attached* **separator**, *which consists of the intersection of adjacent nodes.*

5. **Compute new parameters:** *Each node and separator in the junction tree has an associated table over the configurations of its constituent variables. These are all a table of ones to start with.*

   For *each node X in the original network,*

   (a) *Choose one node Y in the junction tree that contains X and all of X's parents,*

   (b) *Multiply $P(X|Parents(X))$ on Y's table.*

6. **Belief updating:** *Evidence is added and propagated using a message passing algorithm.*

The moralizing step (Step 1) is so called because an arc is added if there is no direct connection between two nodes with a common child; it is "marrying" the parent nodes. For our metastatic cancer example, S and B have C as a common child; however, there is no direct connection between them, so this must be added. The resultant moral graph is shown in Figure 3.11(a).

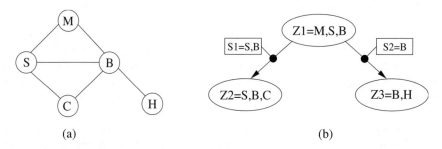

(a)   (b)

FIGURE 3.11: Applying the junction tree algorithm to the metastatic cancer BN gives (a) the moral graph and (b) the junction tree.

As we have seen, the size of the CPT for a new compound node is the product of the number of states of the combined nodes, so the size of the CPT increases exponentially with the size of the compound node. Different triangulations (Step 2) produce different clusters. Although the problem of finding an optimal triangulation is NP complete, there are heuristics which give good results in practice. No arcs need to be added to the moral graph for the metastatic cancer example, as it is already triangulated.

A **clique** of an undirected graph is defined as a set of nodes that are all pairwise linked; that is, for every pair of nodes in the set, there is an arc between them. A **maximal clique** is such a subgraph that cannot be increased by adding any node. In our example the maximal cliques can be read from the moral graph, giving three new compound nodes, $Z_1 = M,S,B$, $Z_2 = S,B,C$ and $Z_3 = B,H$ (Step 3). The junction tree, including the separators (Step 4), is shown in Figure 3.11(b).

Note that sometimes after the compound nodes have been constructed and connected, a junction *graph* results, instead of a junction tree. However, because of the triangulation step, it is always possible to remove links that aren't required in order to form a tree. These links aren't required because the same information can be passed along another path; see Jensen and Nielsen (2007) for details.

The metastatic cancer network we've been using as an example is not sufficiently complex to illustrate all the steps in the junction tree algorithm. Figure 3.12 shows the transformation of a more complex network into a junction tree (from Jensen, 1996).

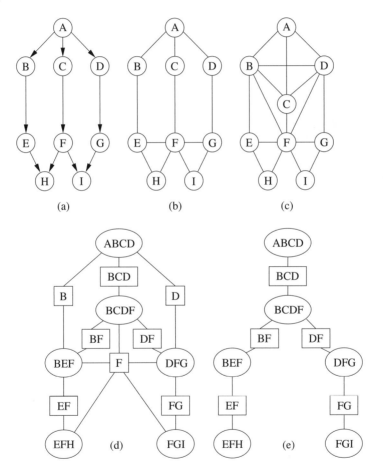

FIGURE 3.12: Example of junction tree algorithm application: (a) the original multiply-connected BN; (b) the moral graph; (c) the triangulated graph; (d) the junction graph; and (e) the junction tree. (From Jensen, F.V. *An Introduction to Bayesian Networks*. UCL Press, London, Copyright (1996). With kind permission of Springer Science and Business Media.)

Finally, Step 5 computes the new CPTs for the junction tree compound nodes. The result of Step 5 is that the product of all the node CPTs in the cluster tree is the product of all the CPTs in the original BN. Any cluster tree obtained using this algorithm is a representation of the same joint distribution, which is the product of all the cluster tables divided by the product of all the separator tables.

Adding evidence to the junction tree is simple. Suppose that there is evidence e about node X. If it is specific evidence that $X = x_i$, then we create an evidence vector with a 1 in the ith position, zeros in all the others. If it is a negative finding for state x_j, there is a zero in the jth position and ones in all the others. If it is virtual evidence, the vector is constructed as in §3.4. In all cases, the evidence vector is multiplied on the table of any node in the junction tree containing X.

Once the junction tree has been constructed (called "**compilation**" in most BN software, including Netica) and evidence entered, belief updating is performed using a message passing approach. The details can be found in (Jensen and Nielsen, 2007). The basic idea is that separators are used to receive and pass on messages, and the computations involve computing products of CPTs.

The cost of belief updating using this junction tree approach is primarily determined by the sizes of the state space of the compound nodes in the junction tree. It is possible to estimate the computational cost of performing belief updating through the **junction tree cost** (Kanazawa, 1992, Dean and Wellman, 1991). Suppose the network has been transformed into a junction tree with new compound nodes $C_1, \ldots, C_i, \ldots, C_n$, The junction tree cost is defined as:

$$\sum_{C_i \in \{C_1, \ldots, C_n\}} \left(k_i \prod_{X \in C_i} |\Omega_X| \right) \tag{3.6}$$

where k_i is the number of separation sets involving clique C_i, i.e., the total number of parents and children of C_i in the junction tree. So the junction-tree cost is the product of the size of all the constituent nodes in the cluster and the number of children and parents, summed over all the compound nodes in the junction tree. The junction tree cost provides a measure that allows us to compare different junction trees, obtained from different triangulations.

While the size of the compound nodes in the junction tree can be prohibitive, in terms of memory as well as computational cost, it is often the case that many of the table entries are zeros. Compression techniques can be used to store the tables more efficiently, without these zero entries.[3]

3.6 Approximate inference with stochastic simulation

In general, exact inference in belief networks is computationally complex, or more precisely, NP hard (Cooper, 1990). In practice, for most small to medium sized net-

---

[3]The Hugin software uses such a compression method.

works, up to three dozen nodes or so, the current best exact algorithms — using clustering — are good enough. For larger networks, or networks that are densely connected, approximate algorithms must be used.

One approach to approximate inference for multiply-connected networks is **stochastic simulation**. Stochastic simulation uses the network to generate a large number of cases from the network distribution. The posterior probability of a target node is estimated using these cases. By the Law of Large Numbers from statistics, as more cases are generated, the estimate converges on the exact probability.

As with exact inference, there is a computational complexity issue: approximating to within an arbitrary tolerance is also NP hard (Dagum and Luby, 1993). However, in practice, if the evidence being conditioned upon is not too unlikely, these approximate approaches converge fairly quickly.

Numerous other approximation methods have been developed, which rely on modifying the representation, rather than on simulation. Coverage of these methods is beyond the scope of this text (see Bibliographic Notes §3.10 for pointers).

3.6.1   Logic sampling

The simplest sampling algorithm is that of logic sampling (LS) (Henrion, 1988) (Algorithm 3.3). This generates a case by randomly selecting values for each node, weighted by the probability of that value occurring. The nodes are traversed from the root nodes down to the leaves, so at each step the weighting probability is either the prior or the CPT entry for the sampled parent values. When all the nodes have been visited, we have a "case," i.e., an **instantiation** of all the nodes in the BN. To estimate $P(X|E)$ with a sample value $P'(X|E)$, we compute the ratio of cases where both X and E are true to the number of cases where just E is true. So after the generation of each case, these combinations are counted, as appropriate.

Algorithm 3.3 Logic Sampling (LS) Algorithm
Aim: *to compute $P'(X|E = e)$ as an estimate of the posterior probability for node X given evidence $E = e$.*

Initialization

- *For each value x_i for node X*
  Create a count variable $Count(x_i,e)$
- *Create a count variable $Count(e)$*
- *Initialize all count variables to 0*

For each round of simulation
1. *For all root nodes*
   Randomly choose a value for it, weighting the choice by the priors.
2. *Loop*
   Choose values randomly for children, using the conditional
   probabilities given the known values of the parents.

   Until all the leaves are reached

3. *Update run counts:*
   If the case includes $E = e$
   $\quad$ $Count(e) \leftarrow Count(e) + 1$

   If this case includes both $X = x_i$ *and* $E = e$
   $\quad$ $Count(x_i, e) \leftarrow Count(x_i, e) + 1$

Current estimate for the posterior probability

$$P'(X = x_i | E = e) = \frac{Count(x_i, E = e)}{Count(E = e)}$$

Let us work through one round of simulation for the metastatic cancer example.

- The only root node for this network is node M, which has prior $P(M = T) = 0.2$. The random number generator produces a value between 0 and 1; any number > 0.2 means the value F is selected. Suppose that is the case.

- Next, the values for children S and B must be chosen, using the CPT entries $P(S = T | M = F)$ and $P(B = T | M = F)$. Suppose that values $S = T$ and $B = F$ are chosen randomly.

- Finally, the values for C and H must be chosen weighted with the probabilities $P(M = | S = T, B = F)$ and $P(H | S = T, B = F)$. Suppose that values $M = F$ and $H = T$ are selected.

- Then the full "case" for this simulation round is the combination of values: $C = F, S = T, B = F, M = F$ and $H = T$.

- If we were trying to update the beliefs for a person having metastatic cancer or not (i.e., $Bel(M)$) given they had a severe headache, (i.e., evidence E that $H = T$), then this case would add one to the count variable $Count(H = T)$, one to $Count(M = F, H = T)$, but not to $Count(M = T, H = T)$.

The LS algorithm is easily generalized to more than one query node. The main problem with the algorithm is that when the evidence E is unlikely, most of the cases have to be discarded, as they don't contribute to the run counts.

3.6.2 Likelihood weighting

A modification to the LS algorithm called likelihood weighting (LW) (Fung and Chang, 1989, Shachter and Peot, 1989) (Algorithm 3.4) overcomes the problem with unlikely evidence, always employing the sampled value for each evidence node. However, the same straightforward counting would result in posteriors that did not reflect the actual BN model. So, instead of adding "1" to the run count, the CPTs for the evidence node (or nodes) are used to determine how likely that evidence combination is given the parent state, and that fractional likelihood is the number added to the run count.

Algorithm 3.4 Likelihood Weighting (LW) Algorithm

Aim: *to compute* $P'(X | E = e)$, *an approximation to the posterior probability for node*

X given evidence $\mathbf{E} = e$, *consisting of* $\{E_1 = e_1, \ldots, E_n = e_n\}$.

Initialization

- *For each value x_i for node X*
    Create a count variable Count(x_i,e)
- *Create a count variable Count(e)*
- *Initialize all count variables to 0*

For each round of simulation

1. *For all root nodes*
    If a root is an evidence node, E_j
        choose the evidence value, e_j
        likelihood($E_j = e_j$) ← $P(E_j = e_j$)
    Else
        Choose a value for the root node, weighting the choice by the priors.

2. *Loop*
    If a child is an evidence node, E_j
        choose the evidence value, e_j
        likelihood($E_j = e_j$) = $P(E_j = e_j|$chosen parent values)
    Else
        Choose values randomly for children, using the conditional probabilities given the known values of the parents.

    Until all the leaves are reached

3. *Update run counts:*
    If the case includes $E = e$
        Count(e) ← Count(e) + $\prod_j$ likelihood($E_j = e_j$)
    If this case includes both $X = x_i$ and $E = e$
        Count(x_i,e) ← Count(x_i,e) + $\prod_j$ likelihood($E_j = e_j$)

Current estimate for the posterior probability

$$P'(X = x_i | E = e) = \frac{Count(x_i, E = e)}{Count(E = e)}$$

Let's look at how this works for another metastatic cancer example, where the evidence is $B = T$, and we are interested in probability of the patient going into a coma. So, we want to compute an estimate for $P(C = T | B = T)$.

- Choose a value for M with prior $P(M) = 0.2$. Assume we choose $M = F$.
- Next we choose a value for S from distribution $P(S|M = F) = 0.20$. Assume $S = T$ is chosen.
- Look at B. This is an evidence node that has been set to T and $P(B = T|M = F) = 0.05$. So this run counts as 0.05 of a complete run.
- Choose a value for C randomly with $P(C|S, B) = 0.80$. Assume $C = T$.
- So, we have completed a run with likelihood 0.05 that reports $C = T$ given $B = T$. Hence, both Count($C = T, B = T$) and Count($B = T$) are incremented.

3.6.3   Markov Chain Monte Carlo (MCMC)

Both the logic sampling and likelihood weighting sampling algorithms generate each sample individually, starting from scratch. MCMC on the other hand generates a sample by making a random change to the previous sample. It does this by randomly sampling a value for one of the non-evidence nodes X_i, conditioned on the current value of the nodes in its Markov blanket, which consists of the parents, children and children's parents (see §2.2.2).

The technical details of why MCMC returns consistent estimates for the posterior probabilities are beyond the scope of this text (see Russell and Norvig, 2010 for details). Note that different uses of MCMC are presented elsewhere in this text, namely Gibbs sampling (for parameter learning) in §6.3.2.1 and Metropolis search (for structure learning) in §9.6.2.

3.6.4   Using virtual evidence

There are two straightforward alternatives for using virtual evidence with sampling inference algorithms.[4]

1. Use a virtual node: add an explicit virtual node V to the network, as described in §3.4, and run the algorithm with evidence $V=T$.

2. In the likelihood weighting algorithm, we already weight each sample by the likelihood. We can set $likelihood(E_j)$ to the normalized likelihood ratio.

3.6.5   Assessing approximate inference algorithms

In order to assess the performance of a particular approximate inference algorithm, and to compare algorithms, we need to have a measure for the quality of the solution at any particular time. One possible measure is the **Kullback-Leibler divergence** between a true distribution P and the estimated distribution P' of a node with states i, given by:[5]

Definition 3.1 Kullback-Leibler divergence

$$KL(P, P') = \sum_i P(i) \log \frac{P(i)}{P'(i)}$$

Note that when P and P' are the same, the KL divergence is zero (which is proven in §7.8). When $P(i)$ is zero, the
convention is to take the summand to have a zero value. And, KL divergence is undefined when $P'(i) = 0$; standardly, it is taken as infinity (unless also $P(i) = 0$, in which case the summand is 0).

---

[4]Thanks to Bob Welch and Kevin Murphy for these suggestions.

[5]Although "Kullback-Leibler distance" is the commonly employed word, since the KL measure is asymmetric — measuring the difference from the point of view of one or the other of the distributions — it is no true distance.

Alternatively, we can put this measure in terms of the updated belief for query node X.

$$KL(Bel(X), Bel'(X)) = \sum_{x_i} Bel(X = x_i) \log \frac{Bel(X = x_i)}{Bel'(X = x_i)} \quad (3.7)$$

where $Bel(X)$ is computed by an exact algorithm and $Bel'(X)$ by the approximate algorithm. Of course, this measure can only be applied when the network is such that the exact posterior *can* be computed.

When there is more than one query node, we should use the marginal KL divergence over all the query nodes. For example, if X and Y are query nodes, and Z the evidence, we should use $KL(P(X, Y|Z), P'(X, Y|Z))$. Often the average or the sum of the KL divergences for the individual query nodes are used to estimate the error measure, which is not exact. Problem 3.11 involves plotting the KL divergence to compare the performance of approximate inference algorithms.

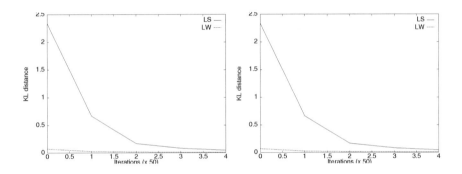

FIGURE 3.13: Comparison of the logic sampling and likelihood-weighting approximate inference algorithms.

An example of this use of KL divergence is shown in Figure 3.13. These graphs show the results of an algorithm comparison experiment (Nicholson and Jitnah, 1996). The test network contained 99 nodes and 131 arcs, and the LS and LW algorithms were compared for two cases:

- Experiment 1: evidence added for 1 root node, while query nodes were all 35 leaf nodes.
- Experiment 2: evidence added for 1 leaf node, while query nodes were all 29 root nodes (29).

As well as confirming the faster convergence of LW compared to LS, these and other results show that stochastic simulation methods perform better when evidence is nearer to root nodes (Nicholson and Jitnah, 1996). In many real domains when the task is one of diagnosis, evidence tends to be near leaves, resulting in poorer performance of the stochastic simulation algorithms.

3.7 Other computations

In addition to the standard BN inference described in this chapter to data — computing the posterior probability for one or more query nodes — other computations are also of interest and provided by some BN software.

3.7.1 Belief revision

It is sometimes the case that rather than updating beliefs given evidence, we are more interested in the most probable values for a set of query nodes, given the evidence. This is sometimes called **belief revision** (Pearl, 1988, Chapter 5). The general case of finding a most probable instantiation of a set of n variables is called **MAP** (maximum aposteriori probability). MAP involves finding the assignment for the n variables that maximizes $P(X_1 = x1, \ldots, X_n = x_n | \mathbf{E})$. Finding MAPs was first shown to be NP hard (Shimony, 1994), and then NP complete (Park, 2002); approximating MAPs is also NP hard (Abdelbar and Hedetniemi, 1998).

A special case of MAP is finding an instantiation of *all* the non-evidence nodes, also known as computing a **most probable explanation (MPE)**. The "explanation" of the evidence is a complete assignment of all the non-evidence nodes, $\{Y_1 = y1, \ldots, Y_n = y_m\}$, and computing the MPE means finding the assignment that maximizes $P(Y_1 = y1, \ldots, Y_n = y_n | \mathbf{E})$. MPE can be calculated efficiently with a similar method to probability updating (see Jensen and Nielsen, 2007 for details). Most BN software packages have a feature for calculating MPE but not MAP.

3.7.2 Probability of evidence

When performing belief updating, it is usually the case that the probability of the evidence, P(E), is available as a by-product of the inference procedure. For example, in polytree updating, the normalizing constant α is just $1/P(E)$. Clearly, there is a problem should $P(E)$ be zero, indicating that this combination of values is **impossible** in the domain. If that impossible combination of values is entered as evidence, the inference algorithm must detect and flag it.

The BN user must decide between the following alternatives.

1. It is indeed the case that the evidence is impossible in their domain, and therefore the data are incorrect, due to errors in either gathering or entering the data.

2. The evidence should not be impossible, and the BN incorrectly represents the domain.

This notion of possible incoherence in data has been extended from impossible evidence to unlikely combinations of evidence. A **conflict measure** has been proposed to detect possible incoherence in evidence (Jensen et al., 1991, Kim and Valtorta, 1995). The basic idea is that correct findings from a coherent case covered by the model support each other and therefore would be expected to be positively correlated. Suppose we have a set of evidence $\mathbf{E} = \{E_1 = e_1, \ldots, E_m = e_m\}$. A conflict measure on $\mathbf{E}$ is

$$C(\mathbf{E}) = \log \frac{P(E_1 = e_1), \ldots, P(E_m = e_m)}{P(E)}$$

$C(\mathbf{E})$ being positive indicates that the evidence may be conflicting. The higher the conflict measure, the greater the discrepancy between the BN model and the evidence. This discrepancy may be due to errors in the data or it just may be a rare case. If the conflict is due to flawed data, it is possible to trace the conflicts.

3.8 Causal inference

There is no consensus in the community of Bayesian network researchers about the proper understanding of the relation between causality and Bayesian networks. The majority opinion is that there is nothing special about a **causal interpretation**, that is, one which asserts that corresponding to each (non-redundant) direct arc in the network not only is there a probabilistic dependency but also a causal dependency. As we saw in Chapter 2, after all, by reordering the variables and applying the network construction algorithm we can get the arcs turned around! Yet, clearly, *both* networks cannot be causal.

We take the minority point of view, however (one, incidentally, shared by Pearl, 2000 and Neapolitan, 1990), that causal structure is what *underlies* all useful Bayesian networks. Certainly not all Bayesian networks are causal, but if they represent a real-world probability distribution, then some causal model is their source.

Regardless of how that debate falls out, however, it is important to consider how to do inferences with Bayesian networks that *are* causal. If we have a causal model, then we can perform inferences which are not available with a non-causal BN. This ability is important, for there is a large range of potential applications for particularly causal inferences, such as environmental management, manufacturing and decision support for medical intervention. For example, we may need to reason about what will happen to the quality of a manufactured product if we adopt a cheaper supplier for one of its parts. Non-causal Bayesian networks, and causal Bayesian networks using ordinary propagation, are currently used to answer just such questions; but this practice is wrong. Although most Bayesian network tools do not explicitly support causal reasoning, we will nevertheless now explain how to do it properly.

3.8.1 Observation vs. intervention

While the causal interpretation of Bayesian networks is becoming more widely accepted, the distinction between causal reasoning and observational reasoning remains

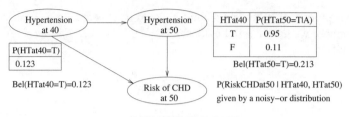

FIGURE 3.14: A causal model linking hypertension at age 40 and 50 with risk of coronary heart disease. Posterior probabilities shown when no evidence has been added.

for many obscure. This is particularly true in application areas where the use of regression models, rather than Bayesian networks, is the norm, since regression models (in ordinary usage) simply lack the capability of modeling interventions.

We illustrate the difference between intervention and observation with a simple example. Figure 3.14 presents a three-variable causal model of coronary heart disease (CHD) risk, which is loosely based upon models of the Framingham heart disease data (e.g., Anderson et al., 1991, D'Agostino et al., 2000). As is normal, each arrow represents a direct and indispensable causal connection between variables.[6] Two contributing factors for CHD are shown: hypertension (HT; elevated blood pressure) at age 40 and HT at age 50. The higher the blood pressure, the greater the chance of CHD, both directly and indirectly. That is, hypertension at 40 directly causes heart disease (in the terms available in this simplified network of three variables!), but also indirectly through HT at 50. In this simplified model, the direct connection between HT at 40 and CHD between 50 and 60 represents all those implicit causal processes leading to heart disease which are *not* reflected in the later HT. Note that P(Risk of CHD at 50 | Hypertension at 40, Hypertension at 50) is given by a 'noisy-or' distribution, see §6.4.2. Figure 3.14 shows the posterior probability distributions computed when no evidence has been added.

Figure 3.15(a) shows the results of observing no HT at age 50. The probability of CHD has decreased from a baseline of 0.052 to 0.026, as expected. But what if we intervene (say, with a medication) to lower blood pressure? The belief propagation (message passing) from Hypertension at 50 to Hypertension at 40, and then down to Risk of CHD (50-60) is all wrong under a **causal intervention**.

Judea Pearl, in his book *Causality* (2000), suggests that we model causal intervention in a variable X simply by (temporarily) cutting all arcs from $Parents(X)$ to X. If you do that with the hypertension example (as in Figure 3.15(b)), then that "fixes" the problem. The probability is reduced by a lesser amount to 0.033. By intervening

---

[6]Being indispensable means that, no matter what other variables within the network may be observed, there is some joint observational state in which the parent variable can alter the conditional probability of the child variable. In case this condition does not hold we have an unfaithful model, in the terminology of §9.7.2.

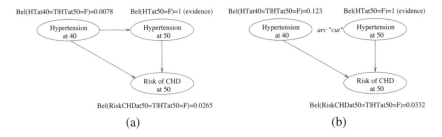

FIGURE 3.15: The hypertension causal model where HT at age 50 is (a) observed as low (b) set to low.

on HT at 50 we have cut the indirect causal path between HT at 40 and CHD, but we have not cut the direct causal path. That is, there are still implicit causal processes leading from HT at 40 to CHD which the proposed intervention leaves intact. Observations of low HT at 50 will in general reflect a lower activation of those implicit processes, whereas an intervention will not. In short, it is better to have low blood pressure at 50 *naturally* than to achieve that by artificial means—and this causal model reflects these facts.

A real-world example of people getting this wrong is in the widespread use of regression models in public health. To assess the expected value of intervention on blood pressure at age 40, for example, regression models of the Framingham data have been used (Anderson et al., 1991, D'Agostino et al., 2000). If those models had exactly the same structure as ours, then (aside from being overly simplistic) there would be no actual problem, since HT at 40 being a root node there is no arc-cutting needed. However, the models actually used incorporate a reasonable number of additional variables, including parents of HT at 40, such as history of smoking, cholesterol levels, etc. By simply *observing* a hypothetical low blood pressure level and computing expected values, these models are being used for something they are incapable of representing.[7] The mis-estimation of effects may well be causing bad public policy decisions.

Pearl's 'cut' method is the simplest way to model causal interventions and often will do the job. There is, however, a more general approach to causal inference, where the Bayesian network modeling can represent a causal intervention that is only probabilistic.

3.8.2 Defining an intervention

In ordinary usage, an intervention represents an influence on some causal system which is extraneous to that system. What kind of influence we consider is not con-

---

[7]In order to be *capable* of representing interventions we require a graphical representation in which the parental effects upon an intervened-upon variable can be cut (or altered). This minimally requires moving from ordinary regression models to path models or structural equation models, and treating these in the ways suggested here.

strained. It may interact with the existing complex of causal processes in the system in arbitrary ways. For example, a poison may induce death in some animal, but it may also interact with an anti-toxin so that it does not. Or again, the action of the poison may be probabilistic, either depending on unknown factors or by being genuinely indeterministic. Also, an intervention may impact on multiple factors (variables) in the system simultaneously or be targeted to exactly one such variable. In the extant literature of both philosophy and computer science there seems to have been an implicit agreement only to consider the very simplest of cases. In that literature, interventions are deterministic, always achieving their intended effect; and their intended effect is always to put exactly one variable into exactly one state. As a consequence, interventions never interact with any other causes of the targeted variable, rather their operation renders the effect of those other parents null. While such a simple model of interaction may be useful in untangling some of the mysteries of causation (e.g., it may have been useful in guiding intuitions in Halpern and Pearl's (2005a, 2005b)) study of token causation, it clearly will not do for a general analysis. Nor will it do for most practical cases. Medical interventions, for example, often fail (patients refuse to stop smoking), often interact with other causal factors (which explains why pharmacists require substantial training before licensing), often impact on multiple variables (organs) and often, even when successful, fail to put any variable into exactly one state (indeterminism!). Hence, we now provide a more general definition of intervention (retaining, however, reference to a single target variable in the system; this is a simplifying assumption which can easily be discharged).

Definition 3.2 *An intervention on a variable C in a causal model M transforms M into the augmented model M' which adds $I_c \to C$ to M where:*

1. *I_c is introduced with the intention of changing C.*

2. *I_c is a root node in M'.*

3. *I_c directly causes (is a parent of) C.*

Interventions are *actions* and, therefore, intentional. In particular, there will be some intended *target distribution* for the variable C, which we write $P^*(C)$. I_c itself will just be a binary variable, reflecting whether an intervention on C is attempted or not. However, this definition does not restrict I_c's interaction with C's other parents, leaving open whether the target distribution is actually achieved by the intervention. Also, the definition does allow variables other than C to be directly caused by I_c; hence, anticipated or unanticipated side-effects are allowed. Figure 3.16 shows a generic fragment of an augmented model.

3.8.3   Categories of intervention

We now develop this broader concept of intervention by providing a classification of the different kinds of intervention we have alluded to above. We do this using two "dimensions" along which interventions may vary. The result of the intervention

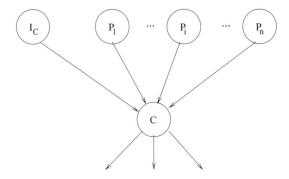

FIGURE 3.16: Fragment of a generic augmented causal model.

is the adoption by the targeted variable of a new probability distribution over its states (even when a single such state is forced by the intervention, when the new probability distribution is degenerate), whether or not this achieved distribution is also the target distribution. To be sure, the new distribution will be identical to the original distribution when the intervention is not attempted or is entirely ineffectual. This special case can be represented

$$P_{M'}(C|\pi_c, \neg I_c) = P_M(C|\pi_c) \tag{3.8}$$

where π_c is the set of the original parents of C.

Dimensions of Intervention

1. The degree of *dependency* of the effect upon the existing parents.

    (a) An entirely independent intervention leads to an achieved distribution which is a function only of the new distribution aimed for by the intervention. Thus, for an independent intervention, we have

$$P_{M'}(C|\pi_c, I_c) = P^*(C) \tag{3.9}$$

    (b) A dependent intervention leads to an achieved distribution which is a function of both the target distribution and the state of the variable's other parents.

An independent intervention on C simply cuts it off from its parents. Dependent interventions depend for their effect, in part, on the pre-existing parents of the target variable. The dependency across the parents, including the new I_c, may be of any variety: linear, noisy-or, or any kind of complex, non-linear interaction. These are precisely the kinds of dependency that Bayesian networks model already, so it is no extension of the semantics of Bayesian networks to incorporate them. Rather, it is something of a mystery that most prior work on intervention has ignored them.

2. Deterministic versus stochastic interventions.

(a) A deterministic intervention aims to leave the target variable in one particular state — i.e., the target distribution is extreme.

(b) A stochastic intervention aims to leave the target variable with a new
distribution with positive probability over two or more states.

A deterministic intervention is by intention simple. Say, get Fred to stop smoking.
By factoring in the other dimension, allowing for other variables still to influence the
target variable, however, we can end up with quite complex models. Thus, it might
take considerable complexity to reflect the interaction of a doctor's warning with
peer-group pressure.

The stochastic case is yet more complex. For example, in a social science study
we may wish to employ stratified sampling in order to force a target variable, say
age, to take a uniform distribution. That is an independent, stochastic intervention.
If, unhappily, our selection into experimental and control groups is not truly random,
it may be that this selection is related to age. And this relation may induce any kind
of actual distribution over the targeted age variable.

Any non-extreme actual distribution will be subject to changes under Bayesian updating, of course, whether it is for a targeted variable or not. For example,
a crooked Blackjack dealer who can manipulate the next card dealt with some high
probability, may intervene to set the next deal to be an Ace with probability 0.95. If
the card is later revealed to be an Ace, then obviously that probability will revised to
1.0.

Most interventions discussed in the literature are independent, deterministic interventions, setting C to some one specific state, regardless of the state of C's other
parents. We can call this sort of intervention Pearlian, since it is the kind of intervention described by Pearl's "do-calculus" (2000). This simplest kind of intervention
can be represented in a causal model simply by cutting all parent arcs into C and
setting C to the desired value.

3.8.4 Modeling effectiveness

There is another "dimension" along which interventions can be measured or ranked:
their effectiveness. Many attempted interventions have only some probability, say
r, of taking effect — for example, the already mentioned fact that doctors do not
command universal obedience in their lifestyle recommendations. Now, even if such
an intervention is of the type that when successful will put its target variable into a
unique state, the attempt to intervene will not thereby cut-off the target variable from
its parents; it is not Pearlian. The achieved distribution will, in fact, be a mixture of
the target distribution and the original distribution, with the mixing factor being the
probability r of the intervention succeeding.

Classifying or ranking interventions in terms of their effectiveness is often important. However, we have not put this scale on an equal footing with the other two
dimensions of intervention, simply because it is conceptually derivative. That is, any
degree of effectiveness r can be represented by mixing together the original with the
target distribution with the factor r. In case the intended intervention is otherwise

independent of the original parents, we can use the equation:

$$P_{M'}(C|\pi_c, I_c) = r \times P^*(C) + (1-r) \times P_M(C|\pi_c) \tag{3.10}$$

This being a function of all the parents of C, it is a special case of dependent interventions.

In practical modeling terms, to represent such interventions we maintain two Bayesian networks: one with a fully effective intervention and one with no intervention. (Note that the first may still be representing a dependent intervention, e.g., one which interacts with the other parents.) There are then two distinct ways to use this mixture model: we can do ordinary Bayesian net propagation, combining the two at the end with the weighting factor to produce new posterior distributions or expected-value computations; or, if we are doing stochastic sampling, we can flip a coin with bias r to determine which of the two models to sample from.

There are many other issues that can arise when updating networks with interventions which take a slightly different twist than when we are simply updating observational networks. In other words, these issues don't make for fundamental distinctions about what types of interventions we need to model, but they have practical consequences in applying these models. For example, although there is a strong tendency in the literature to consider only interventions which are straightforwardly *known* to have been attempted or not (in contrast with their effectiveness being presumed to be total), there are many cases where the actual attempt to intervene is itself unknown, but part of the modeling problem. Thus, in a competitive industrial situation we may wish to consider whether or not we should upgrade the quality of our ingredients in a manufacturing process. From the above, we know basically how we can model this, even if it is to simplify with a Pearlian intervention model. There is no difficulty about setting that intervention variable and reasoning causally; but the model presumably wants a corresponding variable for our competitor, who can also upgrade its quality, and yet we may not directly know whether it will or has done so. We may, in fact, need to apply partial observation methods (e.g., virtual nodes; see §3.4) to intervention variables for some such cases.

3.8.5   Representing interventions

Any Bayesian network tool can be used to implement interventions just by generating the augmented model manually, as in §3.8.2.[8] However, manual edits are awkward and time consuming, and they fail to highlight the intended causal semantics. Hence, we have developed a program, the *Causal Reckoner* (Hope, 2008), which runs as a front-end to the BN tool Netica (Norsys, 2010).

The *Causal Reckoner* makes Pearlian interventions as easy as observing a node and implements more sophisticated interventions via a pop-up, and easy to use, GUI. The mixture modeling representation of effectiveness (§3.8.4) is implemented via a

---

[8] Alternatively, the decision nodes of Chapter 4 can be used to model Pearlian interventions, since their use implies the arc-cutting of such interventions. However, that is an abuse of the semantics of decision nodes which we don't encourage.

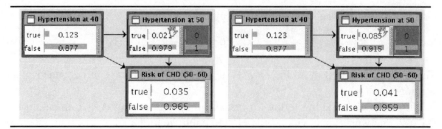

FIGURE 3.17: The hypertension causal model: (a) with 90% effective interventions and (b) 60% effective interventions. (With kind permission of Springer Science+Business Media (Korb et al., 2004) [Fig. 3].)

slider bar, and the target distribution is set by gauges. The full scope of possible interventions is not yet implemented (e.g., causally interactive interventions), as this requires arbitrary replacement of a node's CPT. Our visualization for basic interventions is shown in Figure 3.17. The node is shaded and a hand icon (for "manipulation") is displayed. We don't show the intervention node, simplifying and saving screen space.

When visualizing less than fully effective interventions, it is useful to report extra information. Figure 3.17(a) shows a 90% effective intervention intended to set low blood pressure at age 50. The target distribution is shown to the right of the node's actual distribution, which is a mixture of the original and target distributions. In the hypertension example, the intervention can be interpreted as a drug which fails in its effect 10% of the time. A drug with a weaker effect is shown in Figure 3.17(b).

Even a fully effective intervention can result in an actual distribution that deviates from the target distribution. This can happen when the intervention is stochastic, since other observational evidence also must be incorporated. Figure 3.18(a) shows the hypertension example given a fully effective stochastic intervention. Take a drug that sets the chance of low blood pressure to 95%, irrespective of other causal influences. This particular drug reduces the chances of CHD from 0.052 to 0.038. But what if the patient gets CHD anyway? Figure 3.18(b) (with an eye indicating that

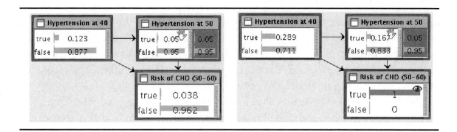

FIGURE 3.18: (a) The hypertension causal model where a stochastic medical intervention has been made. In (b) an observation has also been entered. (With kind permission of Springer Science+Business Media (Korb et al., 2004) [Fig. 4].)

CHD has been observed) shows that under this scenario, it is less likely that the drug *actually* helped with hypertension, since people with hypertension are more susceptible to CHD than others.

In short, the *Causal Reckoner* provides a GUI for mixing observations and interventions seamlessly. We can take existing networks in any domain and investigate various intervention policies quickly, without the trouble of creating new nodes and manually rewriting arbitrarily large CPTs.

3.9 Summary

Reasoning with Bayesian networks is done by updating beliefs — that is, computing the posterior probability distributions — given new information, called evidence. This is called probabilistic inference. While both exact and approximate inference is theoretically computationally complex, a range of exact and approximate inference algorithms have been developed that work well in practice. The basic idea is that new evidence has to be propagated to other parts of the network; for simple tree structures an efficient message passing algorithm based on local computations is used. When the network structure is more complex, specifically when there are multiple paths between nodes, additional computation is required. The best exact method for such multiply-connected networks is the junction tree algorithm, which transforms the network into a tree structure before performing propagation. When the network gets too large, or is highly connected, even the junction tree approach becomes computationally infeasible, in which case the main approaches to performing approximate inference are based on stochastic simulation. In addition to the standard belief updating, other computations of interest include the most probable explanation and the probability of the evidence. Finally, Bayesian networks can be augmented for causal modeling, that is for reasoning about the effect of causal interventions.

3.10 Bibliographic notes

Pearl (1982) developed the message passing method for inference in simple trees. Kim and Pearl (1983) extended it to polytrees. The polytree message passing algorithm given here follows that of Pearl (1988), using some of Russell's notation (2010).

Two versions of junction tree clustering were developed in the late 1980s. One version by Shafer and Shenoy (1990) (described in Jensen and Nielsen, 2007) suggested an elimination method resulting in a message passing scheme for their so-called join-tree structure, a term taken from the relational data base literature. The other method was initially developed by Lauritzen and Spiegelhalter (1988) as a two stage method based on the running intersection property. This was soon refined to a message passing scheme in a junction tree (Jensen, 1988, Jensen et al., 1990) described in this chapter.[9]

The junction tree cost given here is an estimate, produced by Kanazawa (1992), of the complexity of the junction tree method.

Another approach to exact inference is cutset conditioning (Pearl, 1986, Horvitz et al., 1989), where the network is transformed into multiple, simpler polytrees, rather than the single, more complex polytree produced by clustering. This approach has not developed as a serious contender to the clustering methods implemented in BN software.

Another exact inference algorithm implemented in some current BN software is variable elimination for updating the belief of a single query node (and a variant, bucket elimination), based on product and marginalization operators; a clear exposition is given by Russell and Norvig (2010). Note that some software (e.g., JavaBayes) also refers to "bucket tree elimination," a somewhat confusing name for what is essentially a junction tree approach.

The logic sampling method was developed by Henrion (1988), while likelihood weighting was developed by Fung and Chang (1989) and Shachter and Peot (1989). Other approximation methods based on stochastic sampling not covered in this book include Gibbs sampling (Geman et al., 1993), self-importance sampling and heuristic-importance sampling (Shachter and Peot, 1989), adaptive importance sampling (Cheng and Druzdzel, 2000), and backward sampling (Fung and del Favero, 1994).

There have been a number of other approximate inference methods proposed. These include state space abstraction (Wellman and Liu, 1994), localized partial evaluation (Draper, 1995), weak arc removal (Kjærulff, 1994) and using a mutual information measure to guide approximate evaluation (Jitnah, 2000). It has also been shown that applying Pearl's polytree algorithm to general networks, as suggested by Pearl (1988), — so-called **loopy propagation** — can be a both fast and accurate approximate method (Murphy et al., 1999). To our knowledge, none of these methods have been implemented in widely available BN software. Hugin implements an approximation scheme that involves setting very small probabilities in the junction tree tables to zero, which in turns allows more compression to take place.

Cooper (1990) showed that the general problem of inference in belief networks is NP hard, while Dagum and Luby (1993) showed the problem of approximating the new beliefs to within an arbitrary tolerance is also NP hard.

Our Causal Reckoner program (Hope, 2008) provides better visualization and intervention features than any other we have seen. Indeed, *Genie* (Druzdzel, 1999) is the only program with similar capabilities that we know of; it has the feature of 'controlling' nodes to perform Pearlian interventions.

---

[9]Thanks to Finn Jensen for providing a summary chronology of junction tree clustering.

3.11   Problems

Message passing

Problem 1

Consider (again — see Problem 2.8, Chapter 2) the belief network for another version of the medical diagnosis example, where *B=Bronchitis*, *S=Smoker*, *C=Cough*, *X=Positive X-ray* and *L=Lung cancer* and all nodes are Booleans. Suppose that the prior for a patient being a smoker is 0.25, and the prior for the patient having bronchitis (during winter in Melbourne!) is 0.05.

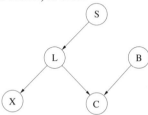

Suppose that evidence is obtained that a patient has a positive X-ray result and a polytree message passing algorithm was to be used to perform belief updating.

1. Write down the π and λ values for the following nodes: *S, B, C, X*.

2. Show the 3 stages of data propagation by the message passing algorithm in terms of the π and λ messages.

3. Suppose that a doctor considered that smoking was a contributing factor towards getting bronchitis as well as cancer. The new network structure reflecting this model is as follows.

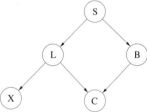

Why is the polytree message passing algorithm unsuitable for performing belief updating on this network?

Virtual / Likelihood evidence

Problem 2

A description of how to enter so-called likelihood evidence in Netica, BayesiaLab and Hugin software is given in Figure 3.19. In §3.4 we only looked at an example

Adding virtual evidence

Suppose you want to add uncertain evidence for node X with values $\{x_1, \ldots, x_n\}$ and that you have worked out that the likelihood ratio vector is $(r_1 : r_2 : \ldots : r_n)$ (as described in §3.4).

BayesiaLab. Right-click on the node's monitor, and choose the "Enter Likelihoods" option. The node's monitor will refresh, showing each state's likelihood with a numeric value (given from 0 to 100), and a horizontal bar.

- To set evidence for a particular state, double-click on the state's numeric value, and enter the new value via the keyboard; or set the value by dragging the edge of the horizontal bar.

- To confirm or discard the entered likelihoods, click on the green or red buttons, respectively, found in the node's monitor.

- To remove virtual evidence, right-click on a node's monitor, and choose the "Remove Likelihoods" option.

Hugin. To add evidence in Hugin, you have the following options.

- With a node selected, select `Network→Enter Likelihood`.
- Right click on belief bar/belief/state entry, and then choose the `Enter Likelihood` option.
- Click on the icon `≡`.

This brings up a window, showing each state with a horizontal bar whose length represents the evidence with a value from 0 to 1. The default for all states is 1.

- To set specific evidence, set a single value x_i to 1.0, and all the others to 0.
- To set negative evidence, set a single value x_j to 0, and leave all the others 1.
- To set likelihood evidence, set each value x_k corresponding to the r_k ratio that you have previously determined.

Netica. Right click on node and select the `likelihood` option. This will bring up a series of windows, the first of which asks you to provide a probability (default is 1) for $P(V|X = x_1)$. Here you should enter the probability r_1. Note that this *is* a probability, unlike everywhere else in Netica, where probabilities are given as values out of 100. You are asked for a series of these probabilities, one for each possible value of X.

If you try to re-enter likelihood evidence for a node, you will be asked if you want to discard previous evidence. If you do not, both the old and new evidence will be incorporated, equivalent to using multiple virtual nodes (assuming each piece of evidence is independent).

FIGURE 3.19: Adding virtual evidence in BayesiaLab, Hugin and Netica.

where we added virtual evidence for a root node; however it can be added for any node in the network. Consider the cancer example from Chapter 2 (supplied as Netica file `cancer.dne`). Suppose that the radiologist who has taken and analyzed the X-ray in this cancer example is uncertain. He thinks that the X-ray looks positive, but is only 80% sure.

- Add this evidence as a likelihood ratio of 4:1.
- Work out the likelihood ratio that needs to be used to produce a new belief of $Bel(X\text{-}ray) = 0.8$
- Add this likelihood ratio as evidence to confirm your calculations.
- Add an explicit virtual node V to represent the uncertainty in the observation. Confirm that adding specific evidence for V gives the same results as adding the likelihood evidence.

Clustering

Problem 3

Consider the BNs for the "Asia" problem described in §2.5.3.

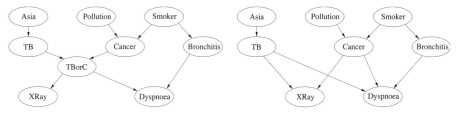

1. Use the Jensen junction tree algorithm (Algorithm 3.2) to construct a junction tree from both these networks, drawing (by hand) the resultant junction trees.

2. Load these networks (`asia1.dne` and `asia2.dne` in the on-line material) into Netica.

   (a) Compile the networks using standard Netica compilation.

   (b) Inspect the junction tree by selecting `Report`→`Junction tree` menu option, and note the elimination ordering (used in Netica's junction tree algorithm) by selecting `Report`→`Elimination Ordering` menu option. How do the junction trees produced by Netica compare to the ones you computed using Algorithm 3.2? What are the junction tree costs of each?

   (c) Re-compile the networks using the `Network`→`Compile Optimizing` menu option. Inspect again the junction trees and the elimination orderings. How much do they differ?

3. Now load the Hugin versions of these networks (`asia.net` and `asia2.net` in the on-line material), and compile them with varying settings of the triangulation heuristic. (See **A Quick Guide to Using Hugin** in Figure 2.11.)

(a) Clique Size

(b) Clique Weight

(c) Fill-in Size

(d) Fill-in Weight

(e) Optimal Triangulation

How do the resultant junction trees differ in structure and corresponding junction tree cost

(a) From each other?

(b) From the those you obtained executing the algorithm by hand?

(c) From those obtained from Netica's standard and optimized compilation?

Approximate inference

Problem 4

The on-line material for this text provides a version of both the Logic Sampling algorithm (Algorithm 3.3) and the Likelihood Weighting algorithm (Algorithm 3.4).

Take an example BN (either provided with the online material, or that you've developed for the problems set in Chapter 2) and do the following.

1. Run the BN software to obtain the exact inference result.
2. Run the LS Algorithm, printing out the approximate beliefs every 10 iterations and stopping when a certain level of convergence has been achieved.
3. Do the same for the LW algorithm.
4. As we have seen, the Kullback-Leibler divergence (§3.6.5) can be used to measure the error in the beliefs obtained using an approximate inference algorithm. Compute and plot the KL error over time for both the LS and LW algorithm.
5. Investigate what effect the following changes may have on the error for the LW algorithm.

   • Vary the priors between (i) more uniform and (ii) more extreme.

   • Vary the location of the evidence (i) root, (ii) intermediate and (iii) leaf.

   • Set evidence that is more or less likely.

Problem 5

As mentioned in the Bibliographic Notes, Hugin implements an approximation scheme that involves setting very small probabilities in the junction tree tables to zero, which in turns allows more compression to take place.

To turn on the approximation scheme, select `Network→Network Properties→Compilation`, then check the approximate optimization box.

As for the previous problem, select an example BN and perform an investigation of the effect of varying the approximation threshold on the belief updating, again using the KL divergence measure.

Causal reasoning

Problem 6

Take Pearl's earthquake example. Suppose there is an intervention on *JohnCalls*.

1. Use ordinary observation and updating. What is *Bel(Burglary)*?
2. Use Pearl's cut-link method. What is *Bel(Burglary)*? this case?
3. Use the augmented model with 0.9 effectiveness. What is *Bel(Burglary)*?

Now add an independent observation for *MaryCalls=T*. Parts 4, 5 and 6 of this problem involve repeating the parts 1, 2 and 3 for this new situation.

4

Decision Networks

4.1 Introduction

By now we know how to use Bayesian networks to represent uncertainty and do probabilistic inference. In this chapter we extend them to support decision making. Adding an explicit representation of both the actions under consideration and the value or **utility** of the resultant outcomes gives us **decision networks** (also called **influence diagrams** by Howard and Matheson, 1981). Bayesian decision networks combine probabilistic reasoning with utilities, helping us to make decisions that maximize the **expected utility**, as described in §1.5.3.

We will begin with utilities and then describe how they are represented together with probabilities in decision networks. Then we present the algorithm for evaluating a decision network to make individual decisions, illustrating with several examples. We are also interested in determining the best sequences of decisions or actions, that is to say, with **planning**. First, we will use a decision network for a "test-then-act" combination of decisions. Then, we introduce **dynamic Bayesian networks** for explicitly modeling how the world changes over time. This allows us to generalize decision networks to **dynamic decision networks**, which explicitly model sequential decision making or planning under uncertainty. We conclude the chapter with a description of **object-oriented Bayesian Decision Networks**, a framework for building large, complex, hierarchical Bayesian decision networks.

4.2 Utilities

When deciding upon an action, we need to consider our **preferences** between the different possible outcomes of the available actions. As already introduced in §1.5.3, **utility theory** provides a way to represent and reason with preferences. A **utility function** quantifies preferences, reflecting the "usefulness" (or "desirability") of the outcomes, by mapping them to the real numbers.

Such a mapping allows us to combine utility theory with probability theory. In particular, it allows us to calculate which action is expected to deliver the most value (or utility) given any available evidence E in its **expected utility**:

$$EU(A|E) = \sum_i P(O_i|E,A)\, U(O_i|A) \qquad (4.1)$$

where

- E is the available evidence,
- A is a non-deterministic action with possible outcome states O_i,
- $U(O_i|A)$ is the utility of each of the outcome states, given that action A is taken,

- $P(O_i|E,A)$ is the conditional probability distribution over the possible outcome states, given that evidence E is observed and action A taken.

The **principle of maximum expected utility** asserts that an essential part of the nature of rational agents is to choose that action which maximizes expected utility. In the following section we will see how to extend Bayesian networks to model such rational decision making.

First, however, we will point out a few pitfalls in constructing utility functions. The simplest, and very common, way of generating utilities, especially for business modeling, is to equate utilities with money. Indeed, in the case of business modeling this will often be satisfactory. Perhaps the most obvious caveat is that there is a time value associated with money, so if some outcome delivers money in the future, it should be discounted in comparison with outcomes delivering money in the present (with the discount being equal to standard interest/discount rates). A slightly more subtle point is that even discounted money is rarely just identical to utility. In fact, the relation between money and utility is represented in Figure 4.1, with the dashed line representing the *abnormal* case where money and utility are identical. The solid line represents the more usual case, which can readily be understood at the extremities: after losing enough money, one is bankrupt, so losing more does not matter; after earning enough money, one is retired, so earning more does not matter. In short, the marginal value of money declines, and this needs to be kept in mind when generating utility functions.

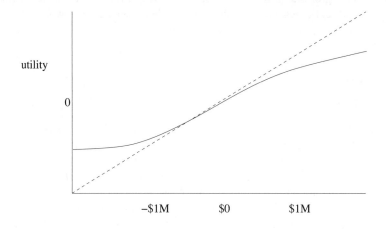

FIGURE 4.1: The utility of money.

Over and above such considerations are others that are possibly less rational.

Thus, cognitive psychologists have discovered that most people are **risk averse**, meaning that they will forgo a significant gain in expected value in order to reduce their uncertainty about the future. For example, many people will prefer to keep $10 in their hand, rather than buy a lottery ticket with an expected value of $20 when the probability of losing the $10 is, say, 0.999. Of course, many others are **risk prone** and will happily part with $1 day after day in the hopes of becoming a millionaire, even when the expected value of the gamble is -95 cents. Such behavior may again be explained, in part, through differences between utility and money. However, there does appear to be residual risk aversion and risk proneness which resists such explanation, that is, which remains even when matters are carefully recast in terms of utilities alone (Bell, 1988).

In generating utilities the choice of unit is arbitrary, as utility functions differing only in scale result in the same decisions. The range of utilities can be set by establishing a scale from the best possible outcome $U(O_{best})$, to some neutral outcome, down to the worst case $U(O_{worst})$. We discuss how to assess utilities further in §10.6.

4.3   Decision network basics

A decision network is an extension of Bayesian networks that is able to represent the main considerations involved in decision making: the state of the world, the decisions or actions under consideration, the states that may result from an action and the utility of those resultant states.

4.3.1   Node types

A decision network consists of three types of nodes, as shown in Figure 4.2.

| Chance | Decision | Utility |

FIGURE 4.2: Decision network node types.

Chance nodes: These have an oval shape and represent random variables, exactly as in Bayesian networks. Each has an associated conditional probability table (CPT), giving the probability of the variable having a particular value given a combination of values of its parents. Their parent nodes can be decision nodes as well as other chance nodes.

Decision nodes: These have a rectangular shape and represent the decision being made at a particular point in time. The values of a decision node are the actions that the decision maker must choose between.[1] A decision node can have chance nodes as a special kind of parent, indicating that evidence about the parent nodes will be available at the time of decision (see §4.3.4). A decision network representing a single decision has only one decision node, representing an isolated decision. When the network models a sequence of decisions (see §4.4), decision nodes can have other decision nodes as parents, representing the order of decisions.

Utility nodes: These have a diamond shape and represent the agent's utility function. They are also called **value nodes**. The parents of a utility node are the variables describing the outcome state that directly affect the utility and may include decision nodes. Each utility node has an associated **utility table** with one entry for each possible instantiation of its parents, perhaps including an action taken. When there are multiple utility nodes, the overall utility is the sum of the individual utilities.

We will now see how these node types can be used to model a decision problem with the following simple example.

4.3.2   Football team example

Clare's football team, Melbourne, is going to play her friend John's team, Carlton. John offers Clare a friendly bet: whoever's team loses will buy the wine next time they go out for dinner. They never spend more than $15 on wine when they eat out. When deciding whether to accept this bet, Clare will have to assess her team's chances of winning (which will vary according to the weather on the day). She also knows that she will be happy if her team wins and miserable if her team loses, regardless of the bet.

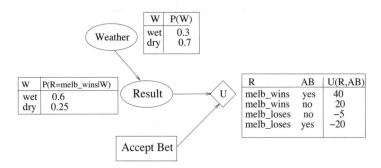

FIGURE 4.3: A decision network for the football team example.

A decision network for this problem is shown in Figure 4.3. This network has one chance node *Result* representing whether Clare's team wins or loses (values

---

[1]Decision nodes are sometimes also called **action** nodes.

{*melb_wins, melb_loses*}), and a second chance node *Weather* which represents whether or not it rains during the match (values {*rain, dry*}). It has a binary decision node *AcceptBet* representing whether or not she bets and a utility node *U* that measures the decision maker's level of satisfaction.

The priors for the *Weather* reflect the typical match conditions at this time of year. The CPT for *Result* shows that Clare expects her team to have a greater chance of winning if it doesn't rain (as she thinks they are the more skillful team).

There are arcs from *Result* and *AcceptBet* to *U*, capturing the idea that Clare's satisfaction will depend on a combination of the eventual match winner and the betting decision. Her preferences are made explicit in the utility table. The numbers given indicate that the best outcome is when her team wins and she accepted the bet (utility = 40) while the next best outcome is her team wins but she didn't bet on the result (utility = 20). When her team loses but she didn't bet, Clare isn't happy (utility = -5) but the worst outcome is when she has to buy the dinner wine also (utility = -20). Clearly, Clare's preferences reflect more than the money at risk. Note also that in this problem the decision node doesn't affect any of the variables being modeled, i.e., there is no explicit outcome variable.

4.3.3 Evaluating decision networks

To evaluate a decision network with a single decision node:

Algorithm 4.1 Decision Network Evaluation Algorithm (Single decision)

1. *Add any available evidence.*

2. *For each action value in the decision node:*

   (a) *Set the decision node to that value;*

   (b) *Calculate the posterior probabilities for the parent nodes of the utility node, as for Bayesian networks, using a standard inference algorithm;*

   (c) *Calculate the resulting expected utility for the action.*

3. *Return the action with the highest expected utility.*

For the football example, with no evidence, it is easy to see that the expected utility in each case is the sum of the products of probability and utility for the different cases. With no evidence added, the probability of Melbourne winning is

$$P(R = melb\_wins) \quad = \quad P(W = w) \times P(R = melb\_wins | W = w) + P(W = d)$$
$$= \quad \times P(R = melb\_wins | W = d)$$

and the losing probability is $1 - P(R = melb\_wins)$. So the expected utility is:

$$
\begin{aligned}
EU(AB = yes) &= P(R = melb\_wins) \times U(R = melb\_wins|AB = yes) \\
&+ P(R = melb\_loses) \times U(R = melb\_loses|AB = yes) \\
&= (0.3 \times 0.6 + 0.7 \times 0.25) \times 40 + (0.3 \times 0.4 + 0.7 \times 0.75) \times -20 \\
&= 0.355 \times 40 + 0.645 \times -20 = 14.2 - 12.9 = 1.3 \\
EU(AB = no) &= P(R = melb\_wins) \times U(R = melb\_wins|AB = no) \\
&+ P(R = melb\_loses) \times U(R = melb\_loses|AB = no) \\
&= (0.3 \times 0.6 + 0.7 \times 0.25) \times 20 + (0.3 \times 0.4 + 0.7 \times 0.75) \times -5 \\
&= 0.355 \times 20 + 0.645 \times -5 = 7.1 - 3.225 = 3.875
\end{aligned}
$$

Note that the probability of the outcomes Clare is interested in (i.e., her team winning or losing) is independent of the betting decision. With no other information available, Clare's decision is to not accept the bet.

4.3.4   Information links

As we mentioned when introducing the types of nodes, there may be arcs from chance nodes to decision nodes — these are called **information links** (Jensen and Nielsen, 2007, p. 305). These links are not involved in the basic network evaluation process and have no associated parameters. Instead, they indicate when a chance node needs to be observed before the decision D is made — but after any decisions prior to D. With an information link in place, network evaluation can be extended to calculate explicitly what decision should be made, *given* the different values for that chance node. In other words, a table of optimal actions conditional upon the different relevant states of affairs can be computed, called a **decision table**. To calculate the table, the basic network evaluation algorithm is extended with another loop (see Algorithm 4.2). We will also refer to this conditional decision table as a **policy**.

Algorithm 4.2 Decision Table Algorithm *(Single decision node, with information links)*

1. *Add any available evidence.*
2. *For each combination of values of the parents of the decision node:*

    (a) For each action value in the decision node:

        i. Set the decision node to that value;

        ii. Calculate the posterior probabilities for the parent nodes of the utility node, as for Bayesian networks, using a standard inference algorithm;

        iii. Calculate the resulting expected utility for the action.

    (b) Record the action with the highest expected utility in the decision table.

3. *Return the decision table.*

To illustrate the use of information links, suppose that in the football team example, Clare was only going to decide whether to accept the bet or not after she heard the weather forecast. The network can be extended with a *Forecast* node, representing the current weather forecast for the match day, which has possible values {*sunny*, *cloudy* or *rainy*}. *Forecast* is a child of *Weather*. There should be an information link from *Forecast* to *AcceptBet*, shown using dashes in Figure 4.4, indicating that Clare will know the forecast when she makes her decision. Assuming the same CPTs and utility table, the extended network evaluation computes a decision table, for the decision node *given* the forecast, also shown in Figure 4.4. Note that most BN software does not display the expected utilities.[2] If we want them, we must evaluate the network for each evidence case; these results are given in Table 4.1 (highest expected utility in each case in **bold**).

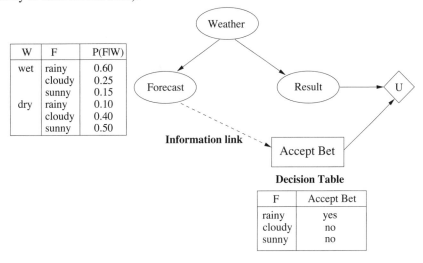

FIGURE 4.4: The football team decision network extended with the *Forecast* node and an information link, with the decision table for *AcceptBet* computed during network evaluation.

TABLE 4.1

Decisions calculated for football team, given the new evidence node *Forecast*.

F	$Bel(W = wet)$	$Bel(R = melb\_wins)$	EU(AB=yes)	EU(AB=no)
rainy	0.720	0.502	**10.12**	7.55
cloudy	0.211	0.324	-0.56	**3.10**
sunny	0.114	0.290	-2.61	**2.25**

---

[2]GeNIe is one exception that we are aware of.

4.3.5 Fever example

Suppose that you know that a fever can be caused by the flu. You can use a thermometer, which is fairly reliable, to test whether or not you have a fever. Suppose you also know that if you take aspirin it will almost certainly lower a fever to normal. Some people (about 5% of the population) have a negative reaction to aspirin. You'll be happy to get rid of your fever, as long as you don't suffer an adverse reaction if you take aspirin. (This is a variation of an example in Jensen, 2001.)

A decision network for this example is shown in Figure 4.5. The *Flu* node (the cause) is a parent of the *Fever* (an effect). That symptom can be measured by a thermometer, whose reading *Therm* may be somewhat unreliable. The decision is represented by the decision node *Take Aspirin*. If the aspirin is taken, it is likely to get rid of the fever. This change over time is represented by a second node *FeverLater*. Note that the aspirin has no effect on the flu and, indeed, that we are not modeling the possibility that the flu goes away. The adverse reaction to taking aspirin is represented by *Reaction*. The utility node, *U*, shows that the utilities depend on whether or not the fever is reduced and whether the person has an adverse reaction. The decision table computed for this network is given in Table 4.2 (highest expected utility in each case in **bold**). Note the observing a high temperature changes the decision to taking the aspirin, but further information about having a reaction reverses that decision.

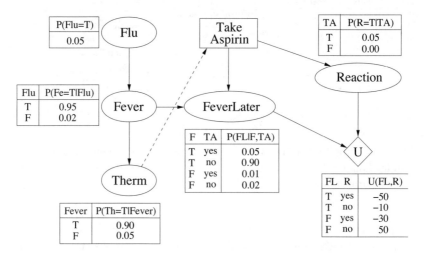

FIGURE 4.5: A decision network for the fever example.

4.3.6 Types of actions

There are two main types of actions in decision problems, intervening and non-intervening. **Non-intervening actions** do not have a direct effect on the chance variables being modeled, as in Figure 4.6(a). Again, in the football team example, making a bet doesn't have any effect on the states of the world being modeled.

TABLE 4.2

Decisions calculated for the fever problem given different values for *Therm* and *Reaction*.

Evidence	*Bel(FeverLater=T)*	EU(TA=yes)	EU(TA=no)	Decision
None	0.046	45.27	**45.29**	no
Therm=F	0.018	45.40	**48.40**	no
Therm=T	0.273	**44.12**	19.13	yes
Therm=T &	0.033	-30.32	**0**	no
Reaction=T				

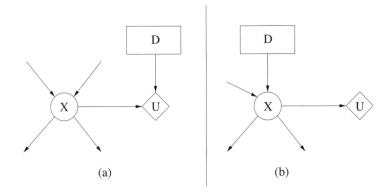

(a)   (b)

FIGURE 4.6: Generic decision networks for (a) non-intervening and (b) intervening actions.

Intervening actions do have direct effects on the world, as in Figure 4.6(b). In the fever example, deciding to take aspirin will affect the later fever situation. Of course in all decision making, the underlying assumption is that the decision will affect the utility, either directly or indirectly; otherwise, there would be no decision to make.

The use of the term "intervention" here is apt: since the decision impacts upon the real world, it is a form of causal intervention as previously discussed in §3.8. Indeed, one can perform the causal modeling discussed there with the standard Bayesian network tools by attaching a parent decision node to the chance node that one wishes to intervene upon. We would prefer these tools to keep decision making and causal modeling distinct, at least in the human-computer interface. One reason is that causal intervention is generally (if not necessarily) associated with a single chance node, whereas the impact of decisions is often more wide-ranging. In any case, the user's intent is normally quite different. Decision making is all about optimizing a utility-driven decision, whereas causal modeling is about explaining and predicting a causal process under external perturbation.

4.4 Sequential decision making

Thus far, we have considered only single decision problems. Often however a decision maker has to select a sequence of actions, or a **plan**.

4.4.1 Test-action combination

A simple example of a sequence of decisions is when the decision maker has the option of running a test, or more generally making an observation, that will provide useful information before deciding what further action to take.

In the football decision problem used in the previous section, Clare might have a choice as to whether to obtain the weather forecast (perhaps by calling the weather bureau). In the lung cancer example (see §2.2), the physician must decide whether to order an X-ray, before deciding on a treatment option.

This type of decision problem has two stages:

1. The decision whether to run a test or make an observation
2. The selection of a final action

The test/observe decision comes *before* the action decision. And in many cases, the test or observation has an associated cost itself, either monetary, or in terms of discomfort and other physical effects, or both.

A decision network showing the general structure for these test-act decision sequences is shown in Figure 4.7. There are now two decision nodes, *Test*, with values {*yes, no*}, and *Action*, with as many values as there are options available. The temporal ordering of the decisions is represented by a **precedence link**, shown as a dotted line.

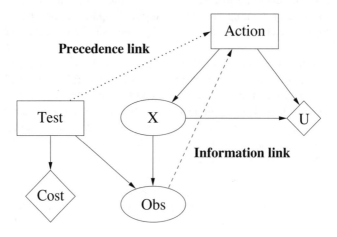

FIGURE 4.7: Decision network for a test-action sequence of decisions.

If the decision is made to run the test, evidence will be obtained for the observation node *Obs*, *before* the *Action* decision is made; hence there is an information link from *Obs* to *Action*. The question then arises as to the meaning of this information link if the decision is made *not* to run the test. This situation is handled by adding an additional state, *unknown*, to the *Obs* node and setting the CPT for *Obs*:

$$P(Obs = unknown|Test = no) = 1$$
$$P(Obs = unknown|Test = yes) = 0$$

In this generic network, there are arcs from the *Action* node to both the chance node *X* and the utility node *U*, indicating intervening actions with a direct associated cost. However, either of these arcs may be omitted, representing a non-intervening action or one with no direct cost, respectively.

There is an implicit assumption of **no-forgetting** in the semantics of a decision network. The decision maker remembers the past observations and decisions, indicated explicitly by the information and precedence links.

Algorithm 4.3 shows how to use the "Test-Action" decision network for sequential decision making. We will now look at an decision problem modeled with such a network and work through the calculations involved in the network evaluation.

Algorithm 4.3 Using a "Test-Action" Decision Network

1. *Evaluate decision network with any available evidence (other than for the Test result).*
   Returns Test decision.

2. *Enter Test decision as evidence.*

3. *If Test decision is 'yes'*
   Run test, get result;
   Enter test result as evidence to network.
   Else
   Enter result 'unknown' as evidence to network.

4. *Evaluate decision network.*
   Returns Action decision.

4.4.2 Real estate investment example

Paul is thinking about buying a house as an investment. While it looks fine externally, he knows that there may be structural and other problems with the house that aren't immediately obvious. He estimates that there is a 70% chance that the house is really in good condition, with a 30% chance that it could be a real dud. Paul plans to re-sell the house after doing some renovations. He estimates that if the house really is in good condition (i.e., structurally sound), he should make a $5,000 profit, but if it isn't, he will lose about $3,000 on the investment. Paul knows that he can get a building surveyor to do a full inspection for $600. He also knows that the inspection report may not be completely accurate. Paul has to decide whether it is worth it to

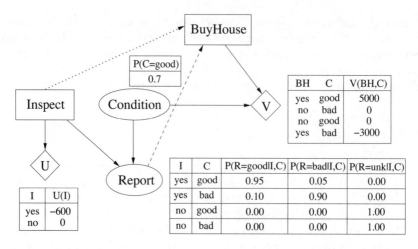

FIGURE 4.8: Decision network for the real estate investment example.

have the building inspection done, and then he will decide whether or not to buy the house.

A decision network for this "test-act" decision problem is shown in Figure 4.8. Paul has two decisions to make: whether to do have an inspection done and whether to buy the house. These are represented by the *Inspect* and *BuyHouse* decision nodes, both {*yes,no*} decisions. The condition of the house is represented by the *Condition* chance node, with values {*good, bad*}. The outcome of the inspection is given by node *Report*, with values {*good, bad, unknown*}. The cost of the inspection is represented by utility node *U*, and the profits after renovations (not including the inspection costs) by a second utility node *V*. The structure of this network is exactly that of the general network shown in Figure 4.7.

When Paul decides whether to have an inspection done, he doesn't have any information about the chance nodes, so there are no information links entering the *Inspect* decision node. When he decides whether or not to buy, he will know the outcome of that decision (either a *good* or *bad* assessment, or it will be *unknown*), hence the information link from *Report* to *BuyHouse*. The temporal ordering of his decisions, first about the inspection, and then whether to buy, is represented by the precedence link from *Inspect* to *BuyHouse*. Note that there is a directed path from *Inspect* to *BuyHouse* (via *Report*) so even if there was no explicit precedence link added by the knowledge engineer for this problem, the precedence could be inferred from the rest of the network structure.[3]

Given the decision network model for the real-estate investment problem, let's see how it can be evaluated to give the expected utilities and hence make decisions.

---

[3] Some BN software, such as Netica and GeNIe, will add such precedence links automatically.

4.4.3 Evaluation using a decision tree model

In order to show the evaluation of the decision network, we will use a **decision tree** representation. The nonleaf nodes in a decision tree are either decision nodes or chance nodes and the leaves are utility nodes. The nodes are represented using the same shapes as in decision networks. From each decision node, there is a labeled link for each alternative decision, and from each chance node, there is a labeled link for each possible value of that node. A decision tree for the real estate investment problem is shown in Figure 4.9.

To understand a decision tree, we start with the root node, which in this case is the first decision node, whether or not to inspect the house. Taking a directed path down the tree, the meaning of the link labels are:

- From a decision node, it indicates which decision is made
- From a chance node, it indicates which value has been observed

At any point on a path traversal, there is the same assumption of "no-forgetting," meaning the decision maker knows all the link labels from the root to the current position. Each link from a chance node has a probability attached to it, which is the probability of the variable having that value *given* the values of all the link labels to date. That is, it is a conditional probability. Each leaf node has a utility attached to it, which is the utility given the values of all the link labels on its path from the root.

In our real estate problem, the initial decision is whether to inspect (decision node I), the result of the inspection report, if undertaken, is represented by chance node R, the buying decision by BH and the house condition by C. The utilities in the leaves are combinations of the utilities in the U and V nodes in our decision network. Note that in order to capture exactly the decision network, we should probably include the report node in the "Don't Inspect" branch, but since only the "unknown" branch would have a non-zero probability, we omit it. Note that there is a lot of redundancy in this decision tree; the decision network is a much more compact representation.

A decision tree is evaluated as in Algorithm 4.4. Each possible alternative scenario (of decision and observation combinations) is represented by a path from the root to a leaf. The utility at that leaf node is the utility that would be obtained if that particular scenario unfolded. Using the conditional probabilities, expected utilities associated with the chance nodes can be computed as a sum of products, while the expected utility for a decision assumes that the action returning the highest expected utility will be chosen (shown with **BOLD**, with thicker arcs, in Figure 4.9). These expected utilities are stored at each non-leaf node in the tree (shown in Figure 4.9 in *underlined italics*) as the algorithm works its way recursively back up to the root node.

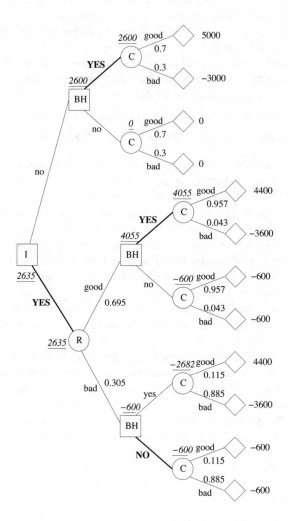

FIGURE 4.9: Decision tree evaluation for real estate investment example.

Algorithm 4.4 Decision Tree Evaluation

1. *Starting with nodes that have only leaves (utility nodes) as children.*

2. *If the node X is a chance node, each outgoing link has a probability and each child has an associated utility. Use these to compute its expected utility*

$$EU(X) = \sum_{C \in Children(X)} U(C) \times P(C)$$

If the node is a decision node, each child has a utility or expected utility attached. Choose the decision whose child has the maximum expected utility and

$$EU(X) = max_{C \in Children(X)}(EU(C))$$

3. *Repeat recursively at each level in the tree, using the computed expected utility for each child.*

4. *The value for the root node is the maximal expected utility obtained if the expected utility is maximized at each decision.*

4.4.4 Value of information

The results of the decision tree evaluation are summarized in Table 4.3 (highest expected utility in each case in **bold**). We can see that the rational decision for Paul is to have the inspection done, as the expected utility (which is the expected profit in this case) is 2635 compared to 2600 if the inspection is not done. Paul's next decision, as to whether to buy the house, will depend on whether he receives a good or bad inspection report. If the report is bad, his decision will be not to buy, with expected utility -600 compared to -2682 for buying. If the report is good, he will go ahead and buy, with expected utility 4055 compared to the same -600 when not buying.

TABLE 4.3 Decisions calculated for the real estate investment problem.

Evidence	Bel(C=good)	EU(I=yes)	EU(I=no)	Decision
None	0.70	**2635**	2600	I=yes
Given *I=no*		EU(BH=yes)	EU(BH=no)	
Report=unknown	0.70	**2600**	0	BH=yes
Given *I=yes*		EU(BH=yes)	EU(BH=no)	
Report=good	0.957	**4055**	-600	BH=yes
Report=bad	0.115	-2682	**-600**	BH=no

This is a situation where additional information may change a decision, as without the test Paul would just go ahead and buy the house. The decision of whether to gather information is based on the value of the information. The difference in the expected utilities with and without the extra information is called the **expected benefit**. In general,

$$EB(Test) = EU(Test = yes) - EU(Test = no)$$

For our real estate example,

$$
\begin{aligned}
EB(\mathit{Inspect}) \quad &= \quad EU(\mathit{Inspect} = yes) - EU(\mathit{Inspect} = no) \\
&= \quad 2635 - 2600 = 35
\end{aligned}
$$

So, even if there is a cost associated with obtaining additional information (such as the inspection fee), if the expected benefit is greater than zero, the price is worth paying.

4.4.5 Direct evaluation of decision networks

This decision tree evaluation method conveys the underlying ideas of evaluating decision networks containing sequential decisions. We start with the final decision and calculate the expected utility for the various options, given the scenario that has been followed to get there, and choose the decision with the maximum expected utility. This is then repeated for the next-to-last decision, and so on, until the first decision is made.

However, expanding a decision network into a decision tree and using Algorithm 4.4 is very inefficient. Various methods have been developed for evaluating decision networks (see Bibliographic notes in §4.9). Many are similar to the inference algorithms described in Chapter 3, involving compilation into an intermediate structure, and propagation of probabilities *and* expected utilities (see Jensen and Nielsen, 2007, Chapter 10 for details). It is possible to avoid the repetitions of the same calculations that we saw using decision tree evaluation, by taking advantage of the network structure. However, all methods face the same complexity problem.

Thus far, we have only looked at sequential decision networks involving just two decisions. There is no theoretical limit to how many sequential decisions can be made. However, this leads us to the general problem of **planning under uncertainty**, which once again is an exponential search problem, as the number of possible plans is the product of the number of actions considered for each plan step.

4.5 Dynamic Bayesian networks

Bayesian and decision networks model relationships between variables at a particular point in time or during a specific time interval. Although a causal relationship represented by an arc implies a temporal relationship, BNs do not explicitly model temporal relationships between variables. And the only way to model the relationship between the current value of a variable, and its past or future value, is by adding another variable with a different name. We saw an example of this with the fever example earlier in §4.3.5, with the use of the *FeverLater* node. In the decision networks we have seen so far, there is an ad hoc modeling of time, through the use of information and precedence links. When making a sequence of decisions that will span a period of time, it is also important to model how the world *changes* during that time. More generally, it is important to be able to represent and reason about changes over time explicitly when performing such tasks as diagnosis, monitoring, prediction and decision making/planning.

In this section we introduce a generalization of Bayesian networks, called **dynamic Bayesian networks** (DBNs),[4] that explicitly model change over time. In the following section we will extend these DBNs with decision and utility nodes, to give **dynamic decision networks**, which are a general model for sequential decision making or planning under uncertainty.

4.5.1 Nodes, structure and CPTs

Suppose that the domain consists of a set of n random variables $\mathbf{X} = \{X_1, \ldots, X_n\}$, each of which is represented by a node in a Bayesian network. When constructing a DBN for modeling changes over time, we include one node for each X_i for each time step. If the current time step is represented by t, the previous time step by $t - 1$, and the next time step by $t + 1$, then the corresponding DBN nodes will be:

- Current: $\{X_1^t, X_2^t, \ldots, X_n^t\}$
- Previous: $\{X_1^{t-1}, X_2^{t-1}, \ldots, X_n^{t-1}\}$
- Next: $\{X_1^{t+1}, X_2^{t+1}, \ldots, X_n^{t+1}\}$

Each time step is called a **time-slice**. The relationships between variables in a time-slice are represented by **intra-slice** arcs, $X_i^t \rightarrow X_j^t$. Although it is not a requirement, the structure of a time-slice does not usually change over time. That is, the relationship between the variables $X_1^t, X_2^t \ldots, X_n^t$ is the same, regardless of the particular T.

The relationships between variables at successive time steps are represented by **inter-slice** arcs, also called **temporal arcs**, including relationships between (i) the same variable over time, $X_i^t \rightarrow X_i^{t+1}$, and (ii) different variables over time, $X_i^t \rightarrow X_j^{t+1}$.

In most cases, the value of a variable at one time affects its value at the next, so the $X_i^t \rightarrow X_i^{t+1}$ arcs are nearly always present. In general, the value of any node at one time can affect the value of any other node at the next time step. Of course, a fully temporally connected network structure would lead to complexity problems, but there is usually more structure in the underlying process being modeled.

Figure 4.10 shows a generic DBN structure, with a sequence of the same static BNs connected with inter-slice arcs (shown with thicker arcs). Note that there are no arcs that span more than a single time step. This is another example of the **Markov assumption** (see §2.2.4), that the state of the world at a particular time depends only on the previous state and any action taken in it.

---

[4]Also called **dynamic belief networks** (Russell and Norvig, 1995, Nicholson, 1992), **probabilistic temporal networks** (Dean and Kanazawa, 1989, Dean and Wellman, 1991) and **dynamic causal probabilistic networks** (Kjærulff, 1992).

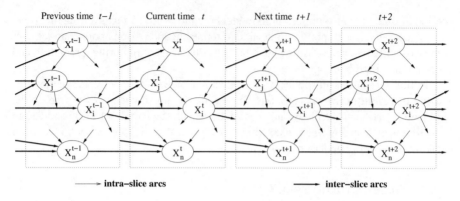

FIGURE 4.10: General structure of a Dynamic Bayesian Network.

The relationships between variables, both intra-slice and inter-slice, are quantified by the conditional probability distribution associated with each node. In general, for node X_i^t with intra-slice parents $Y_1^t, \ldots, Y_m^t$ and inter-slice parents X_i^{t-1} and Z_1^{t-1}, $\ldots, Z_r^{t-1}$, the CPT is

$$P(X_i^t | Y_1^t, \ldots, Y_m^t, X_i^{t-1}, Z_1^{t-1}, \ldots, Z_r^{t-1}).$$

Given the usual restriction that the networks for each time slice are exactly the same and that the changes over time also remain the same (i.e., both the structure and the CPTs are unchanging), a DBN can be specified very compactly. The specification must include:

- Node names
- Intra-slice arcs
- Temporal (inter-slice) arcs
- CPTs for the first time slice t_0 (when there are no parents from a previous time)

- CPTs for $t + 1$ slice (when parents may be from t or $t + 1$ time-slices).

Some BN software packages provide a facility to specify a DBN compactly (see §B.3).

Figure 4.11 shows how we can use a DBN to represent change over time explicitly in the fever example. The patient's flu status may change over time, as there is some chance the patient will get better, hence the $Flu^t \rightarrow Flu^{t+1}$ arc. Taking aspirin at time t, indicated by A^t, may produce a reaction at the next time step $(A^t \rightarrow React^{t+1})$ and it may reduce the fever $(A^t \rightarrow Fever^{t+1})$, although the subsequent fever status depends on the earlier fever status (represented by $Fever^t \rightarrow Fever^{t+1}$). A person's reaction to aspirin is consistent, hence the arc $React^t \rightarrow React^{t+1}$.

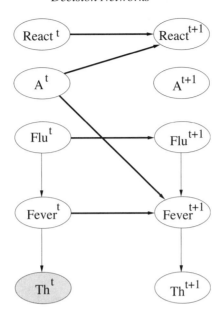

FIGURE 4.11: DBN for the fever example. (Shaded node indicates evidence added.)

4.5.2 Reasoning

Given evidence about a set of nodes, $\mathbf{E}_{\{1,t\}}$, from the first time slice up to and including the current time-slice t, we can perform belief updating on the full DBN, using standard BN inference algorithms. This means obtaining new posterior distributions for all the non-evidence nodes, including nodes in the $t + 1$ and later time-slices. This updating into the future is called **probabilistic projection**.

However, this type of DBN gets very large, very quickly, especially if the interval between time slices is short. To cope, in most cases the DBN is not extended far into the future. Instead, a fixed size, sliding "window" of time slices is maintained. As the reasoning process moves forward with time, one older time slice is dropped off the DBN, while another is added.

Figure 4.12 shows the progress of a simple two time-slice DBN. This structure can be considered a generic DBN, consisting of state node X, its corresponding observation node *Obs*, where evidence is added, and an action node A that will affect the state node at the next time step.[5]

This use of a fixed window means that every time we move the window along, the previous evidence received is no longer directly available. Instead, it is summarized taking the current belief for (root) nodes, and making these distributions the new priors. The DBN updating process is given in Algorithm 4.5. Note that the steps of this DBN updating algorithm are exactly those of a technique used in classical

---

[5]Note that this is a standard BN node representing a random variable, *not* a decision/action node from a decision network.

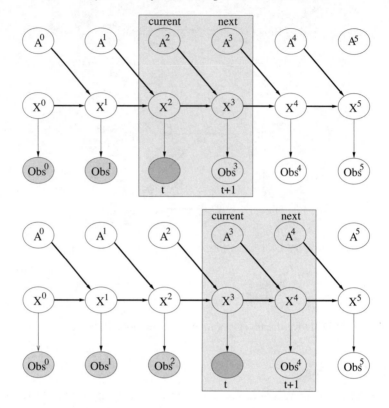

FIGURE 4.12: DBN maintained as a sliding "window" of two time-slices. (Shading indicates evidence node.)

control theory, called a **Kalman Filter** (Kalman, 1960) (see §4.9).

Algorithm 4.5 DBN Updating Process

1. **Sliding:** *Move window along.*
2. **Prediction:**

    (a) *We already know* $Bel(X_{t-1}|\mathbf{E}_{\{1,t-1\}})$, *the estimated probability distribution over* X_{t-1}.

    (b) *Calculate the predicted beliefs,* $\widehat{Bel}(X_t|\mathbf{E}_{\{1,t-1\}})$,

3. **Rollup:**

    (a) *Remove time-slice* $t-1$.

    (b) *Use the predictions for the* t *slice as the new prior by setting* $P(X)$ *to* $\widehat{Bel}(X_t|\mathbf{E}_{\{1,t-1\}})$.

4. **Estimation:**

   (a) Add new observations $\mathbf{E}_t$.

   (b) Calculate $Bel(X_t|\mathbf{E}_{\{1,t\}})$*, the probability distribution over the current state.*

   (c) Add the slice for $t + 1$.

4.5.3 Inference algorithms for DBNs

Exact clustering algorithms can be applied to DBNs, particularly if the inference is restricted to two time-slices.[6] Unfortunately, there normally is a cluster containing all the nodes in a time slice with inter-slice connections, so the clusters become unwieldy. The intuitive reason for this is that even if there are no intra-slice arcs between two nodes, they often become correlated through common ancestors. A version of the junction tree algorithm has been developed for DBNs (Kjærulff, 1995). This method takes advantage of the DBN structure by creating a junction tree for each time slice and performing updating for time slices up to and including the current time-slice using inter-tree message passing. For probabilistic projection into future time-slices, a sampling method is effective as there is no evidence yet for these future time slices.

If the DBN clusters get too large, approximate algorithms using stochastic simulation (described in §3.6) are usually suitable. As we discussed in §3.6.5, stochastic simulation methods are more effective when the evidence is at the root nodes, while the typical DBN structure involves the modeling of one or more state variables, each of which is observed with a possibly noisy sensor. This structure is shown in Figure 4.12; we can see that in these models evidence is at the leaves.

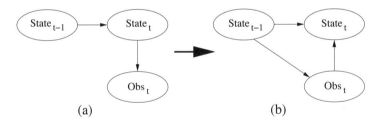

     (a)   (b)

FIGURE 4.13: (a) Original network; (b) new structure after arc reversal process.

One solution to this problem with stochastic simulation is to reverse the arcs to evidence nodes, as proposed by Shachter (1986), and then use the stochastic simulation algorithms. This **arc reversal** method (see §8.3.1.1) ensures that the evidence is at the root nodes, while maintaining the same overall joint probability distribution. It requires the addition of arcs to maintain the conditional independencies. A simple

---

[6]This does not mean that the beliefs being maintained are exact, of course; since past evidence is being summarized, the beliefs are inexact.

example is shown in Figure 4.13. The disadvantage is that we get a more complex structure that does not model the causal relationships.

Many other approximate algorithms for DBN inference have been proposed, including variations of stochastic simulation (e.g., Kanazawa et al., 1995), filtering methods (e.g., Murphy, 2002) and ignoring weak dependencies in the stochastic process (e.g., Boyen and Koller, 1998, Jitnah, 2000, Kjærulff, 1994). Unfortunately, most of these are not implemented in BN software packages.

4.6   Dynamic decision networks

Just as Bayesian networks can be extended with a temporal dimension to give DBNs, so can decision networks be extended to give **dynamic decision networks** (DDNs). Not only do they represent explicitly how the world changes over time, but they model general sequential decision making.

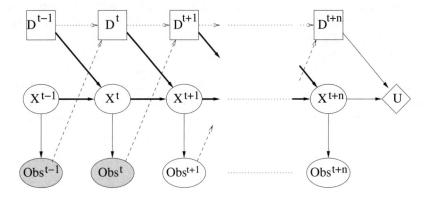

FIGURE 4.14: A generic DDN for a sequence of n decisions, to maximize expected utility at time $t + n$.

Figure 4.14 shows a generic DDN structure, for making a sequence of n decisions $D^t, D^{t+1}, \ldots, D^{t+n}$, from time t into the future. The temporal sequencing of the decision is represented by the precedence link (shown as a dotted line). The single chance node X determines the utility, while *Obs* is the observation node for which evidence will be added before each subsequent decision; this sequencing is represented by the information link (shown as a dashed line). Note that in Figure 4.14 the decision making will maximize the expected utility in n time steps. Another alternative would be to have a utility node at each time slice, in which case the decision making will maximize the cumulative expected utility from time t to $t + n$.

The DDN structure for the fever example is shown in Figure 4.15.

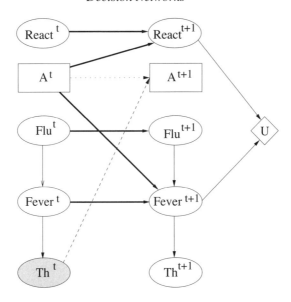

FIGURE 4.15: DDN structure for the fever example.

4.6.1 Mobile robot example

The robot's task is to detect and track a moving object, using sonar and vision sensor information, given a global map of the office floor environment. The robot must also continually reassess its own position (called localization) to avoid getting lost. At any point in time, the robot can make observations of its position with respect to nearby walls and corners and of the target's position with respect to the robot.

This deceptively simple, yet interesting, problem of a mobile robot that does both localization and tracking[7] can be modeled with a DDN as follows. The nodes S_T and S_R represent the locations of the target and the robot, respectively. The decision node is M, representing the robot's movement actions options. The nodes O_R and O_T represent the robot's observations of its own and the target's location, respectively. The overall utility is the weighted sum over time of the utility at each step, U^t, which is a measure of the distance between the robot and its target.

The DDN using these nodes is shown in Figure 4.16. The temporal arcs $S_T^t \rightarrow S_T^{t+1}$ indicate that the target's next position is related to its previous position; the actual model of possible movement is given implicitly in the CPT for S_T^{t+1}. The temporal arcs $S_R^t \rightarrow S_R^{t+1}$ and $M^t \rightarrow S_R^{t+1}$ indicate that the robot's own location depends on its previous position and the movement action taken. Both observation nodes have only intra-slice connections. The robot's observation of own position O_R^t depends only on its actual position ($S_R^t \rightarrow O_R^t$), while its ability to observe the position of the target will depend on both the target's actual location plus its own position (represented by $S_R^t \rightarrow O_T^t$ and $S_T^t \rightarrow O_T^t$), as it has to be close enough for its sensors to detect the target.

---

[7]A slightly simplified version of the one presented by Dean and Wellman, 1991.

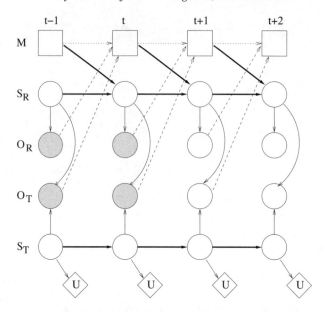

FIGURE 4.16: A dynamic decision network for the mobile robot example. (From Dean, T. and Wellman, M.P. *Planning and Control*. Morgan Kauffman Publishers, Copyright Elsevier (1991). With permission.)

4.7 Object-oriented Bayesian networks

Applying BNs to real-world domains requires modeling large, complex, systems. We now have algorithms to do efficient inference in large scale Bayesian and decision networks. But how should we *build* such systems? The **object-oriented** (OO) software engineering paradigm provides a framework for the large-scale construction of robust, flexible and efficient software (Schach, 2008), while OO database systems provide tools for managing large amounts of complex data (Banerjee et al., 1987). In this section we describe **object-oriented Bayesian Decision Networks** (OOBNs), a framework for building large, complex, hierarchical Bayesian decision networks. There is, as yet, no commonly accepted formal definition of an OOBN, so here we present them informally, with some examples to illustrate the modeling concept, and refer the reader to other sources for different formal definitions (see § 4.9).

4.7.1 OOBN basics

An ordinary BN is made up of ordinary nodes, representing random variables. An OOBN **class** is made up of both ordinary nodes and **objects** which are **instances** of other classes. Thus an object may **encapsulate** multiple sub-networks, giving a composite and hierarchical structure.

Objects are connected to other nodes via some of its own ordinary nodes, called its **interface** nodes. If the interface nodes are a proper subset of all the nodes in a class, it means that the rest of the nodes are not visible to the outside world; this provides so-called **information hiding**, another key OO concept. A class can be thought of as a self-contained "template" for an OOBN object, described by its name, its interface and its hidden part.

Interface nodes may represent either ordinary nodes or decision nodes[8] while hidden nodes may be objects, ordinary nodes, decision nodes or utility nodes. Finally, interface nodes are divided into **input** nodes and **output** nodes. Input nodes are the root nodes within an OOBN class, and when an object (instance) of that class becomes part of another class, each input node may be mapped to a single node (with the same state space) in the encapsulating class.[9] The output nodes are the only nodes that may become parents of nodes in the encapsulating class. However the arcs to the input nodes and from the output nodes may not introduce any directed cycles; the underlying BN must still be a DAG.

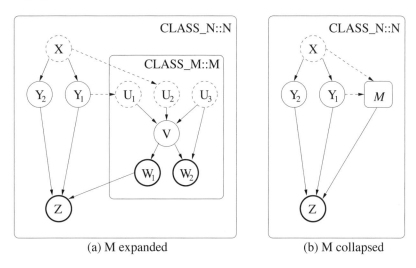

(a) M expanded   (b) M collapsed

FIGURE 4.17: A simple OOBN example.

Figure 4.17(a) illustrates these definitions, showing a simple example with two classes. Following Kjærulff and Madsen (2008), we represent an object as rectangles with curved corners, input nodes as dashed ovals, output nodes as bold ovals; hidden nodes are represented with the same ovals as "ordinary" BN nodes. We introduce the use of a dashed undirected arc to represent the mapping between an input node and the corresponding node in the encapsulating class.[10] We also label each object

---

[8]Koller and Pfeffer's OOBNs (1997a) are more general, with the interface nodes able to be objects. Here we stick to the simpler form, as implemented in Hugin.

[9]These work like actual and formal parameters in OO programming methods.

[10]Elsewhere in the literature, ordinary directed arcs are used, but we feel this is confusing because it does not show visually that it is just a mapping, with no CPT involved and no inference across this arc.

with *CLASS::Object* (or more simply as *Object* if the class is obvious from the object name). Figure 4.17(b) depicts the same OOBN example, but without showing the network structure within the encapsulated object; we say object M is "fully expanded" (or, **unfolded**) in (a), while in (b) it is "collapsed." We note Hugin shows only the interface nodes when expanding an encapsulated object, while the hidden nodes and structure are not shown; we refer to this as a "partially expanded" object.

In this example, *CLASS_N* contains one input node X, two hidden nodes Y_1 and Y_2, a single object M (an instance of *CLASS_M*) and one output node Z. *CLASS_M* has three input nodes U_1 and U_2 and U_3, a hidden node V, and two output nodes W_1 and W_2. X and Y_1 have been mapped to the M1's input nodes U_1 and U_2 which implies they must correspond in "type"; for example, in Hugin, they must have the same number of states, the same state labels and be of the same type. Note that node names can be the same at different levels of the encapsulated object hierarchy, as so-called "scope" rules apply. So in this example, the input nodes in M could have been named X, Y_1 and U_3. If there is any doubt about which node X we are referring to, we can use its "long" name, e.g., *Object.X*. This example also shows that interface nodes need not be linked to the encapsulating class, e.g., U_3 and W_2. Note that W_2 could be a parent of Y_2 or Z but not of Y_1, as that would introduce a cycle.

In an OOBN class, all ordinary nodes, other than input nodes, have parents that are also ordinary nodes from within that class, or are output nodes in an encapsulated object. Naturally, there is a CPT associated with each, quantifying the uncertainty in that relationship. When specifying a class, all input nodes are given default priors. When an object is used within an encapsulating class, if an input node is connected to an appropriate corresponding node in the encapsulating class, it effectively "becomes" that node, and the default priors are no longer relevant. In the Figure 4.17, U_2 will take on X's prior, while U_1 takes Y_1's CPT; U_3 retains its default prior.

OOBNs allow us to view the network at different levels of **abstraction**, which is particularly useful during construction. They can be built "bottom-up," that is, starting with the most detailed level, which in practice means starting with classes that consist only of ordinary nodes, or "top-down," starting with the most abstract description, or alternating between bottom-up and top-down development. OOBNs also allow reuse of subnetworks; once we have a class for a particular type of entity in a network, it can be reused within another object, with as many instances as needed.

4.7.2   OOBN inference

An object-oriented network N has an equivalent **flat** (or unfolded) BN, obtained by recursively unfolding the object nodes of N; Figure 4.18 shows the flat BN obtained from unfolding object M from Figure 4.17. Ordinary BN inference algorithms (see Chapter 3 can then be applied to the flat BN; Hugin uses this method.

Koller and Pfeffer (1997a) described an OOBN inference algorithm that uses the structural information encoded in the OOBN, such as the encapsulation of variables within an object and the reuse of model fragments in different contexts to obtain computational efficiencies; however, this is not used in any generally available BN software that support OOBNs.

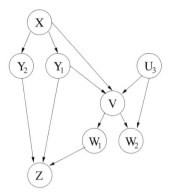

FIGURE 4.18: The BN obtained by recursively unfolding the OOBN in Figure 4.17.

4.7.3 OOBN examples

Example 1: Orchard Farming *Mr Bunce runs a farm, which also contains a small orchard. When his fruit trees are distressed, either through lack of water, or when suffering from a disease, they lose their leaves. His management options include spraying to kill a particular fungal disease and providing more regular water by irrigating, particularly when there is a drought. Water costs more during a drought, while the overall condition of the trees allows him to estimate the expected return from the crop.*[11]

Figure 4.19 shows an OOBN for this example. At the highest level of abstraction, the *BunceFarm* object contains an *Orchard* which is influenced by a combination of Mr Bunce's farm management actions and the *Climate*. The *ORCHARD* class contains multiple tree objects, as well as a node which summarizes the condition of all the trees. The returns from the fruit crop are modelled in the utility table for the *Returns* node.

The *TREE* class is the least abstract in this example, containing only ordinary nodes, showing that a tree's condition, indicated by whether it loses its leaves, depends on whether it is diseased and receiving enough water. The action *Spray* can eliminate the disease, while water – provided by rainfall, not modelled explicitly here – is reduced during *Drought* and can be boosted by *Irrigation*.[12]

Drought is an input node of the *TREE* class, coming from the *Climate* object originally while the *Spray* and *Irrigation* input nodes in each tree object come originally from the *BunceFarm* object, all via the *Orchard* object. Note that while in this example the *CLIMATE* class contains only the single ordinary node *Drought*, it could be extended into a more complex sub-model for predicting drought.

Note that the costs associated with the *Spray* and *Irrigation* decisions are modelled with utility nodes included at the highest level, i.e., where *Spray* and *Irri-*

---

[11]This is an extended version of Kjaerulff and Madsen's (2008) Apple Tree OOBN example.

[12]Note that for sequential decision making, there would be a precedence link between with two action nodes, *Spray* and *Irrigation*.

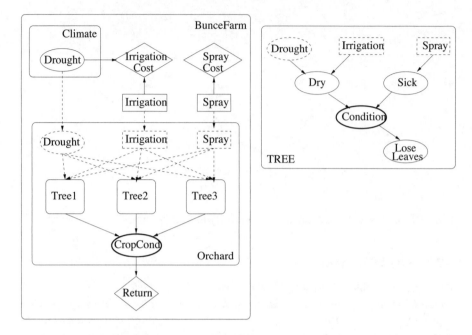

FIGURE 4.19: An OOBN for the Farm Orchard example (top left) together with the TREE class.

gation are not input nodes. Obviously there will be trade-offs between spraying and irrigating to improve the crop condition and hence return, and the costs of those actions. These will depend on the paramaterization of the model, for example the probability of drought, P(*Climate.Drought*), how effective the spraying is, $P(Sick = No|Spray = Yes)$, how likely the trees are to contract the disease if not sprayed, $P(Sick = Yes|Spray = No)$, how well the trees cope with lack of water $P(Condition|Dry = Yes, \_)$, and so on. The parametrization is left as an exercise for the reader.

Example 2: The Car Accident *The amount of damage done in a car accident will depend on a range of factors, including various features of the car, its current value and the speed it was travelling at. These in turn will vary according to the weather, the speed-limit and features of the driver.*[13]

Figure 4.20 shows the *CAR* and *SITUATION* classes for this example. The car *Owner* object is an instance of the *PERSON* class, which has the output nodes *Age* and *Income*. The *CAR* class contains contains other objects – *Owner, Engine, Steering, Tyres* and *Brakes* – that are instances of other classes not shown here. The *CAR* class has hidden nodes *Type, Age, Original_Value, Maintenance* and *Mileage* and output nodes *Max–Speed, Current–Value, Steering–Safety* and *Braking–Power*. The

---

[13]This is a modified version of an example from Koller and Pfeffer (1997a).

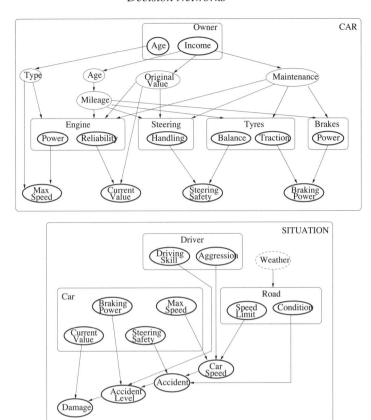

FIGURE 4.20: The *CAR* and *SITUATION* classes for the car accident OOBN example (adapted from Koller and Pfeffer, 1997a[Fig. 1]).

SITUATION class contains, of course, a *Car* object, as well as objects (*Driver* and *Road*), which are instances of other classes also not shown here. The interface to the *SITUATION* class consists of a *Weather* input node (which could be connected to a complex weather model) and several output nodes.

4.7.4 "is-A" relationship: Class inheritance

Capturing relationships is an important part of the object oriented paradigm. The encapsulation of objects within other objects allows us to model "part-of" relationships. There is another key relationship in the OO paradigm, namely the "is-a" relationship, with subclasses that inherit the properties of other classes. An instance of a subclass is an instance of its parent class (called the superclass), giving an "is-a" hierarchy. We follow Kjærulff and Madsen (2008) by defining inheritance as the ability of an instance to take its interface definitions from other instances. So if C_1 is an OOBN network class with input variables I_1 and output variables O_1, then class C_2 may be specified as a subclass of C_1 if and only if $I_1 \subseteq I_2$ and $O_1 \subseteq O_2$. While the hidden nodes and structure are not, in this definition, passed from class to subclass, in practice they may be very similar, with the structure or parameters adjusted to reflect a different, more specialized situation.

For example, in the Farm Orchard example above, we could develop a hierarchy of the Tree class, as in Figure 4.21, which would allow us to model that a particular disease affects only some kinds of fruit tree, or that different species respond differently to the stress of inadequate water.

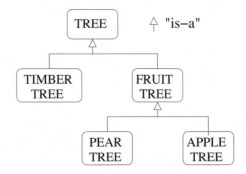

FIGURE 4.21: A possible hierarchical classification of tree using standard UML (Fowler, 2003) OO notation.

Similarly, in the Car Accident example, the *Driver* object in the *SITUATION* class could be an instance of a more detailed subclass of the *PERSON* class, allowing many of its nodes to be inherited rather than redefined.

Unfortunately, to date (to our knowledge), there is no publicly available OOBN software that supports class inheritance.

4.8 Summary

In order to make decisions, we must be able to take into account preferences between different outcomes. Utility theory provides a way to represent and reason with preferences. The combination of utility theory and probability theory gives a framework for decision making, where a rational being should make choices that maximize her or his expected utility. Extending Bayesian networks with decision nodes and utility nodes gives us decision networks (also called influence diagrams). These include an explicit representation of ordering between information and decisions, and decisions and decisions, and allow us to make isolated or sequential decisions. By adding a temporal dimension to Bayesian networks, we get dynamic Bayesian networks (DBNs), which allow us to explicitly model and reason about changes over time. These in turn can be extended with utility and decision nodes, to give dynamic decision networks (DDNs). DDNs run into complexity problems when used for general planning, in which case special purpose planning representations and algorithms may be more useful. Object-oriented Bayesian decision networks (OOBNs) are generalisations of decision networks, based on a hierarchical component structure. They facilitate network construction and reuse by allowing the representation of commonalities of both structure and parameters.

4.9 Bibliographic notes

Influence diagrams were originally developed as a compact representation of decision trees for decision analysis. The basic concepts were developed by a group at SRI (Miller et al., 1976), and were formally introduced by Howard and Matheson (1981).

Jensen and Nielsen (2007) give a detailed presentation of one method for evaluation decision network, and summarizes other approaches. The term "DBN" was first coined by Dean and Kanazawa (1989); other early DBN research in the AI community was undertaken by Nicholson (1992) and Kjærulff (1992).

The chapter by Kevin Murphy on DBNs for Michael Jordan's forthcoming text book, *An Introduction to Probabilistic Graphical Models*, gives a very comprehensive survey of DBNs, their connections to hidden Markov models (HMMs) and state space models, and exact and approximate inference algorithms (including filtering approaches). Kjaerulff (1995) describes the dHugin computation scheme for DBN inference, which includes window expansion and reduction, forward and back interjunction tree propagation and forward sampling for probabilistic projection.

OOBNs appeared in various forms from different researchers from the late 1990s, including (Koller and Pfeffer, 1997a, Laskey and Mahoney, 1997, Bangso and Wuillemin, 2000, Neil et al., 2000). Koller and Friedman (2009) give a very full formal description of one variation, while Kjærulff and Madsen (2008) and Jensen and Nielsen (2007) describe the flavour of OOBNs implemented in Hugin.

4.10 Problems

Utility

Problem 1

You have 20 pounds of cheese. We offer to make a gamble with you: tossing a fair coin, if it comes up heads, we'll give you 130 more pounds of cheese; if it comes up tails, we'll take your 20 pounds of cheese. What is the expected value of the gamble? Assuming that you maximize your expected utility and that you refuse this gamble, what can you infer about your utility function — that is, how many utiles (basic units of your utility) is a pound of cheese worth?

Problem 2

You have 20 utiles worth of chocolate. We offer to make a gamble with you: tossing some coin, if it comes up heads, we'll give you 130 more utiles worth of chocolate; if it comes up tails, we'll take your 20 utiles worth of chocolate. Assuming that you maximize your expected utility and that you refuse this gamble, what can you infer about the probability you give to the coin landing heads?

Modeling

Problem 3

Robert is trying to decide whether to study hard for the Bayesian Artificial Intelligence exam. He would be happy with a good mark (e.g., a High Distinction) for the subject, but he knows that his mark will depend not only on how hard he studies but also on how hard the exam is and how well he did in the Introduction to Artificial Intelligence subject (which indicates how well prepared he is for the subject).

Build a decision network to model this problem, using the following steps.

- Decide what chance nodes are required and what values they should take.
- This problem should only require a single decision node; what will it represent?
- Decide what the casual relationships are between the chance nodes and add directed arcs to reflect them.
- Decide what chance nodes Robert's decision may effect and add arcs to reflect that.
- What is Robert's utility function? What chance nodes (if any) will it depend on? Does it depend on the decision node? Will a single utility node be sufficient? Update the decision network to reflect these modeling decisions.
- Quantify the relationships in the network through adding numbers for the CPTs (of chance nodes) and the utility table for the utility node. Does the number of parameters required seem particularly large? If so, consider how you might reduce the number of parameters.
- Once you have built your model, show the beliefs for the chance nodes and the expected utilities for the decisions before any evidence is added.
- Add evidence and see how the beliefs and the decision change, if at all.
- If you had an information link between the evidence node and the decision node, view the decision table.
- Perform your own decision tree evaluation for this problem and confirm the numbers you have obtained from the software.

Problem 4

Think of your own decision making problem involving reasoning with evidence and uncertainty. Write down a text description of the problem, then model it with a decision network using similar steps to the previous problem. Make the problem sufficiently complex that your network has at least 3 chance nodes (though a single decision node and a single utility node will be enough for a first modeling effort, unless you feel particularly adventurous).

Problem 5

Julia's manufacturing company has to decide whether to go ahead with the production of a new product. Her analysis indicates that the future profit will depend on a combination of the quality of the product and the market demand for it. Before the final decision is made, she has two other possible courses of action. One is to undertake further product development, to make the product quality better before it goes into production. The other is to do more market research to determine the likely demand. She could also choose to do both. Of course both product development and market research will cost money. And she has to be careful about delaying too long, as she knows that the first product of this kind that becomes available will corner the market.

Build a decision network to model this sequential decision problem.

1. Decide what chance nodes, decision nodes (hint: there should be three) and utility node(s) you need.

2. Add the main network links, then the information and precedence links. Note that you need to investigate two possible orderings for the pre-production decisions.

3. The problem description does not provide much in the way of quantitative information for parameterizing your model; so choose initial parameters that seem reasonable.

4. What is the *expected benefit* of additional marketing? And of further product development?

Dynamic Bayesian networks (DBNs)

Problem 6

Peter wants to do simple monitoring of his local weather. Suppose that the main weather types are: sunny, cloudy, showers, rain and snow. He knows that the weather changes from one day to the next in certain patterns, depending on the season. He also has access to the local weather forecast for the following day, and he has been monitoring this for long enough to know that the local forecasters get it right about 70% of the time (though slightly worse for cloudy and showery days).

Build a DBN to model this problem. You should use a BN software package that specifically supports DBNs (see Appendix B).

Dynamic decision networks (DDNs)

Problem 7

Cate brought some shares for $1000. Each day, she has to decide whether to sell, or keep them. Her profit will depend on the price she gets for them. Each morning, she gets the previous day's closing price, and she rings her sister Megan and gets her opinion as to whether the price will go up or down today. She also knows that most of the time, the price of these shares only move within a certain range from one day to the next.

Build a dynamic decision network to model this problem.

Object-oriented Bayesian decision networks (OOBNs)

These modeling exercises should be done using a BN software package that supports OOBNs, such as Hugin or AgenaRisk. See our **Quick Guides to OOBNs in Hugin** in Figure 4.22 and also Appendix B.

Problem 8

Parameterize the Orchard Farming OOBN example given in Section 4.7.3. Investigate the trade-off between the spray and irrigation costs and the return from the crop. See how the decision changes when the probability of drought increase, the disease mutates to become highly virulent, or the tree species is better at handling drought.

Problem 9

Implement the Fever DDN (see Figure 4.15) as an OOBN.

Problem 10

Suppose a student's final grade for a subject is determined by the difficulty of the subject and the student's innate intelligence (ignoring the impact of the student's effort). Build an OOBN to model the situation where each student takes multiple subjects, and each subject has multiple students, but not all students take each subject.

- Assuming the same prior for each student, but different priors for the difficulty of each subject, show how adding the students' grades changes the posteriors for each student's intelligence.

- Assuming you start with no information about each subject's difficulty, but a good estimate of each student's intelligence from their university entrance score, show how adding the students' grades changes the posteriors for each subject's difficulty.

A Quick Guide to OOBNs in Hugin

Classes: OOBN classes are stored in `.oobn` files. Each class is usually stored in a separate `.oobn` file. When a `.oobn` file is opened successfully, classes for all nested instances are also opened, each in a separate window. If Hugin can't find class files for all nested instances of other classes, an error message is given.

Adding an instance: To add an instance of a class, click on ⊞, or select `Edit→ Instance Node Tool` (in edit mode). This brings up a list of all classes already opened; select, drop and drag to create the new instance. By default, Hugin calls the new instance ClassName_1, ClassName_2, etc.

To connect encapsulating class nodes to interface nodes, links are added and connected in the standard way (the instance must be expanded). For links into input nodes, Hugin does a consistency check (state types and names must match exactly). This check is done either when adding link or when compiling; visit `Network menu→Network Properties→ OOBN` to change this setting.

Viewing: An instance is selected with a left click; its rectangular frame is then bolded.

A selected instance can be expanded or collapsed either by a right click then select `Instance Tool→Expand/Collapse` or `View→Expand/Collapse Instance Nodes`. When expanded, Hugin provides 2 layouts of interface nodes, switched using `Instance Tool→Rotate Interface`: (1) horizontal layout, input nodes across the top output nodes at bottom; (2) vertical layout, input nodes down the left, output nodes on the right (typically used for DBNs). The layout can be switched using right click then select `Instance Tool→Rotate Class`.

Hugin has an auto fit capability that moves nodes when expanding or collapsing; visit `Network menu→Network Properties→ OOBN` to change setting.

Within an instance, the relative order of interface nodes can be changed by a left click on the interface node.

To view internal structure of an instance, right click then select `Instance Tool→Show Class`, which brings the window for that class to the front, allowing inspection of hidden part of instance.

The `View→Show Component Tree` option shows graphically the hierarchy of objects, with the selected class at the root.

Editing an instance: Right click on a selected instance, then select `Node Properties`, which brings up a new window. The name or label of the instance can be changed here.

Saving and closing: An OOBN class is usually saved as a separate file. Multiple OOBN classes *can* be saved in the same file (`File→Save all classes in a single file as...`), but this does not support class re-use. An OOBN can be saved as a `.net`; this will save only as an unfolded BN.

Hugin OOBN Tutorials: See also via Hugin Help "Introduction to Object Oriented networks" and "How to build an Object-Oriented BN".

FIGURE 4.22: A quick guide to OOBNs in Hugin.

Problem 11

Implement the Orchard Farm OOBN. (NB: an unparameterized Hugin version is provided online). Investigate how Mr Bunce's decision about irrigating the orchard changes when (a) the prior for *Drought* changes and/or (b) the cost of irrigating changes.

Problem 12

Implement the Car Accident OOBN, then extend it (along the lines of Example 4.3 in (Koller and Pfeffer, 1997a)) with subclasses to mode:

- a fuel-injected engine (a subclass of *ENGINE*), which is less reliable and is more likely to provide a high power level;
- a sport-car (a subclass of *CAR*) with a fuel-injected engine, a higher original value, and has a new output node *Acceleration* that is a function of the engine power and the original value.

5

Applications of Bayesian Networks

5.1 Introduction

In the previous three chapters, we have seen in detail how Bayesian and decision networks can be used to represent and reason with uncertainty. We have used simple illustrative examples throughout, which, together with some of the modeling problems set as exercises, give some indication of the type and scope of the problems to which Bayesian networks may be applied. In this chapter we provide more detailed examples of some BN applications which we have personally developed. By moving beyond elementary examples we can provide a better insight into what is involved in designing and implementing urBayesian models and the range of applications for which they can be useful.

We begin with a very brief survey of BN applications developed by others, especially medical applications and environmental applications, in order to show some of the variety of uses and BN models which are to be found in the literature. The enormous increase over the past 10 years in the number of journal research publications about BNs, or using BNs, is shown in Table 5.1.[1]

We then describe five of our own applications:

1. The first is a medical BN application namely predicting risk of cardiovascular heart disease, built by adapting epidemiological models in the literature.

2. The second application models native fish abundance in a water catchment region, an example of how BNs are being used for ecological and environmental applications.

3. Next is a game playing application, specifically poker, a domain that is rich in uncertainty from a number of sources: physical randomization through shuffling, incomplete information through the opponent's hidden cards and limited knowledge about the opponent's strategies. In this application, the BN is used to compute the probability of winning, with a decision network combined with randomization strategies in a hybrid system to make betting decisions.

4. The fourth application is ambulation monitoring and fall detection, which illustrates the use of DBNs in medical monitoring and also includes typical sensor modeling within a Bayesian network.

---

[1]Note that these citations are limited to journal papers with "Bayesian networks" in title, abstract or keywords and using a single citation index, thus missing many relevant papers using alternative names for BNs or appearing in conferences or non-indexed journals. It is the *trend* in publication numbers across the range of application areas that is important.

TABLE 5.1
Number of BN journal publications (using Current Content search) from 2000-2009, by discipline. "Subjects" here are groupings of CC subjects. Eng = Engineering (includes robotics, automation and control); CS/IT = Computer science and information technology; BioSci = "Biological Sciences" includes biology, zoology, ecology, environmental science, agriculture; MedSci = Medical and veterinary science; PhysSci = "Physical sciences" includes chemistry, physics; Other includes social science, law, business and economics, behavioral sciences, etc.

Subject	2000	2001	2002	2003	2004	2005	2006	2007	2008	2009
Eng	36	23	35	55	61	71	86	81	138	169
CS/IT	21	11	20	32	39	52	44	44	67	84
BioSci	1	7	3	15	11	11	15	16	25	23
MedSci	5	5	11	13	12	20	15	22	24	29
PhysSci	4	5	6	11	9	15	18	20	31	25
Other	0	2	1	3	1	5	7	8	18	13
Total	67	53	76	129	133	174	185	191	303	343

5. Finally, we look at an argument generation application, where BNs for both normative domain modeling and user modeling are embedded in a wider architecture integrating the BNs with semantic networks and an attention focusing mechanism.

5.2 A brief survey of BN applications

5.2.1 Types of reasoning

As we discussed in Chapter 2, a key reasoning task using BNs is that of updating the posterior distribution of one or more variables, given evidence. Depending on what evidence is available, the probabilistic belief updating can be performing one of a number of tasks, such as:

1. **Diagnosis**. Example: *Which illness do these symptoms indicate?*

2. **Monitoring/control**. Example: *Is the patient's glucose level stable, or does extra insulin need to be given?*

3. **Forward prediction**.

   Such predictions may be factual (that is, based on evidence) or hypothetical (such as predicting the effect of an intervention). For example, when considering the question *Will the patient survive the proposed operation?* a factual prediction might be

   Based on the results of the X-ray, MRI and blood tests, will the patient survive the proposed operation?

   while an example hypothetical prediction might be

If the patient is given anticoagulant X, will the patient's chances of survival be improved?

As we saw in the previous chapter, once we extend the network with decision and utility nodes, we can use it for decision making and planning. Examples include *Which treatment plan carries the least risk of failure?* and *Which sequence of actions will maximize the patient's quality of life?*

5.2.2 Medical Applications

Medicine has undoubtedly been the most popular application area to date for Bayesian networks. As a complex domain where much knowledge is implicitly held by experienced medical practitioners, it has long been a target of expert systems. A connection between BN research and medical applications was established in the 1980s by researchers such as David Heckerman and Eric Horvitz who were in the Stanford joint Ph.D./M.D. program. The appeal of BNs for this application area lies in their ability to explicitly model causal interventions, to reason both diagnostically and predictively and the visual nature of the representation, which facilitates their use in explanation.

We have already looked at some simple medical reasoning examples in the preceding chapters. The simplest tree-structured network for diagnostic reasoning is the so-called naive Bayes model, as shown in Figure 5.1(a), where *Disease (D)* node has values for each of the diseases under consideration, while the *F* nodes represent "findings," which encompass both symptoms and test results. This network reflects two unrealistic assumptions: that the patient can have only a single disease and that the symptoms are independent of each other given the disease.

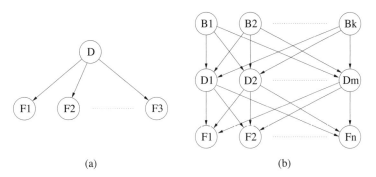

(a)   (b)

FIGURE 5.1: Generic BN structures for medical diagnosis: (a) naive Bayes model; (b) multiply-connected network.

A more realistic, but more complex, model is the multiply-connected network of Figure 5.1(b). Here there is a Boolean node for each disease under consideration, while the *B* nodes represent background information such as the age and sex of the patient, whether they are a smoker, or have been exposed to pollution. However, in practice, this structure is likely to be too complex, requiring us to specify probabilistically the combined effect of every disease on each finding.

The network structure in Figure 5.1(b) is essentially that developed in the QMR-DT project (Shwe et al., 1991, Middleton et al., 1991), a probabilistic version of the frame-based CPCS knowledge base for internal medicine. The QMR-DT network had this two-level structure. The problem of complexity was ameliorated by making it a binary noisy-or model (see §6.4.2). This was done by assuming the effect of a disease on its symptoms and test results is independent of other diseases and independent of other findings. One version of the QMR-DT network (described by Pradhan et al., 1994) had 448 nodes and 908 arcs, including 74 background nodes (which they called "predisposing factors") needing prior probabilities, while the remaining nodes required probabilities to be assessed for each of their values. In total more than 600 probabilities were estimated, a large but not an unreasonable number given the scope of the application. Performing exact inference on networks of this size is generally not feasible. Initial work on the application of likelihood weighting to this medical diagnosis problem is described in Shwe and Cooper (1991), while other approximate methods for QMR-DT are presented in Jaakkola and Jordan (1999).

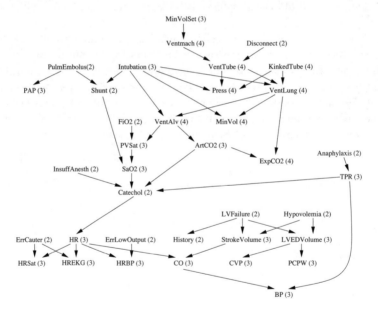

FIGURE 5.2: The ALARM BN for ICU monitoring: an early medical BN (1989).

The ALARM network for monitoring patients in intensive care (Beinlich et al., 1992), shown in Figure 5.2, is a sparsely connected BN consisting of 37 nodes and 42 arcs (the number of values for each node is shown next to the node name). This network is often used as a benchmark in the BN literature. Clearly, it does not map neatly into the generic medical diagnosis structure given above, and it does not provide a template for building other medical monitoring or diagnostic BNs.

TABLE 5.2 A selection of Medical BN applications.

Citation	Topic (System Name)	Task	Type
Clinical			
Andreassen et al. (1989)	neuromuscular disorder (MUNIN)	diagnosis	
Cowell et al. (1991)	adverse drug reaction	prediction	
Heckerman (1991)	lymph-node disease (PATHFINDER IV)	diagnosis	
Dagum and Galper (1993)	sleep apnea	modeling	DBN
Andreassen et al. (1994)	glucose prediction, insulin adjustment	treatment	DBN
Kahn et al. (1997)	breast cancer (MammoNet)	diagnosis	
Díez et al. (1997)	echocardiography (DIAVAL)	???	
Onisko et al. (1998)	liver disorder	diagnosis	
Galán et al. (2002)	nasopharyngeal cancer (NasoNet)	modeling	T, N-OR
van der Gaag et al. (2002)	oesophageal cancer	diagnosis	
Burnside et al. (2004)	mammography	diagnosis	
Tucker et al. (2005)	visual field deterioration	modeling	C
Kline et al. (2005)	venous thromboembolism	prediction	
Maglogiannis et al. (2006)	patient monitoring system	risk analysis	
Shiratori and Okude (2007)	anesthetic practice	modeling	OOBN
Nagl et al. (2008)	breast cancer	prognosis	OOBN
Charitos et al. (2009) Visscher et al. (2009)	pneumonia in ICU patients	diagnosis	DBN+
Smith et al. (2009)	prostate cancer radiation therapy	Treatment	H
Promedas (Promedas)	internal medicine (PROMEDAS)	diagnosis	L
Large Data Sets / Epidemiology			
Wang et al. (1999)	breast cancer	Modeling	L
Getoor et al. (2004)	tuberculosis epidemiology	modeling	L
Acid et al. (2004)	emergency medical service data	modeling	L
Twardy et al. (2006) Nicholson et al. (2008)	cardiovascular heart disease epidemiology	modeling	L
Stajduhar et al. (2009)	censored survival data	prediction	C
Flores et al. (2010)	cardiovascular heart disease	modeling	L,E

Note: L = learned from data, E = elicited, T = explicit time modelled (not DBN), H = hybrid, NOR = noisy-OR, C = classifier.

Since these early medical BNs, many others have been developed. A representative (but by no means exhaustive) selection is given in Table 5.2. These applications can be divided roughly into two categories: those at the clinical level, and those at the epidemiological level. Some focus on a single medical condition, others address a system within the human body. Some are stand-alone BNs, while others are embedded in a broader decision support system. Finally, some networks are built using medical expertise (directly or using information from the medical literature), some are learned from data, or hybrid methods are used. We look at these learning methods in Parts II and discuss how to apply them to "knowledge engineer" BNs in Part III.

Even while basic "static" BNs are growing in popularity for computer modeling, the more powerful BN varieties – dynamic BNs, decision networks and OOBNs – are not yet in wide use, even when they appear apt. For example, Smith et al. (2009) combine BNs (providing probabilities) with multi-attribute preferences and Markov model simulations, but fail to put these together in dynamic decision networks. Given the power of BDNs, plus the existence of "appropriate software, tutorials and proponents...the dissemination and penetration of the influence diagram formalism has been surprisingly limited... the reasons for the slow progress remain unclear" (Pauker and Wong, 2005). At least one clear reason, though, is a simple lack of familiarity with the more powerful kinds of BN and their capabilities.

TABLE 5.3 A selection of Environmental and Ecological BN applications.

Citation	Topic (System Name)	Task	Type
FAUNA & FLORA			
Kuikka et al. (1999)	Baltic Cod	NRM	DN
Marcot et al. (2001)	Fish and wildlife population viability	NRM	
Borsuk et al. (2002, 2004)	Trout populations in Switzerland	modeling	
Pollino et al. (2007)	Native fish abundance	NRM	
Pullar & Phan (2007)	Koala populations near urban environments	modeling	H
Smith et al. (2007)	Endangered mammal habitat	prediction	
Wilson et al. (2008)	Amphibian populations	monitoring	
Pollino et al. (2007)	Endangered Eucalypt species	DSS	
LAND USE			
Bacon et al. (2002)	Land use change and management	NRM	
Aitkenhead et al. (2009)	Land cover	prediction	H
White (2009)	Peatlands in Victorian high plains	modeling	
WATER			
Jensen et al. (1989)	Biological processes in water purification	modeling	
Varis et al. (2002, 2006)	Senegal River, Cambodian Lake	NRM	
Chee et al. (2005)	Environmental flows in Victorian rivers	Monitoring	
Bromley et al. (2005)	Water resources	NRM	OOBN
Pollino & Hart (2006)	Mine-derived heavy metals in PNG rivers	ERA	
Reichert et al. (2007)	River rehabilitation	DSS	
deSanta et al. (2007)	Planning a large aquifer in Spain	NRM	
Henriksen et al. (2007)	Groundwater protection	Modeling	
Ticehurst et al. (2007)	Sustainability of coastal lakes in NSW	ERA	
Barton et al. (2008)	Nutrients in Norwegian river basin	DSS	
AGRICULTURE			
Jensen (1995)	Fungicides for mildew in wheat	DSS	
Kristensen et al. (2002)	Growing barley without pesticide	DSS	
Otto & Kristensen (2004)	Infection in swine herds	modeling	
Jensen et al. (2009)	Causes of leg disorder in finisher herds	modeling	OOBN
Ettema et al. (2009)	Claw and digital skin disease in cows	estimation	
Steeneveld et al. (2010)	Mastitis	DSS	NB
OTHER			
Tighe et al. (2007)	Climate change pressures	NRM	
Voie et al. (2010)	White phosphorous from munitions use	risk analysis	

Note: DSS = decision support system, ERA = environmental risk assessment, NRM = natural resource management, DN = decision network, OOBN = object-oriented BN, H = hybrid BN with other modeling, NB = naive bayes.

5.2.3 Ecological and environmental applications

Our natural environment is complex and changing. Natural resource management is rife with uncertainty – scientists don't fully understand the causal mechanisms, there is physical randomness, and often we cannot directly observe the important variables. Decision makers must balance environmental outcomes such as sustainability and conservation of threatened species with social and economic outcomes. They must involve their stakeholders – farmers, scientists, conservationists, business and the public – in the decision making process. And they must be able to justify their policy frameworks and how they allocate resources. Thus, not surprisingly, Bayesian networks have been enthusiastically adopted as a tool for ecological and environmental modeling and decision making.

For example, in a report looking at environmental water flow assessments, Hart and Pollino (2009) write "Whether an [environmental] assessment is focused on conservation, assessing risk, or aimed at integrating information across disciplines, complexity, trade-offs, and uncertainty are common features. Within each of these frameworks, BNs have proved particularly useful for focusing issues by clearly structuring the formulation of a problem within a participatory-style and transparent process." In many cases, the process of *building* a BN leads to an improved understanding of the environmental system, as well as providing a useful a decision support tool.

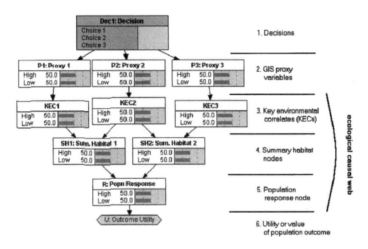

FIGURE 5.3: A general structure suggested by Marcot et al. (2001)[Fig.1] for evaluating population viability outcomes of wildlife species. [Reprinted from Using Bayesian belief networks to evaluate fish and wildlife population viability under land management alternatives from an environmental impact statement. Marcot et al. *Forest Ecology and Management* 153(1-3):29–42. Copyright (2001), with permission from Elsevier.]

Early environmental BNs include Varis' decision networks for fisheries management (e.g., Varis, 1997) while Ellison (1996) advocated BNs as a general framework for ecological research and environmental decision making. Bruce Marcot's work was particularly influential, with his description of general BNs structure for evaluating wildlife species viability (see Figure 5.3) providing an accessible domain specific modeling template for ecologist wanting to use BN technology. Practical guidelines for using BNs in this application area, such as Cain's (2001) and Marco et al.'s (2006), have also played an important role. That there is now a section on Bayesian networks in the Encyclopedia of Ecology (Borsuk, 2008) demonstrates how mainstream Bayesian network technology has become in this field.

A selection of environmental, ecological and agricultural BN applications is given in Table 5.3. Again, as with medical applications, the majority of BNs in this area have been ordinary BNs. Figure 5.4 shows one example, a BN modeling the impact of land use and climate change on peat land in Victoria's Bogong high plains (White, 2009). Given the importance of modeling the decision trade-offs, the temporal and dynamic nature of ecological systems, and the complex of the interacting sub-systems, wider adoption of decision networks, dynamic Bayesian networks and OOBNs seems inevitable.

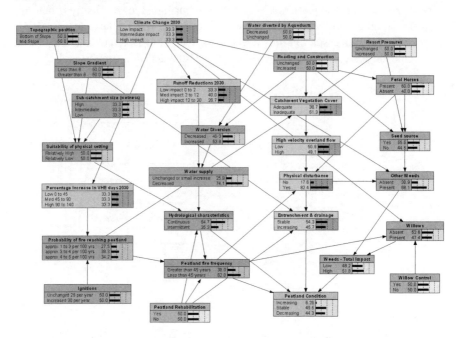

FIGURE 5.4: BN modeling the impact of land use and climate change on peat land in Victoria's Bogong high plains (White, 2009).

5.2.4   Other applications

BNs have now been successfully used in an enormously wide range of application areas, across engineering, the sciences, the military, education, business and finance, and, not least, computing. A small but representative selection of BN applications is given in Tables 5.4 and 5.5.

Vista (Horvitz and Barry, 1995) is an early decision network application for the NASA Mission Control Center in Houston. It uses Bayesian networks to interpret live telemetry and provide advice on possible failures of the space shuttle's propulsion systems. Vista recommends actions of the highest expected utility, taking into account time criticality. The GEMS (Generator Expert Monitoring System) (Morjaria et al., 1993) project is an example of a system where BNs succeeded when a rule-based system did not (Kornfeld, 1991).

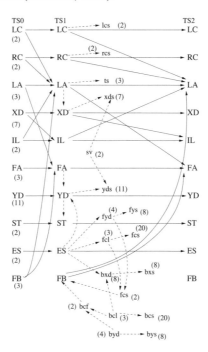

FIGURE 5.5: The BATmobile BN.

BNs are now a well-established reasoning engine for **intelligent tutoring systems (ITS)**. While the educational domains vary widely (e.g., Physics, decimals, medicine), there are common features to them all. The BN is always used to assess a student's knowledge: knowledge of a fact, understanding or misunderstanding a concept, knowledge of which equation to apply, how to perform a mental computations, etc. The BN starts with a prior distribution, which may be uniform, based on a general student population, tailored to a cohort, or student-specific using pre-testing. The student's actions within the ITS (e.g., answers to questions) are then entered as evidence into the BN, which then propagates through to the "knowledge node of interest", updating the profile of the student's knowledge state. A selection of ITS applications are included in Table 5.5, and we use an ITS for decimals as a case study in Chapter 11.

Oil price forecasting is an application that has been modeled with an ordinary BN (Abramson and Finizza, 1991) and with a DBN (Dagum et al., 1992). Other early DBN applications included power plant operation (Katz et al., 1998), robot navigation and map learning (Dean and Wellman, 1991) (see also §4.6.1), monitoring robot vehicles (Nicholson and Brady, 1992a,b) and traffic monitoring in both (Pynadeth and Wellman, 1995) and the BATmobile project for monitoring an automated vehicle traveling on a freeway (Forbes et al., 1995). A fragment of the BATmobile is

TABLE 5.4 A selection of other BN applications (PART I).

Citation	Topic (System name)	Task	Type
Reliability/systems			
Morjaria et al. (1993)	Power Plants (GEMS)	monitoring	
Katz et al. (1998)	Power plant	monitoring	DBN
Horvitz & Barry (1995)	Space shuttle propulsion (VISTA)	DSS	DN
Lee and Lee (2006)	Nuclear waste disposal	risk assessment	
Weber & Jouffe (2006)	Complex system reliability	modeling	DOOBN
Weber et al. (2007)	Model-based fault diagnosis	modeling	DBN
Sahin et al. (2007)	Airplane engine faults	diagnosis	H
Oukhellou et al. (2008)	Transportation (railway)	diagnosis	
Widarsson and Dotzauer (2008)	Thermal engineering boiler leakage	Early Warning	
Montani et al. (2008)	Reliability engineering (Radyban)	analysis	DBN
Neil et al. (2008)	Dependable Systems	modeling	H
General engineering			
Weidl et al. (2005)	Monitoring and root-cause analysis	DSS	OOBN
Naticchia et al. (2007)	Roofpond equipped buildings	design	
Rivas et al. (2007)	Roofing slate quality	evaluation	
Renninger et al. (2002)	Mechatronic systems	modeling	OOBN
Ramesh et al. (2003)	Machine tools	modeling	H
Recognition			
Zweig (2003)	Speech recognition	recognition	
Montero & Sucar (2006)	Video conferencing using gestures	recognition	DN
Emergency			
Ozbay & Noyan (2006)	Accident incident clearance times	estimation	
Cheng et al. (2009)	Firespread in buildings	modeling	
Detection/tracking/surveillence			
Mohan et al. (2007)	Object detection and tracking	modeling	DBN?
Stassopolou et al. (2009)	Web robot detection	modeling	
Jiang et al. (2010)	Spatial event surveillance	modeling	
Gowadia et al. (2005)	Intrusion detection (PAID)	modeling	
Robotics			
Dean & Wellman (1991)	Robot navigation and map learning	modeling	DBN
Nicholson et al. (1992a, 1992b)	Robot vehicle	monitoring	DBN
Pynadeth & Wellman (1995)	Traffic	monitoring	DBN
Forbes et al. (1995)	Automated vehicle (BATmobile)	monitoring	DBN
Zhou & Sakane (2007)	Mobile robot	localization	

Note: DSS = decision support system, DBN = dynamic Bayesian network, DN = decision network,
(D)OOBN = (dynamic) object-oriented BN, H = hybrid BN with other modeling.

TABLE 5.5 A selection of other BN applications (PART II).

Citation	Topic (System name)	Task	Type
Military			
Howard & Stumptner (2005)	Situation assessment		OOBN
Falzon (2006)	Centre of gravity analysis	planning	
Laskey & Mahoney (1997)	Situation assessment		OOBN
Finance/enterprises			
Dagum et al. (1992)	Oil price forecasting	prediction	DBN
Adusei-Poku (2005)	Foreign exchange	Risk management	
Neil et al. (2005)	Operational losses	modeling	
Bonafede and Giudici (2007)	Enterprise risk assessment	risk assessment	
Gupta & Kim (2008)	Customer retention	DSS	H
Jensen et al. (2009)	Employee performance and costs	evaluation	
Software development			
Tang et al. (2007)	Change impact in architecture design	modeling	
de Melo & Sanchez (2008)	Software maintenance project delays	prediction	
Bioinformatics			
Tang et al. (2005)	New pharmaceutical compounds	risk	
Bradford et al. (2006)	ProteinProtein Interfaces	prediction	
Allen & Darwiche (2008)	Genetic linkage (RC_Link)	analysis	
Forensics/law			
Dawid et al. (2006)	DNA identification and profiling	modeling	OOBN
Biedermann et al. (2005a,b)	Investigation of fire incidents	evaluation	
Vicard et al. (2008)	Paternity	estimation	OOBN?
Zadora (2009)	Glass fragments	evaluation	H
User modeling			
Horvitz et al. (1998)	Software users (Lumiere)	???	
Albrecht et al. (1998)	Adventure game players	prediction	DBN
Intelligent tutoring system (ITS)/education			
Conati et al. (1997), (2002)	Physics (ANDES)	ITS	
Nicholson et al. (2001)	Decimals	ITS	
Xenos (2004)	Student behaviour	assessment	
Other			
Weymouth et al. (2007)	Forecasting Fog	prediction	
Abramson et al. (1996)	Forcasting severe summer hail	prediction	
	(Hailfinde)		
Cooper et al. (2006)	Biosurveillance outbreaks	diagnosis	
Fernández & Salmerón (2008)	Computer chess program	games	
	(BayesChess)		

Note: DSS = decision support system, DBN = dynamic Bayesian network, DN = decision network,
(D)OOBN = (dynamic) object-oriented BN, H = hybrid BN with other modeling.

shown in Figure 5.5. The number of values for each node is shown in parentheses, the inter-slice arcs are shown in thicker solid lines, while the intra-slice arcs are thinner. The intra-slice arcs for $t + 1$ are the same as for t and hence omitted. The observation nodes are shaded.

A more recent DBN application is (Weber et al., 2007)'s model-based fault diagnosis, while Mohan et al. (2007) develop a different kind of temporal network for evidence fusion for use in object detection and tracking. Montero and Sucar (2006) describe an automated video conference system that provides observations from a gesture recognition system to a dynamic decision network, to select dynamically which camera view to show to the audience.

Hailfinder is a well-known system for **forecasting** severe summer hail in northeastern Colorado (Abramson et al., 1996). The BN is shown in Figure 5.6. Again, the number of values for each node is shown next to the node name in parentheses.

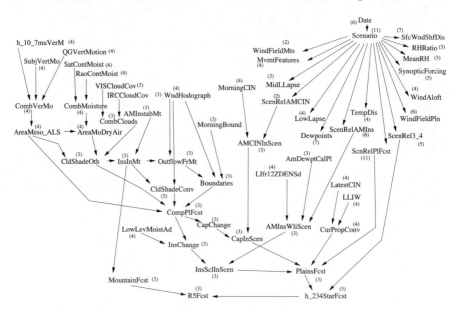

FIGURE 5.6: The Hailfinder BN.

OOBN applications are less numerous to date. This is due partly to their later development (late 1990s, compared to the 1980s for ordinary BNs and decision networks, and the late 1980s for DBNs), and partly because they are not supported in most of the BN software packages. Examples of OOBNs (in addition to the medical and environmental applications described in the previous subsections) include medical and monitoring and root-cause analysis (Weidl et al., 2005), military situation assessment (Howard and Stumptner, 2005), reliability analysis (Langseth, 2002), and DNA analysis in forensics (e.g., Dawid et al., 2006, Vicard et al., 2008).

There are many examples where a BN is combined with other methods in a 'hybrid' application. Examples include agent-based systems (Gowadia et al., 2005), linking with SEM (Gupta and Kim, 2008), combining with particle swarm optimization (Sahin et al., 2007) and support vector machines (Ramesh et al., 2003).

5.3 Cardiovascular risk assessment

In this section we present BN models for predicting coronary heart disease (CHD) that are based on epidemiological models from the medical literature.[2] The epidemiological models were: (1) a regression model from the Australian Busselton study (Knuiman et al., 1998) and (2) a simplified "points-based" model from the German PROCAM study (Assmann et al., 2002). We also adapted both BNs for evaluation on the raw Busselton data.

There are several reasons to convert existing regression models to BNs. First, BNs provide a clear graphical structure with a natural causal interpretation that most people find intuitive to understand. Clinicians who are uncomfortable using or interpreting regression models should be much more amenable to working with the graphical flow of probability through BNs. Second, BNs provide good estimates even when some predictors are missing, implicitly providing a weighted average of the remaining possibilities. Third, BNs clearly separate prior distributions from other model parameters, allowing easy adaptation to new populations. Fourth, BNs can easily incorporate additional data, including (subjective) expert knowledge.

5.3.1 Epidemiology models for cardiovascular heart disease

CHD comprises acute coronary events including myocardial infarction (heart attack) but excluding stroke. Improved predictions of cardiovascular deaths would allow for better allocation of health care resources and improved outcomes. Improved assessment and better explanation and visualization of risk in a clinical setting may also help persuade the patient to adopt preventive lifestyle changes.

We selected two studies which presented regression models for predicting CHD. The first is the Busselton study (Group, 2004), which collected baseline data every three years from 1966 to 1981 and has resulted in many research papers. We used the Cox proportional hazards model of CHD from (Knuiman et al., 1998). The model has the form of a regression: each risk factor has a coefficient, and the risk factors are assumed to act independently. Therefore the structure of the model is a series of independent predictors leading to the target variable. The second study is the *Pro*spective *Ca*rdiovascular *M*ünster study, or PROCAM, which ran from 1979 to 1985, with followup questionnaires every two years. PROCAM has generated several CHD models; we use the "simplified" model in (Assmann et al., 2002). Concerned that clinicians would not use a full logistic regression model, they converted it into

---

[2]These were developed during a broader research project (Twardy et al., 2006, Nicholson et al., 2008). Further details of the BNs, their evaluation, and how they were embedded in a decision support took, TakeHeart II, are given as a knowledge engineering case study in Section 11.5.

a points-based system where the clinician need only add the points from each risk factor, and read the risk off a graph. Therefore the authors had already discretized the continuous variables, making it straightforward to translate to a BN.

5.3.2 The Busselton network

5.3.2.1 Structure

Knuiman et al. (1998) described separate CHD risks for men and women. Rather than having two networks, we made Sex the sole root node in a single network, which then allows us to assess risk across the whole population. The study's other predictor variables (e.g., *Smoking*) are conditional on *Sex*. This effectively gives separate priors for men and women (although those priors do not always differ much).

The full Busselton network is given in Figure 5.7, showing the prior distribution for females. Since the predictors determine the risk score, there is an arc from each predictor variable to the *Score* variable, which is then transformed into the 10-year risk of CHD event. Each state of the 10-year risk node represents a range of percentages. In the example, for a female patient with no other information given, the computed posterior probabilities are 0.749 of a 0-10% risk of a CHD event in the next 10 years, 0.155 probability of a *10-20%* risk, and so on. This risk assessment over percentiles is in turn reduced to a yes/no prediction of a coronary heart disease event, represented by the target node *CHD10* (highlighted in the lower right of the figure); for this example, this risk is 9.43%. Note that this is high because the average age for the survey Busselton population is 58.9 years.

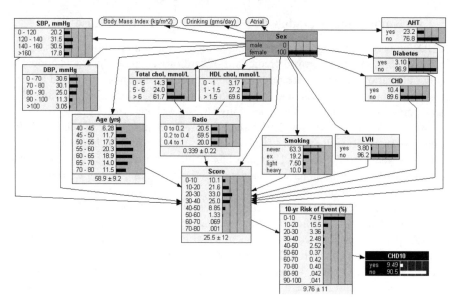

FIGURE 5.7: The Busselton network. Reproduced with permission from Wiley (Nicholson et al., 2008)[Fig.3.1].

Note that the clinician should not care about the *Score* and *Ratio* nodes, as they are just calculating machinery. When embedding the BN in the clinical tool, we use a cleaner view of the Busselton network (depicted in Twardy et al., 2005), where these have been absorbed away. This is a network transform, called **node absorption** (Kjærulff and Madsen, 2008, §7.2.3) that removes nodes from a network in such a way that the full joint probability of the remaining nodes remains unchanged. We left 10-year risk node in because it gives a much better idea about the uncertainty of the estimate.

5.3.2.2 Parameters and discretization

Knuiman et al. (1998) reported summary statistics for their predictors which become the priors for our population: one set for men and another for women. We generated parametric or multi-state priors from their summary statistics (Table 1 in Knuiman et al., 1998). The priors for the multi-state variables just became the CPTs of the corresponding BN node, while the priors for continuous variables were specified using Netica's equation facility, assuming they were Gaussian distributions. To match their model, we also had to create a node for the HDL/Total *Ratio*, as the child of *HDL* and *Total* Cholesterol.

Seven predictors (including the computed *BMI* and *Ratio*) are continuous variables. Table 5.6 summarizes the discretization levels we chose. A detailed discussion of these choices is given in Section 11.5.

TABLE 5.6
Discretization levels for each variable.

Variable	Ranges
Age	0, 40, 45, 50, 55, 60, 65, 70, 75, 80, ∞
SBP	0, 100, 120, 140, 160, ∞
DBP	0, 70, 80, 90, 100, ∞
Chol	0, 5, 6, ∞
HDL	0, 1, 1.5, ∞
Ratio	0, 0.2, 0.4, 1
BMI	0, 20, 25, 30, ∞
Smoking	0, 0, 1, 15, ∞
Alcohol	0, 0, 0, 20, ∞

Score: *Score* is a *continuous* variable which is the weighted sum of all the predictor scores. The weights correspond to the Cox proportional hazards regression and are taken from (Knuiman et al., 1998, p.750). We have separate equations for men and women. Letting the binary variables (e.g., *AHT*, *LVH*, *CHD*) be 1 for "yes" and 0 for "no", these equations are given in Figure 5.8.

Risk: Knuiman et al. did not provide the equation for their risk curve. Simply using their data points would make the model fragile under a new discretization, so instead we fit a curve to the data. Not knowing the precise parametric form of the Cox model likelihood curve, we fit Weibull and logistic curves. As Figure 5.9 shows, the fit was good, but both curves underestimated the risk for low scores. We use the Weibull because it better accommodates the flat top, especially the (70, 99%) point.

Male:	$0.53 \times Age + 0.055 \times SBP$
	$- 56.26 \times Ratio + (0.43 \times AHT)$
	$+ (3.83 \times LVH) + (11.48 \times CHD)$
	$+ (3.2$ if $Smoking \geq 15$, or 3.01 if $15 > Smoking \geq 1)$
	$+ 10$
Female:	$0.846 \times Age + 0.122 \times DBP$
	$- 33.25 \times Ratio + (0.5.86 \times AHT)$
	$+ (2.97 \times LVH) + (7.85 \times DIABETES)$
	$+ (7.99 \times CHD)$
	$+ (6.16$ if $Smoking \geq 15$, or 2.2 if $15 > Smoking \geq 1)$
	$+ 10$

FIGURE 5.8: Equations for *Score* in the Busselton BN (Nicholson et al., 2008)[pp.38-39].

5.3.3 The PROCAM network

5.3.3.1 Structure

For PROCAM, *CHD10* is a weighted sum of the eight risk factors. The full PRO-CAM BN is shown in Figure 5.10(a). There is one root node for each risk factor, which of which has in turn one child, which is the associated scoring node. For example, there is an arc from the *Smoking* root node to its associated score node *SmokingScore*. The eight score nodes are all parents of the combined "Procam Score"node, which is then a parent of the 10 year risk node. Although the scoring scheme was designed to be simpler than logistic regressions, the extra scoring nodes complicate the BN.

Figure 5.10(b) shows the essential structure, with the intermediate score nodes "absorbed" away, as described above: eight nodes converging on a final score. It also shows a hypothetical case where the most likely outcome (0.929 probability) is a 10-20 % risk of a CHD event.

There are some structural differences between the Busselton and PROCAM models. Most obviously, this PROCAM model omits *Sex*, predicting for males only. It also omits *DBP*, *AHT*, *CHD*, and *LVH*, but includes *Family History*. Instead of the ratio of *HDL* to *Total cholesterol*, it uses *HDL*, *LDL*, and *Triglycerides* individually.

5.3.3.2 Parameters and discretization

Assmann et al. (2002) reported summary statistics for their predictors. We generated parametric or multi-state priors as appropriate. Table 5.7 lists the predictors in order of importance in their Cox model, followed by the summary statistics reported in (Assmann et al., 2002, Table 2). For completeness, the equation for the BN's CPDs follows. The priors for the multi-state variables were entered as tables. The priors for continuous variables were specified using Netica's equation facility. Except for *Triglycerides*, they were assumed to be Gaussian distributions.

Age: Assmann et al. (2002) excluded patients younger than 35 or older than 65 at

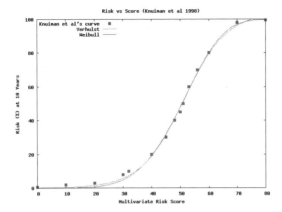

FIGURE 5.9: Risk versus score. Least squares fit on data points are from Knuiman et al.'s Figure 3. Reproduced with permission from Wiley (Nicholson et al., 2008)[Fig.3.2].

TABLE 5.7
Variables and summary statistics from Assmann et al. (2002), Table 2, and corresponding (Netica) BN equations. Reproduced with permission from Wiley (Nicholson et al., 2008)[Table 3.3].

	Summary Stats	BN Equation
Age	46.7 ± 7.5 years	P(Age |) = NormalDist(Age,46.7,7.5)
LDL cholesterol	148.5 ± 37.6 mg/dL	P(LDL |) = NormalDist(LDL,148.5,37.6)
Smoking	31.1% yes	
HDL cholesterol	45.7 ± 11.9 mg/dL	P(HDL |) = NormalDist(HDL, 45.7, 11.9)
Systolic BP	131.4 ± 18.4 mm Hg	P(SBP |) = NormalDist(SBP, 131.4, 18.4)
Family history of MI	16.1% yes	
Diabetes mellitus	6.7% yes	
Triglycerides	126.2* ± 65.9 mg/dL	P(Tri |) = LognormalDist(Tri, log(126.2), 0.45)

(* = Geometric mean)

baseline. We keep their exact range for now. In contrast, Knuiman et al. (1998) used an age range of 40–80, which is significantly older, especially as CHD risk goes up dramatically with age.

Triglycerides: Since the authors reported a geometric mean for Triglycerides, we infer the data must have a lognormal distribution, meaning that log(Triglycerides) would have a normal distribution. A look at the Busselton data confirmed the assumption, as does the common practice of having a LogTriglycerides variable. We then parameterized a lognormal distribution given a geometric mean m and a standard deviation s (see Twardy et al., 2005 for details), ending up with a lognormal equation.

The discretization came from Assmann et al. (2002), reproduced in Table 5.8. *Age* is in 5-year bins, as in our Busselton network. *LDL* and *HDL* are measured in mmol/L instead of mg/dL and discretized more finely. The corresponding breaks for *HDL* would be (0, 0.9, 1.15, 1.4, ∞), compared to the levels we adopted in the

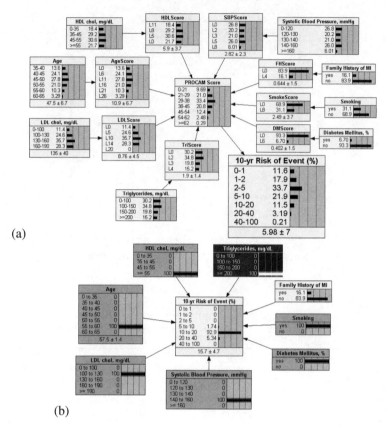

(a)

(b)

FIGURE 5.10: (a) The full PROCAM network and (b) with score nodes "absorbed". Reproduced with permission from Wiley (Nicholson et al., 2008)[Fig.3.3].

Busselton network: (0, 1, 1.5, ∞). *SBP* inserts an extra level at 130, compared to Busselton's (0, 120, 140, 160, ∞).

5.3.3.3   Points

We made a straightforward translation of the discretized predictor variables into scores, by assigning point values to each of the levels of the predictors. In the BN, this means specifying the relationship between the *Score* node and its corresponding predictor variable, as in Table 5.9.[3]

---

[3]The conditional 'if x then y else z' is written 'x ? y : z'. The terms 'L0' ... 'L16' are state names assigned to have the corresponding numeric values 0 ... 16; this is an artifact of the way Netica handles discrete variables that should have numeric values.

TABLE 5.8

Discretization levels for each variable. Values from Assmann et al. (2002), Table 3 and as implemented in Netica. Reproduced with permission from Wiley (Nicholson et al., 2008)[Table 3.4].

Variable	Levels (in paper)	Levels in network
Age	35, 40, 45, 50, 55, 60, 65	(35, 40, 45, 50, 55, 60, 65)
LDL	$< 100, 100, 130, 160, 190, \geq 190$	(0, 100, 130, 160, 190, INFINITY)
HDL	$< 35, 35, 45, 55, \geq 55$	(0, 35, 45, 55, INFINITY)
Tri	$< 100, 100, 150, 200, \geq 200$	(0, 100, 150, 200, INFINITY)
SBP	$< 120, 120, 130, 140, 160$	(0, 120, 130, 140, 160, INFINITY)

TABLE 5.9

Equations defining scores for each predictor. Reproduced with permission from Wiley (Nicholson et al., 2008)[Table 3.5].

AgeScore	*Age<40 ? L0 : Age<45 ? L6 : Age<50 ? L11 : Age<55 ? L16 : Age<60 ? L21 : L26*
LDLScore	*LDL<100 ? L0 : LDL<130 ? L5 : LDL <160 ? L10 : LDL<190 ? L14 : L20*
SmokeScore	*Smoking==yes ? L8 : L0*
HDLScore	*HDL<35 ? L11 : HDL<45 ? L8 : HDL<55 ? L5 : L0*
SBPScore	*SBP < 120 ? L0 : SBP < 130 ? L2 :*
FHScore	*FamHist==yes ? L4 : L0*
DMScore	*Diabetes == yes ? L6 : L0*
TriScore	*Tri<100 ? L0 : Tri<150 ? L2 : Tri<200 ? L3 : L4*
SBP	*SBP<140 ? L3 : SBP<160 ? L5 : L8*

5.3.3.4 Target variables

The PROCAM score is a *continuous* variable which is a simple sum of the individual predictor scores. The work comes in translating this to probabilities. All these risk models have a sigmoid component to convert the risk score to a real risk, thus we can fit a logistic equation to the "risk of acute coronary events associated with each PROCAM score" (Assmann et al., 2002, Table 4). We used the Verhulst equation

$$v(x) = \frac{N_0 K}{N_0 + (K - N_0)exp(-r(x - N_1))}$$

The corresponding graph is shown in Figure 5.11. The fit is good over the the data range (20..60) and tops out at a 70% risk of an event in the next 10 years, no matter how high the PROCAM score goes. As there is always some unexplained variability, our domain experts consider this reasonable (Liew and Rogers, 2004). However, it conflicts with the Busselton study risk curve (Knuiman et al., 1998, p.750, Figure 3) shown in Figure 5.9, which goes up to 100%. Knuiman (2005) says his curve used the standard Cox formula involving baseline survival and defends the high maximum risk, noting that age is a very strong determiner and the Busselton study included people who were up to 80 years old at baseline. In contrast, the PROCAM study contrast, included only those up to 65 at baseline. For now we presume that our fit is reasonable, showing a difference between this PROCAM study and the Busselton study. To define the risk in Netica, we make "Risk" a child of "Score" and set its equation: `Risk (Score) = (30 * 71.7) / (30 + (71.7 - 30) * exp(-0.104 * (Score - 59.5)))`. Table 4 in Assmann et al. (2002) tops out at ≥ 30, and is much finer near the bottom, so we defined the following levels: $(0, 1, 2, 5, 10, 20, 40, 100)$.

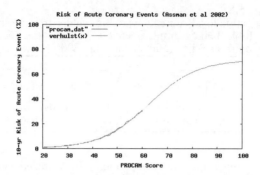

FIGURE 5.11: Logistic curve fitted to Table 4 in Assmann et al. (2002). Reproduced with permission from Wiley (Nicholson et al., 2008)[Fig.3.5].

5.3.4   Evaluation

We wanted to know how well this simple PROCAM model predicts the Busselton data. To do this adapted both the data and the network so that all corresponding variables match, producing the a third network, PROCAM-adapted, which was had essentially the PROCAM structure with the Busselton data priors.

    We evaluated the three networks – Busselton, PROCAM and PROCAM-adapted – using the Busselton epidemiological data. plus other BNs learned from (using the CaMML BN learner, described in Chapter 9), using two metrics. Details of the evaluation are provided in Twardy et al. (2006); a summary is given in knowledge engineering case study in Section 11.5.

5.4   Goulburn Catchment Ecological Risk Assessment

The objective of the Goulburn Catchment (Victoria, Australia) Ecological Risk Assessment (ERA) project was to develop a model to determine the effects of alternative management actions and associated environmental conditions on the native fish community in the Goulburn Broken Catchment, Victoria, Australia. There is evidence that native fish communities have declined substantially since barriers (dams and weirs) were placed in the catchment (Cadwallader, 1978, Erskine, 1996, Gippel and Finlayson, 1993). Although the major factors for the decline have been identified previously (Erskine, 1996), changes to fish communities have not been measured consistently over time.

    BNs were used because they met the project's needs: explicit representation of the causal interactions in the model; representation of uncertainty; ability to combine domain information from domain experts and the monitoring data; ability to incorporate information from multiple scales; and provision of model output that is directly applicable to risk management.

Here we describe the problem domain together with the network structure. Further details of the BN and its parameterization and evaluation, are given as a knowledge engineering case study in Section 11.4.

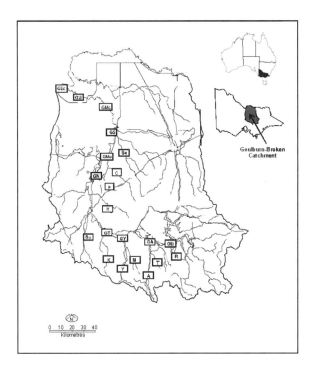

FIGURE 5.12: Location of Goulburn Catchment, showing major system changes. (Reprinted from Parameterisation of a Bayesian network for use in an ecological risk management case study. Pollino et al. *Environmental Modeling and Software* 22(8), 1140–1152. Copyright (2007), with permission from Elsevier.)

5.4.1 Background: Goulburn Catchment

The main stem of the Goulburn Catchment, the Goulburn River, is the largest tributary of the Murray-Darling Basin in the State of Victoria (Australia). The lowland Goulburn River extends from Eildon to its confluence with the Murray River at Echuca (see Figure 5.12). Many rivers and creeks enter the 436 km lowland stretch of the Goulburn River. The headwaters of the Goulburn River flow into Lake Eildon. Water released from Lake Eildon is delivered 218 km downstream to Goulburn Weir. From Goulburn Weir, outflows are to the lower Goulburn River and three irrigation channels. There is evidence that native fish communities in the Goulburn Catchment have declined over the past 100 years (see Pollino et al., 2004). The four major factors identified as influencing native fish abundance and diversity in the Goulburn River are water quality, flow alterations, instream habitat and biological interactions. Although the processes and interactions between these factors and their link to native fish decline are broadly understood, quantitative models to assist in environmental management did not exist.

5.4.2   The Bayesian network

One version of the BN for this application developed by the main domain expert, is shown in Figure 5.13.

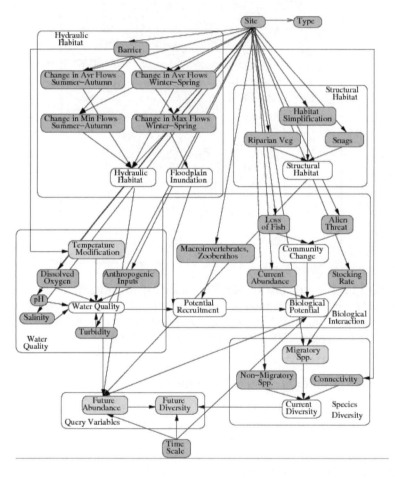

FIGURE 5.13: Prototype Goulburn Catchment ERA BN. (Reprinted from Parameterisation of a Bayesian network for use in an ecological risk management case study. Pollino et al. *Environmental Modeling and Software* 22(8), 1140–1152. Copyright (2007), with permission from Elsevier.)

5.4.2.1 Structure

The BN structure is based on a comprehensive conceptual model developed by domain experts. It consists of five interacting components – water quality, hydraulic habitat, structural habitat, biological potential and species diversity – and has two query variables: Future Abundance and Future Diversity, of the native fish communities. The query variables were also established in collaboration with domain experts and reflect the pragmatic management need to assess what conditions are required to establish sustainable native fish communities in the Goulburn Catchment. Given that no native fish recruitment data was available, these surrogate endpoints were chosen. Clearly, the Goulburn Catchment is a highly complex system with multiple factors (including human-related activities) interacting and influencing fish communities. The empirical relationships within and between chemical, physical and biological system components had not been previously characterized.

TABLE 5.10
ERA Data: Number of cases and missing data for each site. (Sites are grouped into geographical regions.)

Region	Site	No. Cases	Missing data %	Fully Missing Variables
Upper	G_Eild	97	20.0%	
Main	G_Alex	20	7.5%	
	G_Yea	5	26.9%	Temp,Turb,MinSummer
	G_Trawool	4	26.9%	MaxWinter,Food,FutureAbund.,FutureDivers.
Upper	Rubi	27	11.0%	
Tributary	Tagg	25	13.7%	Fishing,Stocking
	Ach	90	32.5%	Food,Fishing
	Murrin	83	26.4%	Fishing
	Yea	24	11.4%	Fishing
	King	15	11.0%	Fishing
	Sunday	8	16.8%	Fishing,FutureAbundance,FutureDiversity
	Hughes	10	18.5%	Salinity,Turb,Fishing
Mid Main	G_LkNag	12	11.5%	Food,FutureAbundance,FutureDiversity
Lower	G_Murch	6	7.7%	FutureAbundance,FutureDiversity
Main	G_Shep	14	10.4%	Food
	G_McCoy	272	26.0%	FutureAbundance,FutureDiversity
	G_Undera	4	11.5%	Turb,FutureAbundance,FutureDiversity
	G_Echuca	3	19.2%	Salinity,Turb,Food,FutureAbund.,FutureDivers.
Lower	Pranjip	8	15.4%	Turb,Food,FutureAbundance,FutureDiversity
Tributary	Crei_Bran	12	18.9%	Salinity,Turb,Fishing
	Castle	6	11.5%	Fishing, FutureAbundance,FutureDiversity
	Sevens	31	14.8%	
Broken	Broken	173	19.4%	
Total		949		

5.4.2.2 Parameterization

We received data from 23 sites which are further aggregated into 6 regions, ranging from 3 to 272 cases, with 949 cases in total (see Table 5.10). For the purpose of parameter learning each case was counted twice to match cases with one- and fiveyear projections of future abundance. In Figure 5.13, the variables for which there was data are indicated by shading, although some sites have data missing for particular variables (see Table 5.10, last column). The number of cases for each site, plus the percentage of data missing, is also shown in Table 5.10, in columns 3 and 4 respectively. Where variables had only limited data available, parameters were initially elicited from domain experts (see § 11.4). These variables are indicated by lighter shading. The CPTs were developed in an iterative manner, combining information from the literature, with expert elicited parameters and data (where available).

5.4.2.3   Evaluation

The domain expert developer conducted a semi-formal model walkthrough with management and ecology experts, with positive feedback. It was recognized by each that more case data was needed, especially with healthy fish populations, to strengthen the model. A sensitivity analysis was also conducted; these results, along with a more detailed description of the knowledge engineering process, are presented in § 11.4.

5.5   Bayesian poker

We will now describe the application of Bayesian networks to a card game, five-card stud poker.[4] Poker is an ideal vehicle for testing automated reasoning under uncertainty. It introduces uncertainty through physical randomization by shuffling and through incomplete information about opponents' hands. Another source of uncertainty is the limited knowledge of opponents, their tendencies to bluff, play conservatively, reveal weaknesses, etc. Poker being a game all about betting, it seems most apt to employ a Bayesian network to compute the odds.

5.5.1   Five-card stud poker

Poker is a non-deterministic **zero-sum game** with imperfect information. A game is zero-sum if the sum of the winnings across all players is zero, with the profit of one player being the loss of others. The long-term goal of all the players is to leave the table with more money than they had at the beginning. A poker session is played in a series of games with a standard deck of 52 playing cards. Each card is identified by its suit and rank. There are four suits: ♣ Clubs, ♢ Diamonds, ♡ Hearts and ♠ Spades. The thirteen card ranks are (in increasing order of importance): Deuce (2),

----

[4]We have worked on this application occasionally since 1993. The version described here is an improved version of that presented in Korb et al. (1999), with some structural changes and the use of utilities to make the betting decision. Details of the system evolution and evaluation are in Chapter 11, as an example of the knowledge engineering process.

Three (3), Four (4), Five (5), Six (6), Seven (7), Eight (8), Nine (9), Ten (T), Jack (J), Queen (Q), King (K) and Ace (A).

In five-card stud poker, after an **ante** (an initial fixed-size bet), players are dealt a sequence of five cards, the first down (hidden) and the remainder up (available for scrutiny by other players). Players bet after each upcard is dealt, in a clockwise fashion, beginning with the best hand showing. The first player(s) may PASS — make no bet, waiting for someone else to open the betting. Bets may be CALLED (matched) or RAISED, with up to three raises per round. Alternatively, a player facing a bet may FOLD her or his hand (i.e., drop out for this hand). After the final betting round, among the remaining players, the one with the strongest hand wins in a "showdown." The strength of poker hand types is strictly determined by the probability of the hand type appearing in a random selection of five cards (see Table 5.11). Two hands of the same type are ranked according to the value of the cards (without regard for suits); for example, a pair of Aces beats a pair of Kings.

TABLE 5.11 Poker hand types: weakest to strongest.

Hand Type	Example	Probability (5-card stud)
Busted	A♣ K♠ J◇ 10◇ 4♡	0.5015629
Pair	2♡ 2◇ J♠ 8♣ 4♡	0.4225703
Two Pair	5♡ 5♣ Q♠ Q♣ K♣	0.0475431
Three of a Kind	7♣ 7♡ 7♠ 3♡ 4◇	0.0211037
Straight (sequence)	3♠ 4♣ 5♡ 6◇ 7♠	0.0035492
Flush (same suit)	A♣ K♣ 7♣ 4♣ 2♣	0.0019693
Full House	7♠ 7◇ 7♣ 10◇ 10♣	0.0014405
Four of a Kind	3♡ 3♠ 3◇ 3♣ J♠	0.0002476
Straight Flush	3♠ 4♠ 5♠ 6♠ 7♠	0.0000134

The basic decision facing any poker player is to estimate one's winning chances accurately, taking into account how much money will be in the pot if a showdown is reached and how much it will cost to reach the showdown. Assessing the chance of winning is not simply a matter of the probability that the hand you have now, if dealt out to the full five cards, will end up stronger than your opponent's hand, if it is also dealt out. Such a pure combinatorial probability of winning is clearly of interest, but it ignores a great deal of information that good poker players rely upon. It ignores the "tells" some poker players have (e.g., facial tics, fidgeting); it also ignores current opponent betting behavior and the past association between betting behavior and hand strength. Our **Bayesian Poker Player** (BPP) doesn't have a robot's sensory apparatus, so it can't deal with tells, but it does account for current betting behavior and learns from the past relationship between opponents' behavior throughout the game and their hand strength at showdowns.

5.5.2   A decision network for poker

BPP uses a series of networks for decision making throughout the game.

5.5.2.1   Structure

The network shown in Figure 5.14 models the relationship between current hand type, final hand type, the behavior of the opponent and the betting action. BPP maintains a separate network for each of the four rounds of play (the betting rounds after two, three, four and five cards have been dealt). The number of cards involved in the current and observed hand types, and the conditional probability tables for them, vary for each round, although the network structure remains that of Figure 5.14.

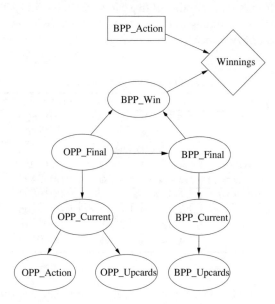

FIGURE 5.14: A decision network for poker.

The node *OPP_Final* represents the opponent's final hand type, while *BPP_Final* represents BPP's final hand type; that is, these represent the hand types they will have after all five cards are dealt. Whether or not BPP will win is the value of the Boolean variable *BPP_Win*; this will depend on the final hand types of both players. *BPP_Final* is an observed variable after the final card is dealt, whereas its opponent's final hand type is observed only after play ends in a showdown. Note that the two final hand nodes are *not* independent, as one player holding certain cards precludes the other player holding the same cards; for example, if one player has four-of-a-kind aces, the other player cannot.

At any given stage, BPP's current hand type is represented by the node *BPP_Current* (an observed variable), while *OPP_Current* represents its opponent's current hand type. Since BPP cannot observe its opponent's current hand type, this must be inferred from the information available: the opponent's upcard hand type, represented by node *OPP_Upcards*, and the opponent's actions, represented by node *OPP_Action*. Note that the existing structure makes the assumption that the opponent's action depends *only* on its current hand and does not model such things as the opponent's confidence or bluffing strategy.

Although the *BPP_Upcards* node is redundant, given *BPP_Current*, this node is included to allow BPP to work out its *opponents* estimate of winning (required for bluffing, see §5.5.4). In this situation, *BPP_Upcards* becomes the observation node as the opponent only knows BPP's upcards.

5.5.2.2 Node values

The nodes representing hand types are given values which sort hands into strength categories. In principle, we could provide a distinct hand type to each distinct poker hand by strength, since there are finitely many of them. That finite number, however, is fairly large from the point of view of Bayesian network propagation; for example, there are already 156 differently valued Full Houses. BPP recognizes 24 types of hand, subdividing busted hands into busted-low (9 high or lower), busted-medium (10 or J high), busted-queen, busted-king and busted-ace, representing each paired hand separately, and with the 7 other hand types as listed in Table 5.11. The investigation of different refinements of hand types is described in §11.2. Note that until the final round, *BPP_Current*, *OPP_Current* and *OPP_Upcards* represent partial hand types (e.g., three cards to a flush, instead of a flush). The nodes representing the opponent's actions have three possible values, *bet/raise, pass/call, fold*.

5.5.2.3 Conditional probability tables

There are four action probability tables $P_i(OPP\_Action|OPP\_Current)$, corresponding to the four rounds of betting. These report the conditional probabilities per round of the actions — folding, passing/calling or betting/raising — given the opponent's current hand type. BPP adjusts these probabilities over time, using the relative frequency of these behaviors per opponent. Since the rules of poker do not allow the observation of hidden cards unless the hand is held to showdown, these counts are made only for such hands, undoubtedly introducing some bias.

The four CPTs $P(OPP\_Upcards|OPP\_Current)$ give the conditional probabilities of the opponent having a given hand showing on the table when the current hand (including the hidden card) is of a certain type. The same parameters were used for $P(BPP\_Upcards|BPP\_Current)$. The remaining CPTs are the four giving the conditional probability for each type of partial hand given that the final hand will be of a particular kind, used for both *OPP_Current* and *BPP_Current*. These CPTs were estimated by dealing out 10,000,000 hands of poker.

5.5.2.4 Belief updating

Given evidence for *BPP_Current*, *OPP_Upcards* and *OPP_Action*, belief updating produces belief vectors for both players' final hand types and, most importantly, a posterior probability of BPP winning the game.

5.5.2.5 Decision node

Given an estimate of the probability of winning, it remains to make betting decisions. Recall that decision networks can be used to find the optimal decisions which will maximize an expected utility. For BPP, the decision node *BPP_Action* in Figure 5.14 represents the possible betting actions *bet/raise, pass/call, fold*, while the utility we wish to maximize is the amount of winnings BPP can accumulate.

5.5.2.6 The utility node

The utility node, *Winnings*, measures the dollar value BPP expects to make based on the possible combinations of the states of the parent nodes (*BPP_Win* and *BPP_Action*). For example, if BPP decided to fold with its next action, irrespective of whether or not it would have won at a showdown, the expected future winnings will be zero as there is no possibility of future loss or gain in the current game. On the other hand, if BPP had decided to bet and it were to win at a showdown, it would make a profit equal to the size of the final pot F_{bet}, minus any future contribution made on its behalf B_{bet}. If BPP bet and lost, it would make a loss equal to any future contribution it made towards the final pot, $-B_{bet}$. A similar situation occurs when BPP decides to pass, but with a differing expected total contribution B_{pass} and final pot F_{pass}. This information is represented in a utility table within the *Winnings* node, shown in Table 5.12.

TABLE 5.12
Poker action/outcome utilities.

BPP_Action	BPP_Win	Utility
Bet	Win	$F_{bet} - B_{bet}$
Bet	Lose	$-B_{bet}$
Pass	Win	$F_{pass} - B_{pass}$
Pass	Lose	$-B_{pass}$
Fold	Win	0
Fold	Lose	0

The amount of winnings that can be made by BPP is dependent upon a number of factors, including the number of betting rounds remaining R, the size of the betting unit B and the current size of the pot C. The expected future contributions to the pot by both BPP and OPP must also be estimated (see Problem 5.10).

The decision network then uses the belief in winning at a showdown and the utilities for each (*BPP_Win, BPP_Action*) pair to calculate the expected winnings (EW) for each possible betting action. Folding is always considered to have zero EW, since regardless of the probability of winning, BPP cannot make any future loss or profit.

5.5.3 Betting with randomization

This decision network provides a "rational" betting decision, in that it determines the action that will maximize the expected utility if the showdown is reached. However, if a player invariably bets strongly given a strong hand and weakly given a weak hand, other players will quickly learn of this association; this will allow them to better assess their chances of winning and so to maximize their profits at the expense of the more predictable player. So BPP employs a mixed strategy that selects an action with some probability based on the EW of the action. This ensures that while most of the time BPP will bet strongly when holding a strong hand and fold on weak hands, it occasionally chooses a locally sub-optimal action, making it more difficult for an opponent to construct an accurate model of BPP's play.

Betting curves, such as that in Figure 5.15, are used to randomize betting actions. The horizontal axis shows the difference between the EW of folding and calling[5] (scaled by the bet size); the vertical axis is the probability with which one should fold. Note that when the difference is zero ($EW(call) - EW(fold) = 0$), BPP will fold randomly half of the time.

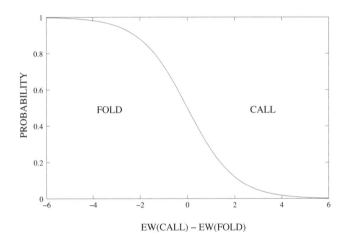

FIGURE 5.15: Betting curve for folding.

Once the action of folding has been rejected, a decision needs to be made between calling and raising. This is done analogously to deciding whether to fold, and is calculated using the difference between the EW of betting and calling.

The betting curves were generated with exponential functions, with different parameters for each round of play. Ideal parameters will select the optimal balance between deterministic and randomized play by stretching or squeezing the curves along

---

[5]More exact would be to compute the differential EW between folding and not folding, the latter requiring a weighted average EW for *pass/call* and for *bet/raise*. We use the EW of calling as an approximation for the latter. Note also that we refer here to calling rather than passing or calling, since folding is not a serious option when there is no bet on the table, implying that if folding is an option, passing is not (one can only pass when there is no bet).

the horizontal axis. If the curves were stretched horizontally, totally random action selection could result, with the curves selecting either alternative with probability 0.5. On the other hand, if the curves were squeezed towards the center, a deterministic strategy would ensue, with the action with the greatest EW always being selected. The current parameters in use by BPP were obtained using a stochastic search of the parameter space when running against an earlier version of BPP.

5.5.4 Bluffing

Bluffing is the intentional misrepresentation of the strength of one's hand. You may over-represent that strength (what is commonly thought of as bluffing), in order to chase opponents with stronger hands out of the round. You may equally well under-represent the strength of your hand ("sandbagging") in order to retain players with weaker hands and relieve them of spare cash. These are tactical purposes behind almost all (human) instances of bluffing. On the other hand, there is an important strategic purpose to bluffing, as von Neumann and Morgenstern pointed out, namely "to create uncertainty in [the] opponent's mind" (von Neumann and Morgenstern, 1947, pp. 188-189). In BPP this purpose is already partially fulfilled by the randomization introduced with the betting curves. However, that randomization occurs primarily at the margins of decision making, when one is maximally uncertain whether, say, calling or raising is optimal over the long run of similar situations. Bluffing is not restricted to such cases; the need is to disguise from the opponent what the situation is, whether or not the optimal response is known. Hence, bluffing is desirable for BPP as an action in addition to the use of randomizing betting curves.

The current version of BPP uses the notion of a "bluffing state." First, BPP works out what its opponent will believe is BPP's chance of winning, by performing belief updating given evidence for *BPP_Upcards*, *OPP_Upcards* and *OPP_Action*. Given this belief is non-zero, it is worth considering bluffing. In which case BPP has a low probability of entering the bluffing state in the last round of betting, whereupon it will continue to bluff (by over-representation) until the end of the round.

5.5.5 Experimental evaluation

BPP has been evaluated experimentally against two automated opponents:

1. A probabilistic player that estimates its winning probability for its current hand by taking a large sample of possible continuations of its own hand and its opponent's hand, then making its betting decision using the same method as BPP;

2. A simple rule-based opponent that incorporated plausible maxims for play (e.g., fold when your hand is already beaten by what's showing of your opponent's hand).

BPP was also tested against earlier versions of itself to determine the effect of different modeling choices (see §11.2). Finally, BPP has been tested against human opponents with some experience of poker who were invited to play via telnet. In all cases, we used BPP's cumulative winnings as the evaluation criterion. BPP performed significantly better than both the automated opponents, was on a par with average amateur humans, but lost fairly comprehensively to an expert human poker player.

We are continuing occasional work on BPP. We have converted it to play Texas Hold'em, which is the game played by the significant computer poker project of Billings and company (see §5.9), and has become the standard for computer poker bot tournaments. Further discussion of BPP's limitations and our ongoing work is given in §11.2.

5.6 Ambulation monitoring and fall detection

Here we present our dynamic belief network (DBN) model for ambulation monitoring and fall detection, an interesting practical application of DBNs in medical monitoring, based on the version described in Nicholson (1996).

5.6.1 The domain

The domain task is to monitor the stepping patterns of elderly people and patients recovering from hospital. Actual falls need to be detected, causing an alarm to be raised. Also, irregular walking patterns, stumbles and near falls are to be identified. The monitoring is performed using two kinds of sensors: foot-switches, which report steps, and a mercury sensor, which is triggered by a change in height, such as going from standing upright to lying horizontally, and so may indicate a fall. Timing data for the observations is also given.

Previous work in this domain performed fall detection with a simple state machine (Davies, 1995), developed in conjunction with expert medical practitioners. The state machine attempts to solve the fall detection problem with a set of *if-then-else* rules. This approach has a number of limitations. First, there is no representation of degrees of belief in the current state of the person's ambulation. Second, there is no distinction between actual states of the world and observations of them, and so there is no explicit representation of the uncertainty in the sensors (Nicholson and Brady, 1992b). Possible sensor errors include:

- **False positives**: the sensor wrongly indicates that an action (left, right, lowering action) has occurred (also called **clutter**, **noise** or **false alarms**).
- **False negatives**: an action occurred but the sensor was not triggered and no observation was made (also called **missed detection**).
- **Wrong timing data**: the sensor readings indicate the action which occurred; however, the time interval reading is incorrect.

5.6.2   The DBN model

When developing our DBN model, a key difference from that state machine approach is that we focus on the causal relationships between domain variables, making a clear distinction between observations and actual states of the world. A DBN for the ambulation monitoring and fall detection problem is given in Figure 5.16. In the rest of this section, we describe the various features of this network in such a way as to provide an insight into the network development process.

5.6.2.1   Nodes and values

When considering how to represent a person's walking situation, possibilities include the person being stationary on both feet, on a step with either the left or right foot forward or having fallen and hence off his or her feet. F represents this, taking four possible values: {*both, left, right, off*}. The Boolean event node *Fall* indicates whether a fall has taken place between time slices. Fall warning and detection relies on an assessment of the person's walking pattern. The node S maintains the person's status and may take the possible values {*ok, stumbling*}. The action node, A, may take the values {*left, right, none*}. The last value is necessary for the situation where a time slice is added because the mercury sensor has triggered (i.e., the person has fallen) but no step was taken or a foot switch false positive was registered.

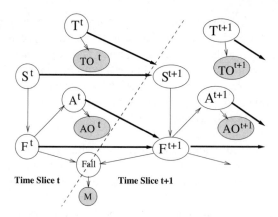

FIGURE 5.16: DBN for ambulation monitoring and fall detection.

There is an observation node for each of the two sensors. The foot switch observations are essentially observations on step actions, and are represented by AO, which contains the same values as the action node. The mercury sensor trigger is represented by the Boolean node M. The time between sensor observations is given by T. Given the problems with combining continuous and discrete variables (see §10.3.7), and the limitations of the sensor, node T takes discrete values representing tenths of seconds. While the fact that there is no obvious upper limit on the time between readings may seem to make it difficult to define the state space of the T node, recall that a monitoring DBN is extended to the next time slice when a sensor observation is made, say n tenths of a second later. If we ignored error in time data, we could add a T with a single value n. In order to represent the uncertainty in the sensor reading, we say it can take values within an interval around the sensor time reading that generates the addition of a new time slice to the DBN. If there is some knowledge of the patient's expected walking speed, values in this range can be added also. The time observation node, TO, has the same state space as T. For each new time slice a copy of each node is added. The possibility of adding further time slices is indicated by the dashed arcs.

5.6.2.2 Structure and CPTs

The CPTs for the state nodes A, F, *Fall* and S are given in Table 5.13. The model for walking is represented by the arcs from F^t to A^t and from F^t, A^t and S^t to F^{t+1}.

We assume that normal walking involves alternating left and right steps. Where the left and right are symmetric, only one combination is included in the table. We have priors for starting on both feet (r) or already being off the ground (s). By definition, if a person finishes on a particular foot, it rules out some actions; for example, if $F^{t+1} = left$, the action could not have been *right*. These zero conditional probability are omitted from the table. The CPT for F^{t+1} for the conditioning cases where $S^{t+1} = stumbling$ is exactly the same as for *ok* except the p and q probability parameters will have lower values, representing the higher expectation of a fall. If there are any variations on walking patterns for an individual patient, for example, if one leg was injured, the DBN can be customized by varying the probability parameters, s, r, p_i, q_i, u, v and w and removing the assumption that left and right are completely symmetric. The fall event node *Fall* has F^t and F^{t+1} as predecessors; a fall only occurs when the subject was on his or her feet to start with ($F^t \neq off$), and finishes off their feet ($F^{t+1} = off$). Note that the fall event node does not really belong to either the t or $t + 1$ time slice; it actually represents the **transition event** between the two.

In this DBN model, the value of walking status node S is determined solely by the time between sensor readings. Also, the T node has no predecessors; its prior can be modified, based on sensor observations over time, to reflect an individual's ordinary walking speed. Note that adding an arc from T^t to T^{t+1} would allow a representation of the expectation that the walking pace should remain fairly constant.

The three observation nodes all have single parents, the variable they are sensing; their CPTs are shown in Table 5.13. Note that the confidence in an observation is given by some value based on a model of the sensor's performance and is empirically obtainable; *pos* is the sensitivity of the positive sensor data and *neg* is the specificity of the negative sensor data (or, 1-*neg* is the probability of ghost data). We make the default assumption that missing or wrong data are equally likely — this need not be the case and can be replaced by any alternative plausible values.

Obviously, this DBN is only one possible model for the ambulation monitoring and fall detection problem and has limitations; for example, it does not handle the cases where both foot switches provide data at the same time or the patient sits down. It also does not handle missing time data.

TABLE 5.13 Ambulation monitoring DBN CPTs.

$P(F_0=left\|right\|)$	$=(1-r-s)/2$	
$P(F_0=both\|)$	$=r$	
$P(F_0=off\|)$	$=s$	
$P(A=left\|F=right)$	$=u$	alternate feet
$P(A=right\|F=right)$	$=v$	hopping
$P(A=none\|F=right)$	$=1-u-v$	stationary
$P(A=\{left\|right\}\|F=both)$	$=w/2$	start with left or right
$P(A=none\|F=both)$	$=1-w$	stationary
$P(A=none\|F=off)$	$=1$	can't walk when off feet
$P(F^{t+1}=left\|F^t=right,A^t=left,S^{t+1}=ok)$	$=p_1$	successful alternate step
$P(F^{t+1}=both\|F^t=right,A^t=left,S^{t+1}=ok)$	$=q_1$	half-step
$P(F^{t+1}=off\|F^t=right,A^t=left,S^{t+1}=ok)$	$=1-p_1-q_1$	fall prob
$P(F^{t+1}=left\|F^t=left,A^t=left,S^{t+1}=ok)$	$=p_2$	successful hop
$P(F^{t+1}=both\|F^t=left,A^t=left,S^{t+1}=ok)$	$=q_2$	half-hop
$P(F^{t+1}=off\|F^t=left,A^t=left,S^{t+1}=ok)$	$=1-p_2-q_2$	fall prob
$P(F^{t+1}=left\|F^t=both,A^t=left,S^{t+1}=ok)$	$=p_3$	successful first step
$P(F^{t+1}=both\|F^t=both,A^t=left,S^{t+1}=ok)$	$=q_3$	unsuccessful first step
$P(F^{t+1}=off\|F^t=both,A^t=left,S^{t+1}=ok)$	$=1-p_3-q_3$	fall prob
$P(F^{t+1}=left\|F^t=left,A^t=none,S^{t+1}=ok)$	$=p_4$	
$P(F^{t+1}=off\|F^t=left,A^t=none,S^{t+1}=ok)$	$=1-p_4$	fall when on left foot
$P(F^{t+1}=right\|F^t=right,A^t=none,S^{t+1}=ok)$	$=p_5$	
$P(F^{t+1}=off\|F^t=right,A^t=none,S^{t+1}=ok)$	$=1-p_5$	fall when on right foot
$P(F^{t+1}=both\|F^t=both,A^t=none,S^{t+1}=ok)$	$=p_6$	
$P(F^{t+1}=off\|F^t=both,A^t=none,S^{t+1}=ok)$	$=1-p_6$	fall when on both feet
$P(F^{t+1}=off\|F^t=off,A^t=left,S^{t+1}=any)$	$=1$	no "get up" action
$P(Fall=T \mid F^{t+1}=off,F^t=\{left\|right\|both\})$	$=1$	from upright to ground
$P(Fall=F \mid F^{t+1}=any,F^t=off)$	$=1$	can't fall if on ground
$P(S^{t+1}=ok\|T^t=t)$	$=1$	if $t \geq y$
$P(S^{t+1}=stumbling\|T^t=t)$	$=1$	if $t < y$
$P(M=T\|Fall=T)$	$=pos_1$	ok
$P(M=F\|Fall=T)$	$=1-pos_1$	missing
$P(M=F\|Fall=F)$	$=neg_1$	ok
$P(M=T\|Fall=F)$	$=1-neg_1$	false alarm
$P(AO=left\|A=left)$	$=pos_2$	ok
$P(AO=right\|A=right)$	$=pos_2$	ok
$P(AO=right\|A=left)$	$=(1-pos_2)/2$	wrong
$P(AO=left\|A=right)$	$=(1-pos_2)/2$	wrong
$P(AO=none\|A=left)$	$=(1-pos_2)/2$	missing
$P(AO=none\|A=right)$	$=(1-pos_2)/2$	missing
$P(AO=none\|A=none)$	$=neg_2$	ok
$P(AO=left\|A=none)$	$=(1-neg_2)/2$	false alarm
$P(AO=right\|A=none)$	$=(1-neg_2)/2$	false alarm
$P(TO=x\|T=x)$	$=pos_3$	ok, $y \neq x$
$P(TO=y\|T=x)$	$=1-pos_3/m-1,$	ok, $y \neq x$

Note: Parameter set used for case-based evaluation results: $s = 0.0$, $r = 0.9$, $u = 0.7$, $v = 0.2$, $w = 0.1$, $p_1 = 0.6$, $q_1 = 0.3$, $p'_1 = 0.5$, $q'_1 = 0.4$ $p_2 = 0.6$, $q_2 = 0.3$, $p'_2 = 0.5$, $q'_2 = 0.4$, $p_3 = 0.6$, $q_3 = 0.3$, $p'_3 = 0.5$, $q'_3 = 0.4$, $p_4 = 0.95$, $p'_4 = 0.85$, $p_5 = 0.95$, $p'_5 = 0.85$, $p_6 = 0.9$, $p'_6 = 0.8$, $pos_1 = 0.9$, $pos_2 = 0.9$, $pos_3 = 0.9$, $neg_1 = 0.95$, $neg_2 = 0.95$.

5.6.3 Case-based evaluation

Let us now look at the behavior of this DBN modeled with the set of parameters given in Table 5.13. For this evaluation, we entered a sequence of evidence that were considered a reasonable series of simulated observations from the sensors. Table 5.14 shows the new beliefs after every new piece of evidence was added. For reasons of space, we left out the initial S_0 node and the T_2 and TO_2 nodes from the model and do not give all the beliefs, especially if they are uniform or otherwise obvious. Probabilities have been rounded to 4 decimal places. The evidence sequence added and the effect on the beliefs was as follows.

No evidence: Beliefs are based on parameters. Belief in an immediate fall is small, bel($Fall_0$ = T)=0.1194, but chance of being off feet in 2 steps is higher, bel(F_0=T) =0.2238.

TO_0 set to t_1: Increased probability that the person is stumbling, bel(S_1 = *stumbling*)=0.9, which slightly increases belief in a fall, bel($Fall_0$ = T) = 0.1828.

AO_0 set to *left*: Foot switch information leads to a change in the belief in the starting state; bel(F_0=*right*) increased 0.2550, reflecting the model of alternate steps.

M_0 set to F: The negative mercury trigger data makes it very unlikely that a fall occurred, bel($Fall_0$=T)=0.0203.

TO_0 set to t_2: "Resetting" of the original timing data makes it less likely the person was stumbling, reducing the belief in a fall, bel($Fall_0$=T) = 0.0098.

M_0 set to T: However, resetting the mercury trigger data makes a fall most probable, bel($Fall_0$=T)=0.6285, although there is still the chance that the sensor has given a wrong reading.

M_1 set to F, TO_1 set to t_4, AO_1 set to *none*: No action, and no mercury trigger data, confirms the earlier fall, bel($Fall_0$=T)=0.7903, since if the person is already on the ground they won't take a left or right step.

5.6.4 An extended sensor model

The DBN described thus far provides a mechanism for handling (by implicitly rejecting) certain inconsistent data. It represents the underlying assumptions about the data uncertainty; however it does not provide an explanation of *why* the observed sensor data might be incorrect. We adapt an idea that has been used in other research areas, that of a moderating or invalidating condition.[6] We can represent a defective sensor by the addition of a sensor status node *SS* or invalidating node (Nicholson and Brady, 1992b). This new node can take the possible values such as {*working, defective*}. Each sensor status node becomes a predecessor of the corresponding observation node, and there is a connection between sensor status nodes across time slices (see Figure 5.17).

---

[6]In the social sciences, the term "moderator" is used for an alternative variable that "moderates" the relationship between other variables (Wermuth, 1987). A similar idea has been used in AI research, e.g., Winston (1977).

TABLE 5.14
Changing beliefs for the ambulation monitoring DBN for an evidence sequence.

Node	Value	Updated beliefs given particular evidence						
		None	$TO_0=t_1$	$AO_0=left$	$M_0=F$	$TO_0=t_2$	$M_0=T$	SET
T_0	t_1	0.25	0.9000	0.9000	0.8914	0.0305	0.0535	0.0616
	t_2	0.25	0.0333	0.0333	0.0361	0.9026	0.8812	0.8736
	t_3	0.25	0.0333	0.0333	0.0361	0.0334	0.0326	0.0323
	t_4	0.25	0.0333	0.0333	0.0361	0.0334	0.0326	0.0323
TO_0	t_1	0.25	**1.0**	**1.0**	**1.0**	0.0	0.0	0.0
	t_2	0.25	0.0	0.0	0.0	**1.0**	**1.0**	**1.0**
F_0	left	0.05	0.05	0.0870	0.0860	0.0856	0.0964	0.0911
	right	0.05	0.05	**0.2550**	0.2717	0.2515	0.2792	0.2767
	both	0.90	0.90	0.6581	0.6422	0.6628	0.6244	0.6322
	off	0.0	0.0	0.0	0.0	0.0	0.0	0.0
A_0	left	0.09	0.09	0.6403	0.6483	0.6453	0.6047	0.5427
	right	0.09	0.09	0.0356	0.0360	0.0359	0.0336	0.0302
	none	0.82	0.82	0.3241	0.3156	0.3188	0.3617	0.4271
AO_0	left	0.1265	0.1265	**1.0**	**1.0**	**1.0**	**1.0**	**1.0**
	right	0.1265	0.1265	0.0	0.0	0.0	0.0	0.0
	none	0.7470	0.7470	0.0	0.0	0.0	0.0	0.0
$Fall_0$	T	**0.1194**	**0.1828**	0.1645	**0.0203**	**0.0098**	**0.6285**	**0.7903**
	F	0.8806	0.8173	0.8355	0.9797	0.9902	0.3715	0.2096
M_0	T	0.1515	0.2053	0.1898	0.0	0.0	**1.0**	**1.0**
	F	0.8485	0.7947	0.8102	**1.0**	**1.0**	0.0	0.0
S_1	ok	0.75	0.1	0.1	0.1086	0.9695	0.9465	0.9383
	stum'g	0.25	**0.9**	0.9	0.8914	0.0305	0.0535	0.0617
F_1	left	0.0638	0.0425	0.2737	0.3208	0.5120	0.1921	0.0340
	right	0.0638	0.0425	0.0168	0.0197	0.0303	0.0114	0.0020
	both	0.7530	0.7322	0.5451	0.6391	0.4478	0.1680	0.1736
	off	0.1194	**0.1828**	0.1645	0.0203	0.0098	0.6285	0.7903
T_1	t_1	0.25	0.25	0.25	0.25	0.25	0.25	0.0326
	t_4	0.25	0.25	0.25	0.25	0.25	0.25	0.9006
TO_1	t_4	0.25	0.25	0.25	0.25	0.25	0.25	**1.0**
A_1	left	0.0950	0.0749	0.0938	0.1099	0.1461	0.0548	0.0035
	right	0.0950	0.0749	0.2222	0.2605	0.3869	0.1451	0.0092
	none	0.8090	0.8502	0.6841	0.6296	0.4670	0.8001	0.9872
AO_1	left	0.1308	0.1137	0.1297	0.1434	0.1741	0.0966	0.0
	right	0.1308	0.1137	0.2389	0.2714	0.3788	0.1734	0.0
	none	0.7383	0.7730	0.6315	0.5851	0.4671	0.7301	**1.0**
$Fall_1$	T	0.1044	0.0975	0.0959	0.1124	0.1099	0.0412	0.0024
	F	0.8956	0.9025	0.9041	0.8876	0.8901	0.9588	0.9976
M_1	T	0.1387	0.1329	0.1315	0.1455	0.1434	0.0850	0.0
	F	0.8612	0.8671	0.8685	0.8545	0.8566	0.9150	**1.0**
S_2	ok	0.75	0.75	0.75	0.75	0.75	0.75	0.9673
	stum'g	0.25	0.25	0.25	0.25	0.25	0.25	0.0327
F_2	left	0.0673	0.0531	0.0898	0.1053	0.1472	0.0552	0.0258
	right	0.0673	0.0531	0.1335	0.1565	0.2291	0.08594	0.0076
	both	0.6415	0.6136	0.5164	0.6055	0.5040	0.1891	0.1740
	off	**0.2238**	0.2802	0.2603	0.1327	0.1197	0.6698	0.7927

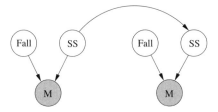

FIGURE 5.17: Extending DBN with sensor status node.

TABLE 5.15
New CPTs with sensor status node added.

$P(M{=}T\|Fall{=}T,SS{=}work)$	= 1	ok
$P(M{=}F\|Fall{=}F,SS{=}work)$	= 1	ok
$P(M{=}F\|Fall{=}T,SS{=}def)$	= ???	missing
$P(M{=}T\|Fall{=}F,SS{=}def)$	= ???	false alarm
$P(SS_{t+1} = def\|SS_t = work)$	= 1 - d	
$P(SS_{t+1} = def\|SS_t = def)$	= X	

The priors for *SS* explicitly represent how likely it is that the sensor is working correctly; in many cases, this can be obtained from data. The CPTs for the *SS* node given in Table 5.15 show how we can model intermittent or persistent faults. The parameter *X* is related to the consistency of the fault. The degradation factor *d* is the probability that a sensor which has been working during the previous time interval has begun to respond defectively. It is based on a model of the expected degradation of the sensor and is a function of the time between sensor readings. It is worth noting that having only the two states doesn't allow us to distinguish between different fault manifestations, which is the source of the query for the CPT numbers when the sensor is defective (see Problem 5.10).

5.7 A Nice Argument Generator (NAG)

Our Nice Argument Generator (NAG) applies Bayesian modeling techniques to a signal problem in artificial intelligence: how to design a computer system which can understand the natural language arguments presented to it and present its own good arguments in reply.[7] This attracted us as an interesting and difficult problem. Furthermore, there is good presumptive reason to believe that Bayesian methods would find a useful role in analyzing and generating arguments, since the considerable majority

---

[7]To be sure, the name of our program emphasizes the generation side of argumentation, neglecting the understanding side; but that was, of course, in view of finding an acronym.

of live arguments in the wild (e.g., on newspaper opinion pages) are inductive and uncertain, unlike mathematical proofs, for example.

In order to argue well one must have a grasp of both the normative strength of the inferences that come into play and the effect that the proposed inferences will have on the audience. NAG is designed to embody both aspects of good argument — that is, to present arguments that are persuasive for an intended audience and also are as close to normatively correct as such persuasiveness allows, which are just the arguments we dub **nice**.

With such a nice system in mind, we incorporated two Bayesian network models within NAG:

- A **normative model**, for evaluating normative correctness
- A **user model**, for evaluating persuasive effect on the user

Currently, both networks are built by hand for each argumentative domain, using the BN software Netica. The normative model should ideally incorporate as many relevant items of knowledge as we can muster and the best understanding of their relationships. The user model should ideally reflect all that we might know of the specific audience that is inferentially relevant: their prior beliefs and their inferential tendencies and biases. We should gather such information about users by building an explicit profile of the user prior to initiating debate, as well as refining the network during debate. As a matter of mundane fact, after building an initial network for the user model, NAG is (thus far) limited to representing some of the simpler cognitive heuristics and biases that cognitive psychologists have established to be widespread, such as the failure to use base rate information in inductive reasoning (Korb et al., 1997).

5.7.1 NAG architecture

Given a normative and a user model, some argumentative context and a goal proposition, NAG is supposed to produce a nice argument supporting the goal. In other words, NAG should produce an argument which will be effective in bringing the user to a degree of belief in the goal proposition within a target range, while simultaneously being "normative" — i.e., bringing the degree of belief in the goal in the normative model within its target. In general terms, NAG does this by building up a subgraph of both the user network and the normative network, called the **argument graph**, which it tests in both the normative and user models. The argument graph is expanded piecemeal until it satisfies the criteria in both models — or, again, until it is clear that the two targets will not be mutually satisfied, when NAG abandons the task.

In somewhat more detail, NAG is composed of three modules as shown in Figure 5.18.

- The **Argument Strategist** governs the argumentation process. In the first in-

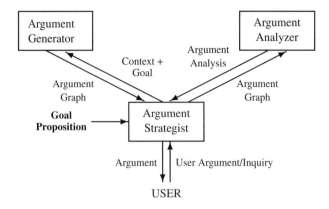

FIGURE 5.18: NAG architecture.

stance it receives a goal proposition and context and then invokes the Generator
to initiate the construction of an argument.[8]

- The **Argument Generator** uses the argumentative context and the goal to construct an initial argument graph.

- The **Argument Analyzer** receives the argument graph from the Strategist and
  tests the effect of the argument on the goal proposition in both the user and the
  normative models, using Bayesian network propagation, with the propositions
  found in the context as the premises (the observed variables) and the goal as
  the query node.

Assuming that the Analyzer discovers that the goal is insufficiently supported in
either the user or the normative model, the Strategist must employ the Generator in
an attempt to expand (or contract) its argument graph. It uses guidance from the Analyzer to do this, such as information about which inferential steps are weak or strong
in either model. If some of the premises are unhelpful, they may be trimmed from
the argument graph, while new nodes that might connect the remaining premises to
the goal may be added. The argument graph will then be returned to the Analyzer
for the next round. This alternating invocation of the Generator and Analyzer continues until either some argument graph is generated which brings the original goal
proposition into the target ranges for strength of belief, the Strategist is unable to fix
a problem reported by the Analyzer, some operating constraint is violated which cannot be overcome (e.g., the overall complexity of the argument cannot be reduced to
an acceptable level) or time runs out. Finally, the Strategist will report the argument
to the user, if a suitable one has been produced.

---

[8]The Strategist may instead receive a user argument, initiating an analysis and rebuttal mode of operation, which we will not describe here.

5.7.2 Example: An asteroid strike

An example might help you get a feel for the process. Consider the Bayesian network shown in Figure 5.19, representing a goal proposition

A large iridium-rich asteroid struck Earth 65-million-years BC. (G)

preceded by the preamble

Around 65-million-years BC the dinosaurs, large reptiles that dominated the Earth for many millions of years, became extinct. (P2) *There was also a massive extinction of marine fauna.* (P3) *At about the same time, a world-wide layer of iridium was laid down on the Earth's crust.* (P1) *It may be that at the same time mass dinosaur migrations led to the spread of disease.* (P4)

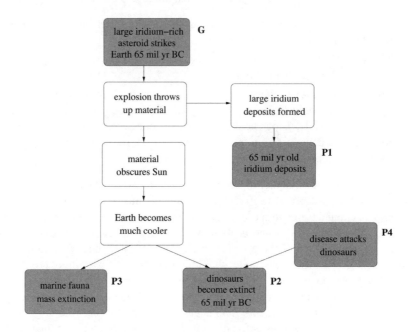

FIGURE 5.19: Asteroid Bayesian network.

The shaded nodes indicate those which are the potential premises in an argument (labeled Pn) and the goal proposition (G). In our example, this Bayesian network represents both the normative and the user models: they do not happen to differ structurally. During the iterative process of building the argument graph in Figure 5.20 the node P4, referring to an alternative explanation for the dinosaur extinction, is dropped because it is unhelpful; the node P3 is dropped because it is unnecessary. It takes two iterations to find the final argument graph, connecting the retained premises with the goal and leading to a posterior belief in the goal node within both target ranges. This argument will then be presented to the user. Note that the arcs in Figure 5.20 are all causal; the final argument proceeds from the existence of the effects to the existence of a cause.

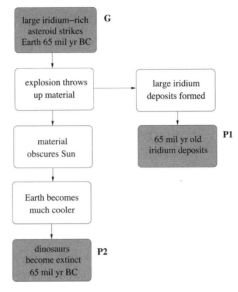

FIGURE 5.20: Asteroid argument graph.

5.7.3   The psychology of inference

*A paradox of inference. A problem may have occurred to the gentle reader: if the premises of the argument are fixed by the argumentative context (excepting the omission of those which turn out to be unhelpful) and if the premises of the argument are identical with those nodes **O** which are observed (i.e., activated for Bayesian network propagation), then the subsequent addition and deletion of nodes in the argument graph is pointless, since Bayesian network propagation rules will guarantee that the goal node G will subsequently take $P(G|\mathbf{O})$ regardless of what ends up in the argument graph. Either the premises immediately and adequately support the goal or they do not, and no further twiddling with any "argument graph" is going to change that.

If this were a correct observation about NAG, then the idea of using NAG to generate any kind of argument would have to be abandoned. Arguments — normative, persuasive or otherwise — are not coextensive with the set of nodes (propositions) which have their conditional probabilities updated given new observations, since that set of nodes, on standard propagation algorithms, is the set of all nodes. What we do with NAG, however, is quite different: we model the psychology of inference. In particular, we model attentional processes during argumentation, and we propagate our Bayesian networks only to the extent that the nodes are being attended to.

Attention. The first fact about cognition, natural or artificial, is that, unlike the presumptions of some philosophers, no cognitive agent has access to infinite reserves of time or inferential power.[9] We employ a constraining mechanism that is simple in concept, namely, attention. There is general agreement that attention serves to apply limited cognitive capacities to problem solving by regulating the flow of information to cognitive processes (e.g., Baars's functions of attention Baars, 1988). We implement such an attentional process through three features:

- An object's **salience**, the extent to which an object is prominent within the set of objects being processed by an agent at some time
- **Recency**, the time elapsed since the cognitive object was last "touched" by a cognitive process
- **Semantic distance**, the degree of semantic or associative relationship between objects

In order to assess semantic distance, we build semantic networks over the Bayesian networks in our user and normative models, as in Figure 5.21. The **semantic networks** represent the semantic relatedness of items directly, in terms of links between nodes and their association strengths. We take the context in which an argument occurs to provide the initial measure of salience: for example, if the user presents an argument to NAG, the propositions in the argument, and any in the preceding discussion, will be marked as salient. Again, when assessing the presentation of an argument, each proposition in the argument graph becomes salient in the order in which the presentation is being considered. Finally, we use activation (Anderson, 1983), spreading from the salient objects (which are clamped at a fixed level of activation) through both the Bayesian and semantic networks, to determine the focus of attention. All items in the Bayesian networks which achieve a threshold activation level while the spreading activation process is iteratively applied will be brought into the span of attention.

This attentional process allows us to decide that some propositions need not be stated explicitly in an argument: if a proposition achieves sufficient activation, without itself being clamped during presentation, it need not be stated.

The attentional focus determines the limits of Bayesian network propagation. The argument graph identifies the premises of the argument — that is, which nodes are to be treated as observational during propagation — but the propagation itself is constrained by the span of attention, which commonly is larger than the argument graph alone. Such an attention-based model of cognition is inherently incomplete: the import of evidence may not be fully realized when that evidence is acquired and only absorbed over time and further argument, as is only to be expected for a limited cognitive agent. Hence, our probability propagation scheme is partial, and the paradox of inference evaded.

5.7.4   Example: The asteroid strike continues

The distinction between activation level and degree of belief needs to be emphasized. A node may have a high or low activation level with any degree of belief attached to

---

[9] See Cherniak's *Minimal Rationality* (1986) for an interesting investigation of the difficulties with such philosophical commitments.

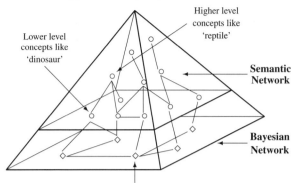

Propositions, e.g., [widespread iridium was deposited 65 mil yr BC]

FIGURE 5.21: NAG combines semantic and Bayesian networks.

it and vice versa. Clamping, for example, effectively means that the node is acting as an observation node during propagation; but observing means being a source in Bayesian network propagation and not necessarily being fully believed (see §2.3.2).

In our asteroid example, if the user model for the asteroid argument differed from the normative model by having a higher prior probability for the disease explanation of saurian death (P4), it would still be activated to the same extent in the initial round, since it is part of the initial argument context. The argument graph of Figure 5.20 might well never be produced, however, because with the disease explanation of a key premise available, the goal node in the user model may not reach the target range with that argument. An argument graph corresponding to the complete Bayesian network of Figure 5.19 could nevertheless reach the target range in both Bayesian networks, since it provides additional evidential support for the asteroidal cause.

5.7.5 The future of argumentation

We have briefly described the NAG concept and architecture and given some idea of what it can do. What it *cannot* do at present is worth noting as well. It was designed around Bayesian networks in concept, so its primary means of interaction is via the networks themselves. Nevertheless, it can handle propositions presented to it in ordinary English, but only when the sentences are preprocessed in various ways. The Bayesian networks themselves, user and normative, must be prepared in advance and by hand. Ideally, of course, we would like to have machine learning software which could generate Bayesian networks for NAG, as well as other software which could generate useful networks from data bases and encyclopedia. As we noted earlier, we would also like to be able to validate and modify user models based upon user performance during argumentation. Extending NAG in these different ways is a long-term, difficult goal.

A more likely intermediate goal would be to employ NAG on the task of explaining Bayesian networks (see §10.4.6.1). It is often, and rightly, remarked that one of the advantages of Bayesian networks over many other AI representations (e.g., neural networks) is that the graphs, both nodes and arcs, have a relatively intuitive semantics. That helps end users to understand what the networks are "saying" and why they are saying it. Our practice suggests that this is quite correct. Despite that, end users still have difficulties understanding the networks and why, say, a target conditional probability has shifted in some particular way. We are planning to utilize the argumentation abilities of NAG in the task of answering such questions as *Why did this network predict that the patient will die under intervention X?* or *Why is intervention A being preferred to intervention B?*

5.8 Summary

In this chapter we have reviewed various BN structures for medical diagnosis applications and surveyed some of the early medical and other applications in the literature. In recent years there have been many new BN and DBN applications, and a full survey is beyond our scope and certainly would be out-of-date quickly. While it is a positive development for BN technology that more applications are being developed commercially, one result is that these are less likely to be published in the research literature or be included in public BN repositories.

We have also described in some detail the development of BN models for several quite different applications. First, we described a BN for cardiovascular risk assessment, based on epidemiological models in the literature. Then we presented a BN for the environmental risk assessment of native fish abundance in a catchment area. Next, we described a game playing application, poker, where the BN was used to estimate the probability of winning and to make betting decisions. Then we saw that a DBN model for ambulation monitoring and fall diagnosis can overcome the limitations of a state-machine approach. Given evidence from sensor observations, the DBN outputs beliefs about the current walking status and makes predictions regarding future falls. The model represents sensor error and is parameterized to allow customization to the individual being monitored. Finally, we looked at how BNs can be used simultaneously for normative domain modeling and user modeling in NAG, by integrating them with semantic networks using an attentional process.

At this stage, it is also worth noting that different evaluation methods were used in each of the example applications, including evaluation on data for the CHD BNs, evaluation by expert and using data for the ERA BN, an experimental/empirical evaluation for the Bayesian poker player, and case-based evaluation for ambulation monitoring DBN. We take up the issue of evaluation in Chapters 7 and 10.

5.9   Bibliographic notes

There have been a number of attempts to collect information about successful BN applications and, in particular, systems in regular use, for example early ones by Russ Greiner and Eugene Santos Jr. More recently, there have been a number of collections of applications published (e.g., Mittal and Kassim, 2007, Pourret et al., 2008, Holmes and Jain, 2008). These and others have been used as a basis for our survey in §5.2.

An interesting case study of PATHFINDER's development is given in (Russell and Norvig, 1995, p. 457). The high level of interest in BN applications for medicine is indicated by the workshop *Bayesian Models in Medicine* at the 2001 European Conference on AI in Medicine (AIME'01). The journal *Artificial Intelligence in Medicine* may also be consulted; there was a 2004 Special Edition on Bayesian networks in biomedicine and health-care, with the edition introduction providing an excellent overview (Lucas et al., 2004). Husmeier et al. have a text (2003) on probabilistic models for medical informatics and bioinformatics.

Our work on BNs for cardiovascular risk assessment has appeared in a series of papers, knowledge engineering from the literature (Twardy et al., 2005), learning from data (Twardy et al., 2006) and the TakeHeart II decision support tool (Hope et al., 2006, Nicholson et al., 2008), while our work on the Goulburn Fish ERA has appeared in Pollino et al. (2005, 2007).

Findler (1977) was the first to work on automated poker play, using a combination of a probabilistic assessment of hand strength with the collection of frequency data for opponent behavior to support the refinement of the models of opponent. Waterman (1970) and Smith (1983) used poker as a testbed for automatic learning methods, specifically the acquisition of problem-solving heuristics through experience. Koller and Pfeffer (1997b) have developed *Gala*, a system for automating game-theoretic analysis for two-player competitive imperfect information games, including simplified poker. Although they use comparatively efficient algorithms, the size of the game tree is too large for this approach to be applied to full poker. Over the past 15 years, Billings et al. have been the leaders in the automation of the poker game Texas Hold'em with their program *Poki* (Billings et al., 2002). In more recent times they have ventured into the commercial world with their poker training software package, Poker Academy Pro, launched in 2003. Our work on Bayesian networks for poker has largely appeared in unpublished honors theses (e.g., Jitnah, 1993, Carlton, 2000, Boulton, 2003, Taylor, 2007), but see also Korb et al., 1999.

The ambulation monitoring and fall detection was an ongoing project, with variations, extensions and implementations described in Nicholson (1996), McGowan (1997), Willis (2000).

NAG has been described in a sequence of research articles, including the first description of the architecture (Zukerman et al., 1996), detailed architectural descriptions (e.g., Zukerman et al., 1998), a treatment of issues in the presentation of arguments (McConachy et al., 1998) and the representation of human cognitive error in the user model (K. B. Korb, 1997).

5.10   Problems

Problems using the example applications

Problem 1

Build the normative asteroid Bayesian network of Figure 5.19 and put in some likely prior probabilities and CPTs. Make a copy of this network, but increase the prior probability for P4; this will play the role of the user model for this problem. Operate both networks by observing P1, P2 and P3, reporting the posterior belief in P4 and G.

   This is not the way that NAG operates. Explain how it differs. In your work has "explaining away" played a role? Does it in the way NAG dealt with this example in the text? Explain your answer.

Problem 2

What are the observation nodes when using the Poker network for estimating BPP's probability of winning? What are the observation nodes when using the network to assess the opponent's estimate of winning chances?

Problem 3

Determine a method of estimating the expected future contributions to the pot by both BPP and OPP, to be used in the utilities in the *Winnings* node. Explain your method together with any assumptions used.

Problem 4

All versions of the Bayesian Poker Player to date have used a succession of four distinct Bayesian networks, one per round. In a DDN model for poker, each round in the poker betting would correspond to a single time slice in the DDN, which would allow explicit representation of the interrelation between rounds of play.

   1. How would the network structure within a single time slice differ from the DN shown in Figure 5.14?
   2. What would the inter-slice connections be?
   3. What information and precedence links should be included?

Problem 5

In §5.6.4, we saw how the use of a sensor status node can be used to model both intermittent and persistent faults in the sensor. In the text this node had only two states, *working* and *defective*, which does not represent the different kinds of faults that may occur, in terms of false negatives, false positives, and so on. Redesign the sensor status modeling to handle this level of detail.

New Applications

The BN, DBN and decision networks surveyed in this chapter have been developed over a period of months or years. It is obviously not feasible to set problems of a similar scope. The following problems are intended to take 6 to 8 hours and require domain knowledge to be obtained either from a domain expert or other sources.

Problem 6

Suppose that a dentist wants to use a Bayesian network to support the diagnosis and treatment of a patient's toothache. Build a BN model including at least 3 possible causes of toothache, any other symptoms of these causes and two diagnostic tests. Then extend the model to include possible treatment options, and a utility model that includes both monetary cost and less tangible features such as patient discomfort and health outcomes.

Problem 7

Build a Bayesian network for diagnosing why a car on the road has broken down. First, identify the most important five or six causes for car breakdown (perhaps, lack of fuel or the battery being flat when attempting to restart). Identify some likely symptoms you can use to distinguish between these different causes. Next you need the numbers. Do the best you can to find appropriate numbers within the time frame of the assignment, whether from magazines, the Internet or a mechanic. Run some different scenarios with your model, reporting the results.

Problem 8

Design a Bayesian network bank loan credit system which takes as input details about the customer (banking history, income, years in job, etc.) and details of a proposed loan to the customer (type of loan, interest rate, payment amount per month, etc.) and estimates the probability of a loan default over the life of a loan. In order to parameterize this network you will probably have to make guestimates based upon aggregate statistics about default rates for different segments of your community, since banks are unlikely to release their proprietary data about such things.

Part II

LEARNING CAUSAL MODELS

These chapters describe how to apply machine learning to the task of learning causal models (Bayesian networks) from statistical data. This remains a hot topic in the data mining community. Modern databases are often so large they are impossible to make sense of without some kind of automated assistance. **Data mining** aims to render that assistance by discovering patterns in these very large databases and so making them intelligible to human decision makers and planners. Much of the activity in data mining concerns rapid learning of simple association rules which may assist us in predicting a target variable's value from some set of observations. But many of these associations are best understood as deriving from causal relations, hence the interest in automated causal learning, or **causal discovery**.

The machine learning algorithms we will examine here work best when they have large samples to learn from. This is just what large databases are: each row in a relational database of N columns is a joint observation across N variables. The number of rows is the sample size.

In machine learning samples are typically divided into two sets from the beginning: a **training set** and a **test set**. The training set is given to the machine learning algorithm so that it will learn whatever representation is most appropriate for the problem; here that means either learning the causal structure (the dag of a Bayesian network) or learning the parameters for such a structure (e.g., CPTs). Once such a representation has been learned, it can be used to predict the values of target variables. This might be done to test how good a representation it is for the domain. But if we test the model using the very same data employed to learn it in the first place, we will reward models which happen to fit noise in the original data. This is called **overfitting**. Since overfitting almost always leads to inaccurate modeling and prediction when dealing with new cases, it is to be avoided. Hence, the test set is isolated from the learning process and is used strictly for testing *after* learning has completed.

Almost all work in causal discovery has been looking at learning from **observational data** — that is, simultaneous observations of the values of the variables in the network. There has also been work on how to deal with joint observations where some values are missing, which we discuss in Chapter 6. And there has been some work on how to infer **latent structure**, meaning causal structure involving variables that have not been observed (also called **hidden variables**). That topic is beyond the scope of this text (see Neapolitan (2003) for a discussion). Another relatively unexplored topic is how to learn from **experimental data**. Experimental data report observations under some set of causal interventions; equivalently, they report joint observations over the augmented models of §3.8, where the additional nodes report causal interventions. The primary focus of Part II of this book will be the presentation of methods that are relatively well understood, namely causal discovery with observational data. That is already a difficult problem involving two parts: searching through the **causal model space** $\{h_i\}$, looking for individual (causal) Bayesian networks h_i to evaluate; evaluating each such h_i relative to the data, perhaps using some score or metric, as in Chapter 9. Both parts are hard. The model space, in particular, is exponential in the number of variables (§8.2.3).

We present these causal discovery methods in the following way. We begin in Chapter 6 with the automated paramaterization of discrete Bayesian networks, i.e., the learning of conditional probability distributions from sample data. The network structures may come from either expert elicitation or the structure learning methods addressed in the succeeding chapters. We start with the simplest case of a binomial root node and build to multinomial nodes with any number of parents. We then discuss learning parameters when the data have missing values, as is normal in real-world problems. In dealing with this problem we introduce **expectation maximization (EM)** techniques and **Gibbs sampling**. In Chapter 7 we survey methods of learning restricted Bayesian network models specifically aimed at prediction tasks. The success of naive Bayes models and their relatives has sustained a strong interest in these techniques over the last two decades. We also consider there the strengths and weaknesses of various techniques for evaluating predictive performance, which has a wider application, including in evaluating Bayesian networks that are not strictly limited to prediction problems.

In Chapter 8 we turn to structure learning, describing methods for the discovery of linear causal models in particular (also called path models, structural equation models). Linear models were the first significant graphical tool for causal modeling, but are not well known in the AI community. Those who have a background in the social sciences and biology will already be familiar with the techniques. In the linear domain the mathematics required is particularly simple, and so this domain serves as a good introduction to causal discovery. It is in this context that we introduce **constraint-based learning** of causal structure: that is, applying knowledge of conditional independencies to make inferences about what causal relationships are possible.

In the final chapter of this part, Chapter 9, we introduce the metric learning of causal structure for discrete models, including our own, **CaMML**. Metric learners search through the exponentially complex space of causal structures, attempting to find that structure which optimizes their metric. The metric usually combines two scores, one rewarding fit to the data and the other penalizing overly complex causal structure. We also discuss how expert knowledge can be seamlessly combined with learning from data. We conclude by examining some difficulties with causal discovery, both theoretical and practical, and evaluate various commonly used methods for evaluating causal discovery programs, and proposing a new alternative of our own.

6

Learning Probabilities

6.1 Introduction

Bayesian network structure may be constructed by hand, presumably during a process of eliciting causal and probabilistic dependencies from an expert (as in Part III), or else they can be learned via causal discovery (Chapters 8 and 9 below), or they can be built using a combination of both approaches (Part III). However the structures are arrived at, they will be useless until parameterized, i.e., the conditional probability tables are specified, characterizing the direct dependency between a child and its parents. Identifying conditional probabilities from expert elicitation is again deferred to Part III; here we consider how to learn these probabilities from data, given an accepted causal structure. (Parameters for linear models are dealt with in Chapter 8, where we simultaneously treat linear structure learning.)

We will first examine how to learn the probability parameters that make up the conditional probability tables of discrete networks when we have non-problematic samples of all the variables. Non-problematic here means that there is no **noise** in the data — all the variables are measured and measured accurately. We will then consider how to handle sample data where some of the variables fail to have been observed, that is when some of the data are incomplete or missing. Finally, we will look at a few methods for speeding up the learning of parameters when the conditional probabilities to be learned depend not just upon the parent variables' values but also upon each other, in **local structure** learning.

6.2 Parameterizing discrete models

In presenting methods for parameterizing discrete networks we will first consider parameterizing **binomial models** (models with binary variables only) and then generalize the method to parameterizing arbitrary discrete models.

6.2.1 Parameterizing a binomial model

The simplest possible Bayesian network is one of a single binomial variable X. Suppose X reports the outcome of a coin toss, taking the values *heads* and *tails*. We wish to learn the parameter value $\Theta = \theta$, which is the probability $P(X = heads)$. Suppose we learn exactly and only that the next toss is *heads*. Then by Bayes' theorem:

$$P(\theta|heads) = \beta P(heads|\theta)P(\theta) \qquad (6.1)$$

(where β is the inverse of the probability of the evidence). $P(heads|\Theta = \theta) = \theta$, so

$$P(\theta|heads) = \beta\theta P(\theta) \qquad (6.2)$$

which multiplication we can see graphically in Figure 6.1. Of course, the observation of heads skews the posterior distribution over $\Theta = \theta$ towards the right, while tails would skew it towards the left.

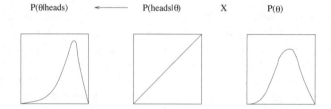

FIGURE 6.1: Updating a binomial estimate in a visual rendition of Bayes' Theorem.

If we get two heads in a row $e =< heads, heads >$ (letting e represent our evidence), then our Bayesian updating yields:

$$P(\theta|e) = \beta\theta^2 P(\theta) \qquad (6.3)$$

In general, evidence e consisting of m heads and $n - m$ tails gives us:

$$P(\theta|e) = \beta\theta^m(1 - \theta)^{n-m}P(\theta) \qquad (6.4)$$

on the assumption that all the coin tosses are **independently identically distributed** (i.i.d.), which means: each toss is independent of every other toss, and each toss is a sample drawn from the very same probability distribution as every other toss.

6.2.1.1   The beta distribution

Equation (6.4) needs a prior distribution over Θ to get started. It doesn't matter what that distribution may be (i.e., the update procedure applies regardless), but it's natural — particularly for automating the learning of Bayesian networks — to choose a distribution that is easy to work with and gives reasonable performance. So, for simplicity we can restrict our prior to the family of **beta distributions** $B(\alpha_1, \alpha_2)$:[1]

$$P(\theta|\alpha_1, \alpha_2) = \beta\theta^{\alpha_1 - 1}(1 - \theta)^{\alpha_2 - 1} \qquad (6.5)$$

---

[1] β in this equation remains the normalization constant, but for the beta distribution (and Dirichlet distribution below) it has a specific form, which is given in Technical Notes §6.7.

To identify a particular beta distribution within the family we must set the hyper-parameters α_1 and α_2 to positive integers. (A **parameter**, when set, takes us from a class of models to a specific model; a **hyperparameter**, when set, takes us from a superclass [or family] of models to a class of models, distinguished from one another by their parameter values.) We identify *heads* with the value 1 and *tails* with the value 0 (as is common), and again *heads* with α_1 and *tails* with α_2. Figure 6.2 displays two beta distributions for fixed small values of α_1 and α_2, namely $< 2,2 >$ and $< 2,8 >$.

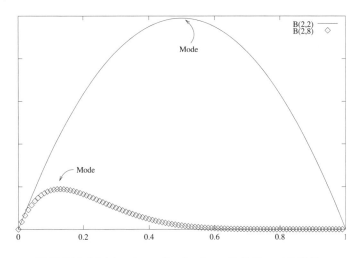

FIGURE 6.2: Two beta distributions: B(2,2) and B(2,8).

The expected value of the beta distribution is:

$$E_{B(\alpha_1,\alpha_2)} = \int \theta P(\theta|\alpha_1,\alpha_2)d\theta = \frac{\alpha_1}{\alpha_1 + \alpha_2} \tag{6.6}$$

Having selected a specific beta distribution, then, by Equation (6.4), after observing evidence e of m heads and $n - m$ tails we will have:

$$P(\theta|e,\alpha_1,\alpha_2) = \beta\theta^{\alpha_1+m-1}(1 - \theta)^{\alpha_2+(n-m)-1} \tag{6.7}$$

Interestingly, we remain within the family of beta distributions, since this has the same form as Equation (6.5), with α_1 replaced by $\alpha_1 + m$ and α_2 replaced by $\alpha_2 + n - m$. A family of distributions which has this property — where Bayesian updating always keeps you within the same family, but with altered hyperparameters — is called a **conjugate family of distributions**.

So, after we have observed m heads and $n - m$ tails, Bayesian updating will move us from $B(\alpha_1,\alpha_2)$ to $B(\alpha_1+m,\alpha_2+n-m)$. Thus, in Figure 6.2 $B(2,2)$ may represent our prior belief in the value of the binomial parameter for the probability of heads. Then $B(2,8)$ will represent our posterior distribution over the parameter after six tails are observed, with the posterior mode around 0.10.

The expected result of the next coin toss will, of course, have moved from $E = \alpha_1/(\alpha_1 + \alpha_2)$ to $E = (\alpha_1 + m)/(\alpha_1 + \alpha_2 + n)$. Selecting the hyperparameters α_1 and α_2 thus fixes how quickly the estimate of θ adjusts in the light of the evidence. When $\alpha_1 + \alpha_2$ is small, the denominator in the posterior expectation, $\alpha_1 + \alpha_2 + n$, will quickly become dominated by the sample size (number of observations) n. When $\alpha_1 + \alpha_2$ is large, it will take a much larger sample size to alter significantly the prior estimate for θ. Selecting a prior beta distribution with $\alpha_1 + \alpha_2 = 4$ in Figure 6.2 represents a readiness to accept that the coin is biased, with the expectation for the next toss moving from 0.5 to 0.2 after only 6 tails. Compare that with $B(10, 10)$ and $B(10, 16)$ in Figure 6.3, when both the posterior mode and posterior mean (expectation) shift far less than before, with the latter moving from 0.5 to 0.38.

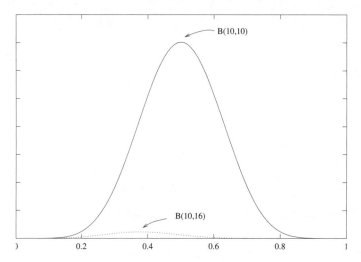

FIGURE 6.3: Two beta distributions: $B(10,10)$ and $B(10,16)$.

In fact, setting the size of $\alpha_1 + \alpha_2$ plays the same role in subsequent learning as would an initial sample of the same size $\alpha_1 + \alpha_2$ (ignoring the fact that our initial beta distribution requires positive hyperparameters). So, an initial selection of values for α_1 and α_2 can be thought of as a kind of "pretend initial sample," leading people to refer to $\alpha_1 + \alpha_2$ as the **equivalent sample size**.

In short, estimating the binomial parameter under these assumptions — i.i.d. samples and a beta prior — just means mixing the prior hyperparameters with the frequency of heads in the sample.

6.2.2   Parameterizing a multinomial model

This process generalizes directly to variables with more than two states — that is, to **multinomial variables** — using another conjugate family of distributions, namely the **Dirichlet** family of distributions. The Dirichlet distribution with τ states is written $D[\alpha_1, \ldots, \alpha_i, \ldots, \alpha_\tau]$ with α_i being the hyperparameter for state i. In direct generalization of the binomial case, the probability of state i is

$$P(X = i) = \frac{\alpha_i}{\sum_{j=1}^{\tau} \alpha_j} \tag{6.8}$$

The learning problem is to estimate the set of parameters $\{\Theta_i = \theta_i : 1 \leq i < \tau\}$. Let $\vec{\theta}$ refer to the vector $< \theta_1, \ldots, \theta_\tau >$. A possible simplification is to assume **local parameter independence**:

$$P(\vec{\theta}) = \prod_{i=1,\ldots,\tau} P(\theta_i) \tag{6.9}$$

that is, that the probability of each state is independent of the probability of every other state. With this assumption, we can update each multinomial parameter in the same way as we did binomial parameters, following Spiegelhalter and Lauritzen (1990). Thus, observing state i moves you from the original distribution $D[\alpha_1, \ldots, \alpha_i, \ldots, \alpha_\tau]$ to $D[\alpha_1, \ldots, \alpha_i + 1, \ldots, \alpha_\tau]$. In this case the equivalent sample size is the original $\sum_{j=1}^{\tau} \alpha_j$.

These techniques provide parameter estimation for BNs with a single node. In order to parameterize an entire network we simply iterate over its nodes:

Algorithm 6.1 *Multinomial Parameterization (Spiegelhalter and Lauritzen method)*

1. For each node X_j

> *For each instantiation of Parents(X_j), assign some Dirichlet distribution for the τ states of X_j $D[\alpha_1, \ldots, \alpha_i, \ldots, \alpha_\tau]$*

2. For each node X_j

> *For each joint observation of all variables $X_1, \ldots, X_k$*
>
> *(a) Identify which state i X_j takes*
>
> *(b) Update $D[\alpha_1, \ldots, \alpha_i, \ldots, \alpha_\tau]$ to $D[\alpha_1, \ldots, \alpha_i + 1, \ldots, \alpha_\tau]$ for the distribution corresponding to the parent instantiation in the observation*

Thus, we have a very simple counting solution to the problem of parameterizing multinomial networks. This solution is certainly the most widely used and is available in the standard Bayesian network tools.

The assumptions behind this algorithm are:

1. Local parameter independence, per Equation (6.9).

2. Parameter independence *across* distinct parent instantiations. That is, the parameter values when the parents take one state do not influence the parameter values when parents take a different state.

3. Parameter independence across non-local states. That is, the states adopted by other parts of the network do not influence the parameter values for a node once its parent instantiation is given.

4. The parameter distributions are within a conjugate family of priors; specifically they are Dirichlet distributed.

The third assumption is already guaranteed by the Markov property assumed as a matter of general practice for the Bayesian network as a whole.[2] The first and second assumptions are more substantial and, frequently, wrong. When they are wrong, the implication is that dependencies between parameter values are not being recognized in the learning process, with the result that the information afforded by such dependencies is neglected. The upshot is that Algorithm 6.1 will still work, but it will work more slowly than methods which take advantage of parameter dependencies to re-estimate the values of some parameters given those of others. The algorithm must painstakingly count up values for each and every cell in each and every conditional probability table without any reference to other cells. This slowness of Algorithm 6.1 can be troublesome because many parent instantiations, especially when dealing with large arity (large numbers of joint parent states), may be rare in the data, leaving us with a weak parameterization of the network. We will examine different methods of taking advantage of parameter dependence in probability learning in §6.4 below.

The fourth assumption, that the parameter priors are Dirichlet distributed, enables the application of the simple Algorithm 6.1 to parameterization. Of course, there are infinities of other possible prior distributions over parameters; but choosing outside of the Dirichlet family requires a different estimation algorithm. The exponential family of distributions, which subsumes the Dirichlet family, admits of tractable estimation methods (DeGroot, 1970). In any case, choosing inaccurate hyperparameters for the Dirichlet is a more likely source of practical trouble in estimating parameters than the initial restriction to the Dirichlet family. The hyperparameters should be selected so as to reflect what is known or guessed about the probability of each state, as revealed in Equation (6.8), as well as the degree of confidence in those guesstimates, expressed in equivalent sample size.

6.3   Incomplete data

Machine learning from real data very commonly has to deal with data of poor quality. One way in which data are often poor is that the measurements are **noisy**, meaning that many of the attribute (variable) values reported are incorrect. In Chapter 9 we will look at some information-theoretic techniques which deal with noise.

Another kind of data poverty is when some attribute values are simply missing in some of the joint observations, that is, when the data samples are incomplete. Thus, when responding to surveys some people may fail to state their ages or their incomes, while reporting their other attributes, as in Table 6.1. This section will present a

---

[2]To be sure, the Markov property does not imply parameter independence from the parameters of descendants, so the third assumption has this stronger implication.

number of techniques for parameterizing a Bayesian network even though some of the data available are corrupted in such a manner.

TABLE 6.1

Example of incomplete data in joint observations of four variables (— indicates a missing value).

Name	Occupation	Income	Automobile
Jones	surgeon	120K	Mercedes
Smith	student	3K	none
Johnson	lawyer	—	Jaguar
Peters	receptionist	23K	Hyundai
Taylor	pilot	—	BMW
—	programmer	50K	BMW

A more extreme kind of incompleteness is simply to be missing all the measurements for some relevant variables. Rectifying that problem — learning causal structure and parameterizing a network lacking all the values for one or more variables — is called **latent variable** (or hidden variable) discovery. The complexities of latent variable discovery will, for the most part, not be addressed in this text; they are under active research in the Bayesian network community. In statistics research there are various known methods for introducing and testing for latent variables, including factor analysis (see, e.g., Loehlin, 1998).

In this section we will first look at the most complete solution to dealing with missing data, namely directly computing the conditional density $P(\theta|\mathbf{Y})$ when the observed variables $\mathbf{Y}$ are a proper subset of all the variables $\mathbf{V}$ (and so, incomplete). This turns out to be an intractable computation in general, so we move on to consider two approximate solutions.

6.3.1 The Bayesian solution

There is an exact Bayesian answer to the question: What is the conditional density $P(\theta|\mathbf{Y})$? — where the observed variables $\mathbf{Y} \subset \mathbf{V}$ are incomplete. We demonstrate the idea of the answer with the simplest possible scenario. Suppose the set of binary variables $\mathbf{Y}$ are observed and the single binary variable X is unobserved, so that the entire set of variables is $\mathbf{V} = \mathbf{Y} \cup X$. Here we are dealing simply with binomial parameterization. By Theorem 1.3 we have, for any binomial parameter, $P(\theta|\mathbf{Y}) = P(\theta|X,\mathbf{y})P(X|\mathbf{y}) + P(\theta|\neg X,\mathbf{y})P(\neg X|\mathbf{y})$. We can see that, in effect, we must compute each possible way of completing the incomplete data (i.e., by observing X or else by observing $\neg X$) and then find the weighted average across these possible completions. Under the assumptions we have been applying above, both $P(\theta|X,\mathbf{y})$ and $P(\theta|\neg X,\mathbf{y})$ will be beta densities, and $P(\theta|\mathbf{Y})$ will be a linear mixture of beta densities. If the set of unobserved attributes $\mathbf{X}$ contains more than one variable, then the mixture will be more complex — exponentially complex, with the number of products being $2^{|\mathbf{X}|}$.

The generalization of this analysis to multinomial networks is straightforward, resulting in a linear mixture of Dirichlet densities. In any case, the mixed densities must be computed over every possible completion of the data, across all joint samples which are incomplete. This exact solution to dealing with incomplete data is clearly intractable.

6.3.2   Approximate solutions

We will examine two approaches to approximating the estimation of parameters with incomplete data: a stochastic sampling technique, called Gibbs sampling, and an iterative, deterministic algorithm, called expectation maximization (EM). Both techniques make the strong simplifying assumption that the missing data are independent of the observed data. That assumption will frequently be false. For example, in a consumer survey it may be that income data are missing predominately when wealthy people prefer to cover up their wealth; in that case, missing income data will be independent of other, observed values only in the unlikely circumstance that wealth has nothing to do with how other questions are answered. Nevertheless, it is useful to have approximate methods that are easy to use; and they are easy to use because they rely upon the independence assumption.

6.3.2.1   Gibbs sampling

Gibbs sampling is a stochastic procedure introduced by Geman et al. (1993), which can be used to sample the results of computing any function f of $\mathbf{X}$, yielding an estimate of $P(f(\mathbf{X}))$. In particular, it can be used to estimate by sampling the conditional distribution $P(\mathbf{X}|e)$ where the evidence e is partial. Here we present the Gibbs sampling algorithm in a simple form for computing the expected value of the function $f(\mathbf{X})$, i.e., $E[f(\mathbf{X})]$. At each step we simply compute $f(\mathbf{x})$ (the algorithm assumes we know how to compute this), accumulating the result in *sum*, and in the end return the average value. To estimate the full distribution of values $P(f(\mathbf{X})|e)$ we only need to alter the algorithm to collect a histogram of values.

Intuitively, Gibbs sampling estimates $P(\mathbf{X})$ by beginning at an arbitrary initial point in the state space of $\mathbf{X}$ and then sampling an adjacent state, with the conditional probability $P(\mathbf{X}|e)$ governing the sampling process. Although we may start out at an improbable location, the probability pressure exerted on the sampling process guarantees convergence on the right distribution (subject to the convergence requirement, described below).

Algorithm 6.2 *Gibbs Sampling Algorithm for the expected value $E[f]$*
Let j index the unobserved variables in the full set of variables $\mathbf{X}$ and limit be the number of sampling steps you wish to take.

   0.   *(a) Choose any legitimate initial state for the joint $\mathbf{X} = \mathbf{x}$; the observed variables take their observed values and the unobserved variables take any value which is allowed in $P(\mathbf{X})$.*

        (b) sum $= 0.0$

        (c) $i = 0$

   1. While $i < limit$ do

(a) *Select the next unobserved X_j*

(b) *Replace its value with a sample from $P(X_j|\mathbf{X}\backslash X_j)$*

(c) *sum $\leftarrow$ sum $+ f(\mathbf{x})$*

(d) *$i \leftarrow i+1$*

2. *Return sum$/i$.*

In the initialization Step (0.a) it does not matter what joint state for $\mathbf{X}$ is selected, so long as the observed variables in $\mathbf{X}$ are set to their observed values and also the convergence requirement (below) is satisfied. In Step (1.b) we are required to compute the conditional probability distribution over the unobserved X_j given what we know (or guess) about the other variables; that is, we have to compute $P(X_j|\mathbf{X}\backslash X_j)$, which is easy enough if we have a complete Bayesian network. The next version of the algorithm (Algorithm 6.3 just below) provides a substitute for having a complete Bayesian network.

Note that this algorithm is written as though there is a *single* joint observation. In particular, Step (0.a) assumes that variables in $\mathbf{X}$ either are observed or are not, whereas across many joint observations some variables will be both observed in some and unobserved in others. To implement the algorithm for multiple joint observations, we need only embed it in an iteration that cycles through the different joint observations.

Now we present another variation of the Gibbs sampling algorithm which specifically computes an approximation for multinomial networks to the posterior density $P(\theta|e)$, where e reports a set of incomplete observations. This algorithm assumes Dirichlet priors.

Algorithm 6.3 *Gibbs Sampling Algorithm for estimating $P(\theta|e)$*
Let k index the sample — i.e., x_{jk} is the value observed for the jth variable in the kth joint observation in the sample. This algorithm is not presented in complete form: where the phrase "per Cooper & Herskovits" appears, this means the probability function $P(\cdot)$ can be computed according to Theorem 9.1 in Chapter 9.

0. *Complete $e \rightarrow e^*$ arbitrarily, choosing any legitimate initial state for the joint samples; set limit to the number of sampling steps; $i = 0$.*

1. *While $i < $ limit do*

   (a) *For each unobserved $X_{jk} \in e^*\backslash e$*

   Reassign X_{jk} via a sample from

   $$P(x_{jk}|e^*\backslash x_{jk}) = \frac{P(x_{jk}, e^*\backslash x_{jk})}{\sum_{x'_{jk}} P(x'_{jk}, e^*\backslash x_{jk})}$$

   per Cooper & Herskovits. Here the denominator sums over all possible values for $X_{jk} = x'_{jk}$.

   This produces a new, complete sample, e^{**}.

(b) Compute the conditional density $P(\theta|e^{**})$ using Algorithm 6.1.

(c) $e^{*} \leftarrow e^{**}; i \leftarrow i+1$.

2. *Use the average value during sampling to estimate $P(\theta|e)$.*

Convergence requirement. For Gibbs sampling to converge on the correct value, when the sampling process is called **ergodic**, two preconditions must be satisfied (see Madigan and York, 1995). In particular:

1. From any state $\mathbf{X} = \mathbf{x}$ it must be possible to sample any other state $\mathbf{X} = \mathbf{x}'$. This will be true if the distribution is positive.

2. Each value for an unobserved variable X_j is chosen equally often (e.g., by round robin selection or uniformly randomly).

6.3.2.2 Expectation maximization

Expectation maximization (EM) is a deterministic approach to estimating θ asymptotically with incomplete evidence, and again it assumes missing values are independent of the observed values; it was introduced by Dempster et al. (1977). EM returns a point estimate $\hat{\theta}$ of θ, which can either be a **maximum likelihood** (ML) estimate (i.e., one which maximizes $P(e|\hat{\theta})$) or a **maximum aposteriori probability** (MAP) estimate (i.e., one which maximizes $P(\hat{\theta}|e)$, taking the mode of the posterior density). First, a general description of the algorithm:

Algorithm 6.4 *Expectation Maximization (EM)*

0. *Set $\hat{\theta}$ to an arbitrary legal value; select a desired degree of precision ε for $\hat{\theta}$; set the update value $\hat{\theta}'$ to an illegally large value (e.g., MAXDOUBLE, ensuring the loop is executed at least once).*

While $|\hat{\theta}' - \hat{\theta}| > \varepsilon$ do:

$\hat{\theta} \leftarrow \hat{\theta}'$ *(except on the first iteration)*

1. **Expectation Step:** *Compute the probability distribution over missing values:*

$$P(e^{*}|e, \hat{\theta}) = \frac{P(e|e^{*}, \hat{\theta})P(e^{*}|\hat{\theta})}{\sum_{e^{*}} P(e|e^{*}, \hat{\theta})P(e^{*}|\hat{\theta})}$$

2. **Maximization Step:** *Compute the new ML or MAP estimate $\hat{\theta}'$ given $P(e^{*}|e, \hat{\theta})$.*

Step 1 is called the expectation step, since it is normally implemented by computing an expected **sufficient statistic** for the missing e^{*}, rather than the distribution itself. A sufficient statistic is a statistic which summarizes the data and which itself contains all the information in the data relevant to the particular inference in question. Hence, s is a sufficient statistic for θ relative to e^{*} if and only if θ is independent of e^* given the statistic — i.e., $\theta \perp\!\!\!\perp e^* | s$. Given such a statistic, the second step uses it to maximize a new estimate for the parameter being learned. The EM algorithm generally converges quickly on the best point estimate for the parameter; however, it is a best estimate locally and may not be the best globally — in other words, like hill climbing, EM will get stuck on local maxima (Dempster et al., 1977). The other limiting factor is that we obtain a point estimate of θ and not a probability distribution.

We now present EM in its maximum likelihood (ML) and maximum aposteriori (MAP) forms.

Algorithm 6.5 *Maximum Likelihood EM*

ML estimation of θ given incomplete e.

0. Set $\hat{\theta}$ arbitrarily; select a desired degree of precision ε for $\hat{\theta}$; set the update value $\hat{\theta}'$ to MAXDOUBLE.

While $|\hat{\theta}' - \hat{\theta}| > \varepsilon$ do:

   $\hat{\theta} \leftarrow \hat{\theta}'$ *(except on the first iteration)*

   1. Compute the expected sufficient statistic for e^*:

$$E_{P(X|e,\hat{\theta})} N_{ijk} = \sum_{l=1}^{N} P(x_{ik}, Parents(X_{ij}) | y_l, \hat{\theta})$$

   N_{ijk} counts the instances of possible joint instantiations of X_i and $Parents(X_i)$, which are indexed by k (for X_i) and j (for $Parents(X_i)$). These expected counts collectively (across all possible instantiations for all variables) provide a sufficient statistic for e^*. For any one N_{ijk} this is computed by summing over all (possibly incomplete) joint observations y_l the probability on the right hand side (RHS). Since in this step we have a (tentative) estimated parameter $\hat{\theta}$ and a causal structure, we can compute the RHS using a Bayesian network.

   2. Use the expected statistics as if actual; maximize $P(e^*|\hat{\theta}')$ using

$$\hat{\theta}'_{ijk} = \frac{E_{P(X|e,\hat{\theta})} N_{ijk}}{\sum_k' E_{P(X|e,\hat{\theta})} N_{ijk'}}$$

Algorithm 6.6 *Maximum Aposteriori Probability EM*

  0. Set $\hat{\theta}$ arbitrarily; select a desired degree of precision ε for $\hat{\theta}$; set the update value $\hat{\theta}'$ to MAXDOUBLE.

While $|\hat{\theta}' - \hat{\theta}| > \varepsilon$ do:

    $\hat{\theta} \leftarrow \hat{\theta}'$ *(except on the first iteration)*

    1. Compute the expected sufficient statistic for e^*:

$$E_{P(X|e,\hat{\theta})}N_{ijk} = \sum_{l=1}^{N} P(x_{ik}, Parents(X_{ij})|y_l, \hat{\theta})$$

    (See Algorithm 6.5 Step 1 for explanation.)

    2. Use the expected statistics as if actual; maximize $P(\hat{\theta}'|e^*)$ using

$$\hat{\theta}'_{ijk} = \frac{\alpha_{ijk} + E_{P(X|e,\hat{\theta})}N_{ijk}}{\sum_{k'} \alpha_{ijk'} + E_{P(X|e,\hat{\theta})}N_{ijk'}}$$

where α_{ijk} is the Dirichlet parameter.

6.3.3 Incomplete data: summary

In summary, when attribute values are missing in observational data, the optimal method for learning probabilities is to compute the full conditional probability distribution over the parameters. This method, however, is exponential in the arity of the joint missing attribute measurements, and so computationally intractable. There are two useful approximation techniques, Gibbs sampling and expectation maximization, for asymptotically approaching the best estimated parameter values. Both of these require strong independence assumptions — especially, that the missing values are independent of the observed values — which limit their applicability. The alternative of actively modeling the missing data, and using such models to assist in parameterizing the Bayesian network, is one which commends itself to further research. In any case, the approximation techniques are a useful start.

6.4 Learning local structure

We now turn to a different kind of potential dependence between parameters: not between missing and observed values, but between different observed values. **Local structure** learning involves learning the relation between a variable and its parents and, in particular, learning any interdependencies between its parameters. Algorithm 6.1 assumed that the different states which a child variable takes under different parent instantiations are independent of each other (i.e., exhibiting parameter independence), with the consequence that when there *are* dependencies, they are ignored, resulting in slower learning times. When there are dependencies between parameters we should like our learning algorithms to take advantage of them.

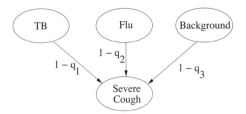

FIGURE 6.4: A noisy-or model.

6.4.1 Causal interaction

One of the major advantages of Bayesian networks over most alternative uncertainty formalisms (such as PROSPECTOR, Duda et al., 1976 and Certainty Factors, Buchanan and Shortliffe, 1984) is that Bayesian networks allow, but do not require, conditional independencies to be modeled. Where there are dependencies, of any complexity, they can be specified to any degree required. And there are many situations with local dependencies, namely all those in which there is at most limited **causal interaction** between the parent variables. To take a simple example of interaction: one might ingest alkali, and die; one might instead ingest acid, and die; but if one ingests both alkali and acid together (to be sure, only if measured and mixed fairly exactly!) then one may well not die. That is an interaction between the two potential causes of death. When two parent causes fully interact, each possible instantiation of their values produces a probability distribution over the child's values which is entirely independent of all their other distributions. In such a case, the full power, and slowness, of the Spiegelhalter and Lauritzen method of learning CPTs (Algorithm 6.1) is required.

The most obvious case of local structure is that where the variables are continuous and the child is an additive linear function of its parents, as in path models. In this case, the magnitude of the child variable is influenced independently by the magnitudes of each of its parents. And the learning problem (solved in Chapter 8) is greatly reduced: one parameter is learned for each parent, so the learning problem is linear in the number of parents, rather than exponential.

We shall now briefly consider three ways of *modeling* local structure in the CPTs of discrete models and of taking advantage of such structure to learn parameterizations faster, namely with noisy-or connections, classification trees and graphs, and with logit models. Each of these model different kinds of non-interactive models.

6.4.2 Noisy-or connections

Noisy-or models are the most popular for dealing with non-interactive binomial causal factors. Figure 6.4 shows a noisy-or model of severe coughing. This model assumes that the illnesses *TB* and *Flu* are independent of each other and that each has a probability $(1 - q_i)$ of causing the coughing which is independent of the other causes and some background (unattributed) probability of *Severe Cough*. Thus, the

q_i parameters of the model can be thought of as the probability of each cause *failing* — it is the "noise" interfering with the cause. Since they are required to operate independently, the CPT relating the three causal factors can be easily computed, on the assumption that the *Background* is always active (*On*), as in Table 6.2.

Thus, to parameterize the noisy-or model we need find only three parameters versus the four in the CPT. Although that may not be an impressive savings in this particular case, once again the simpler model has one parameter per parent, and so the task grows linearly rather than exponentially.

TABLE 6.2
CPT for *Severe Cough* generated from noisy-or parameters.

Severe Cough	TB, Flu	$TB, \neg Flu$	$\neg TB, Flu$	$\neg TB, \neg Flu$
F	$q_1 q_2 q_3$	$q_1 q_3$	$q_2 q_3$	q_3
T	$1 - q_1 q_2 q_3$	$1 - q_1 q_3$	$1 - q_2 q_3$	$1 - q_3$

Algorithm 6.1 can be readily adapted to learning noisy-or parameters: first, learn the probability of the effect given that all parent variables (other than *Background*) are absent (q_3 in our example); then learn the probability of each parent in the absence of all others, dividing out the *Background* parameter (q_3). Since all causal factors operate independently, we are then done.

6.4.3 Classification trees and graphs

A more general technique for learning local structures simpler than a full CPT is to apply the mature technology of **classification tree** and graph learning to learning the local structure.[3] Classification trees are made up of nodes representing the relevant attributes. Branches coming off a node represent the different values that attribute may take. By following a branch out to its end, the leaf node, we will have selected values for all of the attributes represented in the branch; the leaf node will then make some prediction about the target class. Any instance which matches all of the selected values along a branch will be predicted by the corresponding leaf node. As an example, consider the classification tree for the acid-alkali ingestion problem in Figure 6.5. This tree shows that the ingestion of *Alkali* without the ingestion of *Acid* leads to a 0.95 probability of *Death*; that is, $P(Death|Alkali, \neg Acid) = 0.95$. Every other cell in the CPT for the *Death* variable can be similarly computed. Indeed, any classification tree which has a node for each parent variable along every branch, and which splits those nodes according to all possible values for the variable, will (at the leaf nodes) provide every probability required for filling in the CPT.

In this case (and many others) the equivalent **classification graph**, which allows branches to *join* as well as split at attribute nodes, is simpler, as in Figure 6.6. Since

---

[3]In much of the AI community these are called decision trees and graphs. However, in the statistics community they are called classification trees and graphs. We prefer the latter, since decision trees have a prior important use in referring to representations used for decision making under uncertainty, as we have done ourselves in Chapter 4.

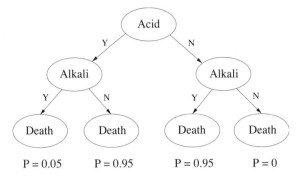

FIGURE 6.5: A classification tree for acid and alkali.

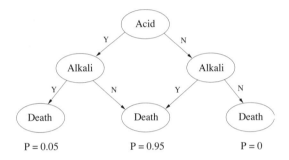

FIGURE 6.6: A classification graph for acid and alkali.

the classification graph combines two branches, it also has the advantage of combining all the sample observations matching the attributes along the two branches, providing a larger sample size when estimating the corresponding parameter.

One of the major aims of classification tree and graph learning is to minimize the complexity of the representation — of the tree (graph) which is learned. More exactly, beginning already with the Quinlan's ID3 (1986), the goal has been to find the right trade-off between model complexity and fit to the data. This **complexity trade-off** is a recurring theme in machine learning, and we shall continue to encounter it through the remainder of this text. Quinlan's approach to it was to use a simple information-theoretic measure to choose at each step in building the tree the optimal node for splitting, and he stopped growing the tree when further splits failed to improve anticipated predictive accuracy. The result was that the least predictively useful attributes were unused. From the point of view of filling in a CPT, this is equivalent to merging CPT cells together when the probabilities are sufficiently close (or, to put it another way, when the probabilities in the cells are not sufficiently distinct for the available evidence to distinguish between them). More recent classification tree and graph learners play the same trade-off in more sophisticated ways (e.g., Oliver, 1993, Quinlan, 1996).

Should the parents-to-child relation have a local dependency structure, so that some CPT entries are strict functions of others, the classification tree can take advantage of that. Current methods take advantage of the local structure to produce smaller trees just in case the probabilities in distinct cells are approximately identical, rather than standing in any other functional relation (Boutilier et al., 1996).

The technology of classification tree learning is, or ought to be, the technology of learning probabilities, and so its fruits can be directly applied to parameterizing Bayesian networks.[4]

6.4.4   Logit models

Log-linear approaches to representing local structure apply logarithmic transformations to variables or their probability distributions before looking for linear relations to explain them. One of this kind is the logit model.

Suppose we have the binomial variables X, Y, Z in the v-structure $X \to Z \leftarrow Y$. A **logit model** of this relationship is:

$$\log \frac{P(Z = 1|x, y)}{P(Z = 0|x, y)} = a + bx + cy + dxy \qquad (6.10)$$

This models any causal interaction between X and Y explicitly. The causal effect upon Z is decomposed into terms of three different orders:[5]

- Order 0 term (a): Identifies the propensity for Z to be true independent of the parents' state.

- Order 1 terms (b, c): Identify the propensity for Z dependent upon each parent independent of the other.

- Order 2 term (d): Identifies the propensity for Z dependent upon the interaction of both parents.

If we have a **saturated** logit model, one where all parameters are non-zero, then clearly we have a term (d) describing a causal interaction between the parents. In that case the complexity of the logit model is equivalent to that of a full CPT; and, clearly, the learning problem is equally complex.

On the other hand, if we have an unsaturated logit model, and in particular when the second-order term is zero, we have a simpler logit model than we do a CPT, resulting in a simpler learning problem. Again, as with classification trees and noisy-or models, learning fewer parameters allows us to learn the parameters with greater precision — or equal precision using smaller samples — than with full CPT learning.

If we are learning a first-order model (i.e., when $d = 0$), some simple algebra reveals the relation between the CPT and the logit parameters, given in Table 6.3.

---

[4]The reason for the reservation here is just that many researchers have thought that the classifications in the leaf nodes ought to be *categorical* rather than probabilistic. We have elsewhere argued that this is a mistake (Korb et al., 2001).

[5]The order refers to the number of variables in a product. For example, in the product dxy d is an order 2 term.

TABLE 6.3
CPT for the first-order logit model.

X	Y	Z
0	0	$\dfrac{\exp(a)}{1+\exp(a)}$
1	0	$\dfrac{\exp(a+b)}{1+\exp(a+b)}$
0	1	$\dfrac{\exp(a+c)}{1+\exp(a+c)}$
1	1	$\dfrac{\exp(a+b+c)}{1+\exp(a+b+c)}$

6.5 Summary

Parameterizing discrete Bayesian networks when the data are not problematic (no data values are missing, the parameter independence assumptions hold, etc.) is straightforward, following the work of Spiegelhalter and Lauritzen leading to Algorithm 6.1. That algorithm has been incorporated into many Bayesian network tools. The Gibbs sampling and EM algorithms for estimating parameters in the face of missing data are also fairly straightforward to apply, so long as their assumptions are satisfied — especially, that missing values are independent of observed values. What to do when the simplifying assumptions behind these methods fail is not clear and remains an area of active research. Another applied research area is the learning of local structure in conditional probability distributions, where classification trees and graphs, noisy-or models and logit models can all be readily employed.

6.6 Bibliographic notes

An excellent concise and accessible introduction to parameter learning is David Heckerman's tutorial (Heckerman, 1998), published in Michael Jordan's anthology *Learning in Graphical Models* (Jordan, 1999). That collection includes numerous articles elaborating ideas in this chapter, including an overview of **Markov Chain Monte Carlo** methods, such as Gibbs sampling, by Mackay (1998) and a discussion of learning local structure by Friedman and Goldszmidt (1999). For more thorough and technically advanced treatments of parameterizing discrete networks, see the texts by Cowell et al. (1999) and Neapolitan (2003).

6.7 Technical notes

We give the formal definitions of the beta and Dirichlet distributions, building upon some preliminary definitions.

Gamma function

The gamma function generalizes the factorial function. Limiting it to integers alone, it can be identified by the recurrence equation:

$$\Gamma(n+1) = \begin{cases} n\Gamma(n) & n > 1 \\ 1 & n = 0, 1 \end{cases} \tag{6.11}$$

If n is a positive integer, then $\Gamma(n+1) = n!$

Beta distribution

$\Theta = \theta$ has the beta distribution with hyperparameters $\alpha_1, \alpha_2 > 0$ if its density function is

$$f(\theta) = \begin{cases} \dfrac{\Gamma(\alpha_1+\alpha_2)}{\Gamma(\alpha_1)\Gamma(\alpha_2)} \theta^{\alpha_1-1}(1-\theta)^{\alpha_2-1} & 0 < \theta < 1 \\ 0 & \text{otherwise} \end{cases} \tag{6.12}$$

From this, we see that the normalization factor β in Equation (6.5) is $\frac{\Gamma(\alpha_1+\alpha_2)}{\Gamma(\alpha_1)\Gamma(\alpha_2)}$.

Dirichlet distribution

$\Theta = \vec{\theta}$ is Dirichlet distributed with hyperparameters $\alpha_1, \ldots, \alpha_\tau > 0$ if its density function is

$$f(\Theta) = \frac{\Gamma(\sum_{i=1}^{\tau} \alpha_i)}{\prod_{i=1}^{\tau} \Gamma(\alpha_i)} \prod_{i=1}^{\tau} \theta_i^{\alpha_i-1} \tag{6.13}$$

when $\theta_1, \ldots, \theta_\tau \geq 0$ and $\sum_{i=1}^{\tau} \theta_i = 1$.

6.8 Problems

Distribution Problems

Problem 1

Suppose your prior distribution for the probability of heads of a coin in your pocket is B(2,2). Toss the coin ten times. Assuming you update your distribution as in §6.2.1, what is your posterior distribution? What is the expected value of tossing the coin on your posterior distribution? Select a larger equivalent sample size for your starting point, such as B(10,10). What then is the expected value of a posterior toss?

Problem 2

Suppose your prior Dirichlet distribution for the roll of a die is $D[2,2,2,2,2,2]$ and that you update this distribution as in §6.2.2. Roll a die ten times and update this. Is the posterior distribution flat? How might you get a flatter posterior distribution?

Experimental Problem

Problem 3

In this problem you will analyze a very simple artificially created data set from the book Web site
    http://www.csse.monash.edu.au/bai.html
which was created by a v-structure process $X \rightarrow Y \leftarrow Z$ — i.e., one with three variables of which one is the child of the other two which are themselves not directly related. However, it will be instructive if you *also* locate a real data set generated by a similar v-structure process and answer the questions for both data sets.

Parameterize the Bayesian network $X \rightarrow Y \leftarrow Z$ from the data set in at least two of the following ways:

- using the algorithm for the full CPT, that is, Algorithm 6.1
- using a noisy-or parameterization
- using a classification tree algorithm, such as J48 (available from the WEKA Web site: http://www.cs.waikato.ac.nz/~ml/weka/)
- using an order 1 logit model

Compare the results. Which parameterization fits the data better? For example, which one gives better classification accuracy?

Since in answering this last question you have (presumably) used the very same data both to parameterize the network and to test it, a close fit to the data may be more an indication of overfitting than of predictive accuracy. In order to test a model's **generalization accuracy** you can divide the data set into a training set and a test set, using only the former to parameterize it and the latter to test predictive (classification) accuracy (see Part II introduction).

Programming Problems

Problem 4

Implement the multinomial parameterization algorithm (6.1). Test it on some of the data sets on the book Web site and report the results.

Problem 5

Implement the Gibbs Sampling algorithm (6.3) for estimating parameters with incomplete data. The missing Cooper & Herskovits computations can be filled in by copying the Lisp function for them on the book Web site, or by skipping forward and implementing Equation (9.3) from Chapter 8. Try your algorithm out using some of the data sets with missing values from the book Web site and report the results. How many sampling steps are needed to get good estimates of the parameters in the different cases? How do you know when you have a good estimate for a parameter?

Problem 6

1. Implement maximum likelihood expectation maximization (Algorithm 6.5).
2. Implement MAP expectation maximization (Algorithm 6.6).

Try your algorithm(s) out using some of the data sets with missing values from the book Web site and report the results. Note that there is Lisp code available on the book Web site for computing the expected sufficient statistics for these particular cases. Use a variety of different convergence tests — values for ε — and report the relation between the number of iterations and ε.

7

Bayesian Network Classifiers

7.1 Introduction

Most of this book is about the use of Bayesian networks for modeling physical or social systems in the real world; and most models of such systems are implicitly or explicitly causal models. That's for the fairly obvious reason that most of the worldly systems we are interested in understanding are in fact causal systems, so modeling them *as* causal systems is natural and intuitive, and it satisfies normal expectations about our models. As we've seen above there are other advantages to a causal approach, such as compactness (Chapter 2), and we'll see below some additional advantages connected with causal discovery. However, there is a substantial and growing body of work using Bayesian networks strictly as predictive models, that is, without any expectation or interest in them as causal, explanatory models. Learning predictive models addresses the problem of **classification**, that is, learning how to classify joint samples of variables (attributes). Here we shall consider Bayesian networks specifically for **supervised classification**, which is learning a classifier using sample data consisting of both attribute values and explicit target class values, where the classes have been identified by some expert or "supervisor".

The popularity of predictive Bayesian networks for supervised classification lies in the combination of generally good predictive accuracy with simplicity and computational efficiency, which is a combination that frequently compares favorably with what is offered by alternatives such as regression models and classification tree learners, such as C4.5 (Quinlan, 1993), at least for many simpler classification problems (for a very brief introduction to classification trees see §7.8).

The predictive use of Bayesian networks does not require that the arcs have any causal interpretation, but only that they indicate probabilistic dependencies, although it still also requires that the Markov property holds — i.e., that missing arcs imply independencies (see §2.2.4). In fact, most such predictive models are strictly incompatible with a causal interpretation of the networks, including the earliest, simplest and best known such model, called the **naive Bayes (NB)** model. NB models have been very widely applied and perhaps are especially known for spam filtering (e.g., Graham, 2003). In this chapter we shall first review these models and some of their close relatives, and then look at how predictive classifiers are evaluated and how they might be evaluated better. The issues involved with classifier evaluation have application beyond problems of classification, since the techniques employed are often used also for validating Bayesian networks developed in other contexts. For methods of parameterizing predictive models see Chapter 6.

7.2 Naive Bayes models

In a supervised classification problem, if the class happens to be the *result* of the factors being identified as attributes, then the naive Bayes model is the *opposite* of a causal model, for the class variable is always the parent of all its attributes. For example, in the mushroom problem from the UCI machine learning archive the target class variable is binary, indicating whether or not a mushroom is edible; there are 22 other variables describing attributes, such as color and shape, which may or may not be predictive of edibility (Asuncion and Newman, 2007). In a naive Bayes model the class variable is made the parent of all attribute variables, and no other arcs are allowed. For example, the naive Bayes model for mushrooms is shown in Figure 7.1

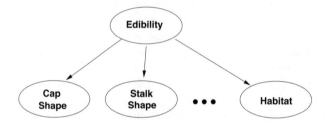

FIGURE 7.1: A naive Bayes mushroom model.

Naive Bayes models have the significant advantage over most alternatives that they are extremely simple. In particular, since each attribute node has exactly one parent, each attribute variable's CPT is minimal, containing the attribute's distribution for each possible parent value. In other words, instead of a joint encoding of all attributes given the class variable, they are encoded independently, as is required by the lack of any arcs *between* attributes. So, for naive Bayes models:

$$P(A_1,\ldots,A_n|C) = \prod_i P(A_i|C) \qquad (7.1)$$

where the A_i are n attributes and C the class variable. This implies that there are far fewer parameters than usual for Bayesian networks, meaning that far less data than usual are needed for such models to be estimated, so they can be learned much faster than other models.

The computation of NB predictive accuracy is likewise simplified, since to compute it for each sample we need to find

$$\text{argmax}_c P(c|a_1, \ldots a_n) \quad = \quad \text{argmax}_c \frac{P(a_i, \ldots, a_n|c)P(c)}{P(a_1, \ldots a_n)} \tag{7.2}$$

$$= \quad \text{argmax}_c \prod_i \frac{P(a_i|c)P(c)}{P(a_1, \ldots a_n)} \tag{7.3}$$

$$= \quad \text{argmax}_c \prod_i P(a_i|c)P(c) \tag{7.4}$$

where argmax_c is finding that value of $C = c$ that maximizes the subsequent expression.

Another significant advantage for NB is the fact that, in contrast with alternative methods, there is a very simple and natural way of dealing with one of the most common kinds of missing data, namely missing attribute values. Alternative classification learners typically have to resort to fairly crude techniques such as modal imputation or deleting samples with missing values or else to computationally expensive techniques such as expectation maximization (see Chapter 6). With naive Bayes models, or any Bayesian network, missing attributes can be handled simply by not setting those variables, but instead setting all and only attribute variables whose values have been observed, then updating the network and reading off the posterior distribution over the target variable. This is as natural as it gets, since attributes are missing precisely because they were unobserved when the data were collected.[1]

The major caveat for NB is that the independence assumptions are very strong, and where they are not justified, at least as some crude approximation, then the predictive accuracy of the naive Bayes model may be quite poor. That, in fact, is why they are called "naive" models (or, even, "idiot Bayes" models). In case of independence violations NB classifiers will typically not be **statistically consistent**, that is, they will fail to converge on the true probability of the class even given unlimited volumes of data. Nevertheless, these models have been effective in prediction across a wide range of problems, and they often perform significantly better given realistic volumes of data than competitive methods, such as C4.5 (Domingos and Pazzani, 1997). So, naive Bayes is a natural first choice for a predictive model for a new problem, at least until evidence arises that it's not working for that problem. When such evidence does arise, it's a further natural choice to relax the independence assumptions one by one — i.e., introduce new arcs to the model in a minimal way — leading in recent years to a slow progression in the complexity of these kinds of models under investigation in the machine learning community.

---

[1]This technique works well for values that are "missing completely at random", that is, when the fact that they have gone missing is not dependent upon either its actual value or upon the values of other attributes or the target variable. In the latter cases, a better job of dealing with the absent data can in principle be done by modeling and estimating the relevant dependencies. On the other hand, given that NB models are ignoring all but first-order dependencies to begin with, the reward may well not be worth the trouble. The types of absent data (together with a proposal to substitute "absent data" for "missing data" and "absence" for the absurd neologism "missingness") are treated in Wen et al. (2009).

7.3   Semi-naive Bayes models

Tree Augmented Naive Bayes (TAN) models are perhaps the first natural step in desimplifying naive Bayes models (Friedman et al., 1997). They relax the independence assumptions of NB by allowing some arcs directly between attributes, in particular allowing a tree amongst the attributes to be constructed, separately from their direct relations with the class variable. For example, Figure 7.2 is a (very simple) TAN for the mushroom case, restricted for simplicity to two attributes.

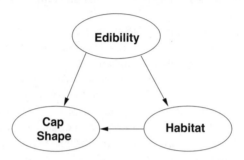

FIGURE 7.2: A TAN mushroom model.

In Friedman et al. (1997) the tree is required to include all attribute nodes (i.e., it is a spanning tree), with the result that all attributes but one (the tree's root) will have two parents, the class variable and one other attribute. As a result, many circumstances where attributes are inter-related even given the class value can be accommodated, so we can expect in such circumstances TAN models will outperform NB models. For (an invented) example, in Figure 7.2, if both *Habitat* and *Cap Shape* are positively related to *Edibility*, but are themselves negatively related,[2] it can easily happen that NB will seriously underestimate the contribution of a positive *Cap Shape* for *Edibility*. The TAN model will not suffer from this kind of error. Of course, it will suffer from analogous interactions of any higher order, and it is possible to go beyond TAN models by further relaxing independence assumptions and allowing, for example, the number of attribute parents α to increase from one to two, or three or any number up to $n-1$; these have been called α-dependence Bayesian classifiers. Unfortunately, the space complexity of the model (the number of its parameters) grows exponentially in α, while the search complexity (the number of potential TAN models that need to be examined) grows rapidly, if polynomially. However, for simple one-dependence TAN models there is a polynomial time algorithm for finding the maximum likelihood model, namely a maximal weighted spanning tree algorithm that runs in time $O(n^2K)$, where K is the sample size (Friedman et al., 1997).

Friedman et al. (1997) performed empirical studies comparing NB, TAN and

---

[2]Where we imagine some ordering of the values of these variables, so we may meaningfully talk of positive and negative relations (correlations).

C4.5, along with other methods, using many of the UC Irvine machine learning data sets (Murphy and Aha, 1995). For those problems, NB and C4.5 performed about the same, while TAN almost dominated both, meaning TAN performed either the about same or, in some cases, better, as measured by predictive accuracy.

7.4 Ensemble Bayes prediction

Friedman et al. (1997) also compared the predictive accuracy of NB and TAN with full causal discovery algorithms (in particular, an MDL discovery algorithm; we address discovery algorithms in Chapter 9). They found NB and TAN outperformed causal discovery when there were a large number of attributes in the data. Causal discovery can potentially return *any* Bayesian network containing the attribute and class variables. If it happens to find the true model, the model which actually generated the data available, then that model (disregarding any noise in the data) will necessarily be the best predictor for the target variable. The prediction in question will be that made by the target variable's Markov blanket, since by definition all other variables are conditionally independent of the target given its Markov blanket. The problem with full causal discovery for prediction, however, is the same as the problem with any kind of feature selection or model selection: frequently the true model is *not* what is learned, but some similar, yet different, model is learned instead. The result may well be that variables which ought to be in the target variable's Markov blanket are not, and so are ignored during prediction, with potentially disastrous consequences for predictive accuracy.

NB and TAN, on the contrary, include *all* the attributes in their predictions, so this source of error is not even possible. To be sure, by being all-inclusive NB and TAN introduce a different potential source of error, namely overfitting. It may be that some variables are directly associated with the target variable only accidentally, due to noise in the available data. With a maximum likelihood parameterization NB and TAN will nevertheless use those accidents to fit the data, to the detriment of generalization error on new test data. To compensate for this one may either relax the parameterization process, so that it is no longer strictly maximizing likelihood (see Chapter 6) or else introduce variable selection, eliminating those attributes from the model which are contributing little to the prediction. In the latter case, again, incorrect variable selection returns us to the problem faced by causal discovery for prediction: variables missing from the target's Markov blanket.

Another response to the problem of incorrectly identifying the Markov blanket, aside from utilizing *all* attributes as predictors, is to move to **ensembles** of predictive models. This means mixing the predictions of some number of distinct models together, using some weighting over the models. For example, in Figure 7.3 there are two alternative (partial) Bayesian networks for the mushroom problem (note that they are not NB models). Figure 7.3(a) shows a Markov Blanket for the target *Edibility* of *Stalk Shape* and *Cap Shape*, while Figure 7.3(b) shows *Cap Shape* and *Habitat*.

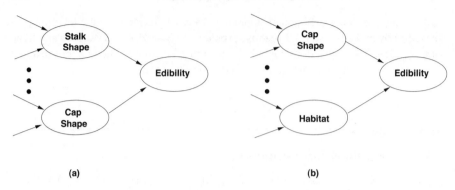

(a)   (b)

FIGURE 7.3: Two BN models explaining (and predicting) edibility.

If the true model or system (i.e., the one generating the data set) has a Markov Blanket of exactly these three attributes, then either model alone will be suboptimal for prediction, whereas a mixture of the two may well do better. Ensembling methods have been widely applied in classification learning and are among the most popular techniques in use. (For some references see §7.7.)

The same opportunity exists for causal discovery. If instead of using causal discovery to find a single "best" model given the data, which, best or not, will very often not have the correct Markov blanket, we can use causal discovery to estimate a posterior distribution over the space of causal models and use that to produce a posterior distribution over the target variable, averaged over *all* the models sampled during the discovery process. At least some of those models are likely to incorporate the correct Markov blanket, and if the posterior distribution over them is reasonable and guided by good data, then the correct blanket should be contributing a large weight to the prediction. Such weighted predictions will very often substantially improve predictive accuracy.

In short, it is to be expected that single model predictors will generally be inferior to ensemble predictors: the latter will (at least often) include the true model, and so the true Markov blanket, amongst its predictor variables, whereas the former will not. Naturally, there is an opportunity to apply ensembling to Bayesian predictive models and not just to full causal discovery. There is, of course, no opportunity for ensemble prediction with NB models themselves, since there is always only one NB model per problem. But NB has many relatives.

Averaged One-Dependence Estimators (AODE) provide just such an ensembling approach to generalizing TAN predictive models (Webb et al., 2005). These are based upon a particular kind of TAN model, namely Super-Parent One-Dependence Estimators, in which the augmenting tree has a single attribute being a "super-parent", i.e., a parent to all other attributes. For TAN models, a restriction to a single common attribute parent is fairly severe, but AODE takes *all* such models and uses them to compute an equal-weighted average prediction over the target class. Combining the simple structure, and the avoidance of search or variable selection, on the one hand, with the averaging of a comprehensive set of predictive models of this kind, on the other, yields a computationally efficient predictor with accuracy frequently superior to both NB and TAN (Webb et al., 2005). In short, AODE selects the class c according to:

$$\text{argmax}_c \sum_{s=1}^{n} P(\mathbf{a}|c, a_s) P(c, a_s)$$

where $\mathbf{a}$ specifies all of the attribute values (i.e., it's a vector identifying values for all attribute variables). In cases where $\mathbf{a}$ happens not to occur in the data, AODE reverts to base rate of the target class in the training set.

Averaged n-Dependence Estimators (AnDE) generalize AODE by expanding the number of super-parents from 1 to n (Webb et al., 2010). Thus, AnDE selects the class c according to:

$$\text{argmax}_c \sum_{s \in S^n} P(\mathbf{a}|c, \mathbf{a_s}) P(c, \mathbf{a_s})$$

where S^n is the set of all subsets of $\{a_i\}$ of size n and $\mathbf{a}$ specifies all of the attribute values. Where $n = 0$ we have NB; where $n = 1$ we have AODE. And, of course, $n = 2$ gives us A2DE, which Webb et al. (2010) report as a reasonable compromise between computational efficiency and predictive accuracy for a variety of problems. They report prohibitive computational costs for $n > 2$ on high-dimensional data sets.

In summary, ensemble prediction simply has to be superior in terms of predictive accuracy across a wide range of problems and a wide range of techniques. AODE is, of course, very often predictively superior to NB and super-parent TAN. Similarly, other ensembling techniques such as boosted trees and random forests have been shown to be superior to NB (Caruana and Niculescu-Mizil, 2006). Where the primary problem is a predictive one, rather than one of modeling and understanding a process, ensemble methods are always worth considering, so long as the computational costs are manageable.

7.5 The evaluation of classifiers

There is no agreed standard for how to assess the performance of classifiers. There are some common practices, but many of them are unjustified, or even unjustifiable. Here we introduce the most commonly used approaches, together with their strengths and weaknesses, and some possible improvements upon them.

7.5.1 Predictive accuracy

Predictive accuracy is far and away the most popular technique for evaluating predictive models, whether they be Bayesian networks, classification trees, or regression models. The predictive accuracy of a BN is assessed relative to a target variable via the algorithm:

Algorithm 7.1 *Predictive Accuracy*

1. *Set* count *to 0.*
2. *For each case in the test data set:*
   (a) *Add any available evidence for the attribute nodes.*
   (b) *Update the network.*
   (c) *Take the predicted value for the target variable to be that which has the highest posterior probability (the modal probability).*
   (d) *If this value is equal to the actual value for that variable in the sample, increment* count.

3. *Return* count *divided by the sample size.*

For example, if a model attempting to relate measured attributes of mushrooms to their edibility reported for a particular mushroom that the probability of it being edible is 0.99, then it would be taken as predicting edibility. Running through the entire sample, we can record the frequency with which the model gets its predictions right. That is its **predictive accuracy**. We could equivalently report its error rate as 100% minus its predictive accuracy.

One difficulty with using accuracy to compare learning methods is that if the sample data have been used to learn the model in the first place (i.e., they *are* the training data) — whether learning a representational structure, or parameters or both — then they cannot safely be reused to gauge the model's predictive accuracy. Otherwise the model is not being tested on its predictions, but just on its memory — how well it has been tuned to the training data. Fit to the training data can be very far from the ability of the model to *predict* new values. When a fixed sample must be used for both training and testing a model, a plausible approach is divide the data into separate training and test sets, for example, in a 90% and 10% split. The model's predictive accuracy on the test set will now provide some estimate of its **generalization accuracy**, how well it can predict genuinely new cases, rather than just rewarding the overfitting of the training data.

No matter how large the training and test sets are, however, a single such test will only provide a *single* estimate of the classifier's predictive accuracy — the relevant sample size is *one!* A sample of size one provides no opportunity to judge the variation in the model's predictive accuracy over similar test sets in the future, and so provides no serious basis for assessing one learner as superior to another. The well-known early study of classification learners, Thrun et al. (1991), for example, suffered from this mistake.

If the training-test procedure can be repeated enough times, then each predictive accuracy measure will provide a separate estimate, and the collection of estimates can be used to assess the variation in the model's performance. For example, by estimating the accuracy of two models on the same training and test sets, it then becomes possible to perform an orthodox difference-of-means statistical significance test to assess one model's performance against the other's (as any statistics reference book will explain; e.g., Spiegel et al., 2008). This classical approach works well enough when new data can be easily collected, or when a given data set is sufficiently large, so that no training or tests sets overlap across the samples required. However, there may be no opportunity to collect enough data for such repeated testing.

There may nevertheless be enough data for **k-fold cross validation**, which reuses the same data set, generating k splits of the data set into non-overlapping training and test sets with proportions (k-1)/k and 1/k respectively. The difference-of-means t test might still be applied. Dietterich (1998) found that an alternative resampling method, two-fold cross validation repeated five times (or, 5x2cv), had more power to discriminate distinct predictive accuracies than the above method. This result was substantially improved upon by Nadeau and Bengio (2003), who noted that Dietterich's technique failed to account for overlapping test sets across its iterations and so underestimated the variance of the classifiers. They provide a corrected statistical test. Improving these kinds of resampling statistics remains an active area of research in machine learning.

Despite these difficulties, predictive accuracy is relatively simple to measure, and for that reason may provide your best initial guide to the merit of a model. We shall consider some alternatives below that in many circumstances are preferable to predictive accuracy. When using accuracy to compare classifiers, at least bear in mind the caveats:

- Absolute accuracy can be very misleading. If the base rate for edible mushrooms is 90% (i.e., 90% of cases in a sample are edible), then accuracy of 91% may sound impressive without *being* impressive. Accuracy should always be relativized to the base rate.

- Accuracy will always vary, not just between classifiers, but between samples by the same classifier. The variance in performance needs to be estimated. An absolute minimum is repeated sampling leading to a test of statistical significance.

These caveats apply just as well to the alternative measures we look at below, after a brief interlude on bias and variance.

7.5.2 Bias and variance

When dealing with prediction, and more particularly with predictive error (the flip side of predictive accuracy), the bias and variance of the predictor are useful analytical concepts. The **mean squared error (MSE)** of a predictor is its expected squared deviation from the true value:

$$MSE(\hat{p}, p) = E[(\hat{p} - p)^2]$$

where $\hat{p}$ is the predicted and p the actual values. This value decomposes into the sum of squared bias and variance:

$$
\begin{aligned}
E[(\hat{p} - p)^2] &= (\text{Bias}(\hat{p}, p))^2 + \text{Variance}(\hat{p}) \\
&= (E[\hat{p}] - E[p])^2 + E[(\hat{p} - E[\hat{p}])^2]
\end{aligned}
$$

The **bias** measures the degree and direction of systematic prediction errors, whereas the **variance** reports the asystematic deviation of predictions from the mean prediction. In the language of target shooting, the bias reflects a continuing tendency of a gun to deviate from the center, which might be corrected by adjusting its sight, while the variance reflects random jiggles that simply require a steadier hand. Both effects contribute to error, and it is possible to reduce error by reducing either bias or variance or both.

On the other hand, very often in machine learning there is a kind of trade-off, where improving predictive error in bias comes at a cost in terms of variance, and vice versa. When learned representations can range from very simple to very complex, then the simpler representations will tend to show little variance when learned from different data sets taken from the same process: since they have lesser representational power than their more complex relatives, they *cannot* vary so greatly. However, they are fairly likely to be biased systematically; in particular, they are likely biased towards representing simple processes. More complex representations do not have that bias. However, since they have high representational capacity they will differ more strongly in the representations they learn over varying data sets. For one thing, they will better represent noise in the data — i.e., they are more likely to overfit their training data. A prominent example of this kind of trade-off is that of artificial neural networks, with single-layer networks being so simple they cannot even represent XOR and other non-linear interactions, whereas networks with open-ended hidden layers potentially can represent every data point exactly, by adding new hidden nodes between their input and output layer, producing maximal variance.

Another good illustration of this phenomenon is semi-naive Bayesian networks and, in particular, AnDEs. A0DEs (Averaged Zero-Dependence Estimators, i.e., naive Bayes models) have notoriously high bias and low variance. Any variance they show is entirely due to fluctuations in their parameter estimates across data sets, since NB models don't involve any variable selection: the full set of attributes are used to predict every time. The bias enters in through ignoring all interdependencies between attributes. AODEs reduce that bias, without greatly increasing variance, since they too involve no variable or model selection, averaging over *all* super-parent one-dependence models. There is nevertheless some increase in variance with AODEs, because they have more parameters needing estimation from the data, with the result that smaller subsets of data are available to estimate individual parameters, thereby increasing the variation in their estimation. A2DE is the next step in this direction: its further relaxed independence assumptions allow for a reduction in bias at the cost of a further increase in need for data for its parameterization. Every increase in the n of AnDE will continue the pushing this tradeoff, up to the point where n is equal to one less than the number of attributes. For a discussion of this particular tradeoff with AnDE see Webb et al. (2010).

When available data are limited, the variation from sample to sample will be very high (in consequence of the Law of Large Numbers). If we employ high-variance classifiers, we will get wildly fluctuating results, and most likely find them useless. When large data sets are available, this problem is greatly diminished, and, contrariwise, we would be well advised to concentrate on avoiding high-bias learners, especially if their biases cannot be overcome even by large data sets, as is the case with naive Bayes. In short, as a rule of thumb we should prefer high-bias, low-variance classifiers when the data are sparse and low-bias, high-variance classifiers when data are readily available.

7.5.3 ROC curves and AUC

Use of the **Receiver Operating Characteristic (ROC)** curve is a popular alternative to predictive accuracy for examining and comparing the predictive performance of classifiers. It lays out such performance along the two dimensions of **false positive** rate (the rate with which negative instances are wrongly classified into the target class) versus the **true positive** rate (the rate with which positive instances are correctly classified into the target class), as in Figure 7.4. A perfect classifier would have

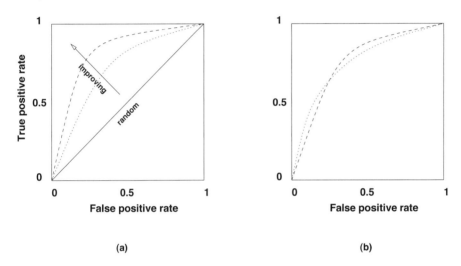

(a)   (b)

FIGURE 7.4: (a) ROC curves for three idealized classifiers; (b) Two ROC curves intersecting.

a false positive rate of 0% and a true positive rate of 100%, putting its performance at the upper left of the square. If a classification learner has performance that is tunable in terms of its false positive rate, then a useful ROC curve can be estimated for it. The solid diagonal in Figure 7.4(a), for example, represents the worst possible ROC curve one might be willing to contemplate, namely random guessing, produced by varying the probability threshold for positively classifying any instance at random from 0 to 1. The dotted curve shows a better classifier, which has a greater true than false positive rate at every point other than the extremes, while the dashed line is better still at every such point. In this case, the dashed line represents a classifier whose predictive performance is strictly dominating its alternatives, and it would be preferred to them.

When comparing classifiers it is possible, of course, that they alternate regions in which they perform best — for example, one might have a higher true positive rate when the false positive rate is constrained below 25% and the other classifier outperforms the first when the false positive rate is greater than 25%, as in Figure 7.4(b).

Area Under the Curve (AUC) is a popular way of summarizing the classification performance displayed in an ROC curve, by integrating the area under the ROC curve. While AUC ignores particular regions along the x axis where one classifier is doing better than another, it averages those performances across the whole range of false positive rates. If there is no specific way of controlling the false positive rate or no preferred range within which the false positive rate should be constrained, then this may be a good way of characterizing overall predictive performance.

If we have precise control over the false positive rates for the different available classifiers, we can adopt a "meta-learning" approach; that is, we can mix and match our classifiers, selecting always that classifier for application to a problem when its performance is best for that problem's particular false positive rate. The appropriate ROC curve in that case is the **convex hull** formed from the constituent classifiers' ROC curves (Provost and Fawcett, 2001).

ROC and AUC provide different views of predictive performance from that of predictive accuracy. AUC, like accuracy, is an average measure of that performance. ROC, on the other hand, gives a finer-grained view of what is happening with the positive classifications across the range of false positive rates. ROC is naturally suited to binomial classification problems, but it can be extended to multinomial classification by iterating through the classes, taking each one in turn as the target. On the other hand, if the classifiers vary in which performs best for the different target classes, it is not clear how such disparate results may be meaningfully combined without reference to something like expected value, which we take up in §7.5.5.

In summary, there are a number of notable problems with the use of ROC curves:

- They require that the classifier be tunable in terms of the false positive rate. This is natural enough for probabilistic classifiers, since the threshold for classifying an instance positively can be varied continuously, yielding (most likely) continuously varying performance in terms of false positives. For other classifiers, however, this may be a problem.

- In typical practice, a single point estimate of the true positive rate is made for each false positive rate. This entirely ignores the variance in the performance of classifiers and so is a poor practice.

- Although multinomial classes can be treated with ROC curves, the results may be unclear. In actual practice, multinomial classes are usually treated as if binomial, with one value alone taken as the value of interest and all other values lumped together. This loses information.

- ROC curves share the weakness of predictive accuracy in focusing on categorical classification, ignoring the confidence (probability) with which instances are claimed to lie within the target class, so long as some threshold is reached. As a result, classifier performance is assessed without regard to either calibration or the expected value of its performance, issues which we address next.

7.5.4   Calibration

A fundamental problem with the evaluative approaches focused upon predictive accuracy is that they entirely disregard the confidence of the prediction. In the mushroom classification problem, for example, a prediction of edibility with a probability of 0.51 may count exactly the same as a prediction of edibility with a probability of 0.99. Now, if *we* were confronted with the first prediction, we might rationally hesitate to consume such a mushroom. The predictive accuracy measure does not hesitate. According to standard practice, any degree of confidence is as good as any other if it leads to the same prediction: that is, all predictions in the end are categorical, rather than probabilistic. Any business, or animal, which behaved this way would soon cease doing so! Predictive accuracy pays no attention to the calibration of the model's probabilities. Bayesian theory suggests a very different orientation to the problem of evaluation.

Bayesian decision making under uncertainty is in large measure founded upon gambling as a metaphor for all decision making, which goes back at least to the work of Frank Ramsey (1931), if not to the origins of formal probability theory given that it arose from efforts to analyse gambling devices. We can view classification as gambling on which class an instance belongs to. There are two fundamental ingredients to gambling success, and we would like any evaluative measure to be maximized when they are maximized:

Property 1: Domain knowledge (first-order knowledge), which can be measured by the frequency with which one is inclined to correctly assert $x_i = T$ or $x_i = F$ — i.e., by predictive accuracy. For example, in sports betting the more often you can identify the winning team, the better off you are.

Property 2: Calibration (meta-knowledge), the tendency of the bettor to put $P(x_i = T) = p$ close to the objective probability (or, actual frequency). That betting reward is maximized by perfect calibration is proven as Theorem 6.1.2 in Cover and Thomas's *Elements of Information Theory* (Cover and Thomas, 2006).[3]

With Property 1 comes a greater ability to predict target states; with Property 2 comes an improved ability to assess the probability that those predictions are in error. These two are not in a trade-off relationship: they can be jointly maximized.

More typically, however, calibration is neglected, both during prediction, by classifiers, and afterwards, in the evaluation of classifiers. In fact, there is a considerable empirical literature showing that people are strongly inclined to be overconfident in their predictions, as illustrated in Figure 7.5 (e.g., Von Winterfeldt and Edwards, 1986). In that figure, the x axis is the objective probability of some event (or, an objective stand-in or estimator of such a probability, such as a long-run frequency), while the y axis is some person's predictive probability for the same event. Perfect

---

[3]This is recognized in David Lewis's "Principal Principle," which asserts that one's subjective probability for an event, conditioned upon the knowledge of a physical probability of that event, should equal the latter (Lewis, 1980).

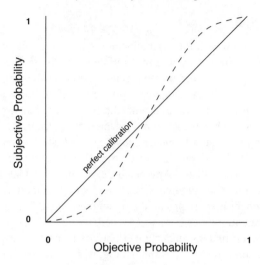

FIGURE 7.5: Overconfidence compared to perfect calibration.

calibration is reflected by the solid diagonal line. Lying on that line means that predictions that are put at 50-50 turn out, on average, to be true 50% of the time, etc. Most people, guessing about most things, tend to show overconfidence, meaning that events they are indifferent about indeed turn out to be true about half the time, but as the underlying event becomes more (or less) probable they tend to get over (or under) enthusiastic about their chances. The literature suggests that experience and expertise diminish, but do not eliminate, this tendency. Our point here, however, is just that good calibration means better predictive performance and ought to result in a higher evaluation. The most common evaluative techniques, employing predictive accuracy or ROC curves, are, however, utterly indifferent to calibration.

When we are using real data to evaluate the performance of classifiers, we often lack a known, or even good, model for the system under study. After all, we presumably are applying machine learning to the problem because we do *not* already know the answer! But this makes real data a difficult tool to use for assessing which machine learning method is best for a domain. Frequently, therefore, when the task at hand is assessing classifiers (or other machine learning techniques) artificial data are used, generated by a known model. In such cases, and certainly for probabilistic prediction problems, Kullback-Leibler divergence (also known as relative entropy; see also §3.6.5) is an ideal measure of how close a learner has come to discovering the true probability distribution underlying the available data. That is, we can use

Definition 7.1 (Kullback-Leibler divergence[4])

$$KL(p,q) = \sum_{x \in X} p(x) \log \frac{p(x)}{q(x)}$$

where p is the true distribution and q the learned distribution. This is just the expected log likelihood ratio:

$$KL(p,q) = E_p \log \frac{p(x)}{q(x)} \qquad (7.5)$$

Kullback-Leibler divergence is a measure which has the two properties of betting reward in reverse: that is, by minimizing KL divergence you maximize betting reward, and vice versa. Consider, the KL divergence from the binomial distribution $B(p = 0.5)$ for various $B(q)$, as in Figure 7.6. Clearly, this KL divergence is minimized when $q = p$. This is proved for all positive distributions in the Technical Notes §7.8. Of course, $q = p$ implies that the model is perfectly calibrated. But it also implies Property 1 above: clearly there is no more domain knowledge to be had (disregarding causal structure, at any rate) once you have exactly the true probability distribution in the model.

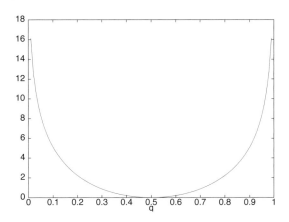

FIGURE 7.6: KL divergence.

KL divergence is the right measure when there is a preferred "point of view" — when the true probability distribution is known, p in Definition 7.1, then KL divergence reports how close a model's distribution q is to the generating distribution. This provides a natural evaluation measure for a model that increases monotonically as the model's distribution diverges from the truth. So, applied to multiple probabilistic prediction problems using multiple classifiers, it provides a natural method for distinguishing good and bad classifier performance, and one that is vastly superior to ROC curves and predictive accuracy. There has been a growing uptake of KL divergence in the empirical literature on classification.[5]

The drawback to KL divergence, as we noted, is just that when assessing models

---

[4]Where by convention $0 \log \frac{0}{0} = 0$, $0 \log \frac{0}{q} = 0$, and $p \log \frac{p}{0} = \infty$.

[5]KL divergence is often referred to as KL "distance", by the way, but since it is asymmetric (the expectation is taken relative to the true distribution) and since it violates the triangle inequality, it is not a true distance measure.

for real-world processes, the true model is necessarily unknown.[6] In consequence, we want some *other* performance measure, sharing with KL the same two properties of being optimized with maximal domain knowledge and calibration, while not depending on some privileged access to the truth.

7.5.5   Expected value

An answer to this need is readily available when data come not just with attribute and class values, but also with the value of making correct classifications and the disvalue of making mistakes. In that case we can employ **cost-sensitive classification** methods. Instead of preferring an algorithm or model which simply has the highest predictive accuracy, the idea is to prefer an algorithm or model with the best weighted average cost or benefit computed from its probabilistic predictions. In other words, the best model is that which produces the highest expected utility for the classification task at hand.

Since classifications are often done with some specific purpose in mind, such as selecting a treatment for a disease, it should come as no surprise that known utilities should be relevant to judging the predictive model. Indeed, it is clear that an evaluation which ignores available utilities, such as predictive accuracy, cannot be optimal in the first place, since it will, for example, penalize a false negative of some nasty cancer no more and no less than a false positive, even if the consequences of the former swamp the latter.

TABLE 7.1
Utilities for binary classification and misclassification.

	$x_i = T$	$x_i = F$
$x_i = T$ predicted	$u_1 > 0$	$u_2 < 0$
$x_i = F$ predicted	$u_3 < 0$	$u_4 > 0$

Here is a simple binomial example of how we can use expected value to evaluate a model. Suppose that the utilities associated with categorically predicting $X = T$ or $X = F$ are as shown in Table 7.1. Then when P is the model's probability function, the model's score would be

- $P(x_i = T)u_1 + P(x_i = F)u_3$ across those cases where in fact $x_i = T$, and
- $P(x_i = T)u_2 + P(x_i = F)u_4$ otherwise.

Its average score would be the sum of these weighted by the true probability of T and F respectively, which is the best estimate of the expected utility of its predictions, if the model is forced to make a choice about the target class.

This expected value measurement explicitly takes into account the full probability distribution over the target class. If a model is overconfident, then for many

---

[6]Another limitation is that KL doesn't tell the full story when applied to the learning of causal models. We will take up this issue in Chapter 9.

cases it will, for example, attach a high probability to $x_i = T$ when the facts are otherwise, and it will be penalized with the negative utility u_2 multiplied with that high probability.

Naturally, the substitution of the expected utility of a classifier for predictive accuracy can completely rewrite the story of which classifier is the best for any given problem. For an example, a very large number of studies of classifiers report predictive accuracy results for one or another method relative to the "German credit scoring" data set from the UCI archives (Asuncion and Newman, 2007; originally from Michie et al., 1994), usually finding that their preferred method scores somewhat better than the default prediction, that the credit applicant will not default on a loan (which has a 70% accuracy). Distributed with the data set, but almost universally ignored, is Table 7.2(a). This table looks like a very simple-minded representation of the costs of misclassification. The cost of 0 for refusing a loan to one who would default makes sense as a baseline choice for the utility scale. Similarly, the cost of 1 for refusing what would have been a good loan establishes the unit for the utility scale, representing the opportunity loss of not making a good loan. The 5-to-1 ratio of bad loans to missed good loans is intuitively an underestimate of the cost of bad business; 10-to-1 seems more plausible (although a real study of this issue need not rely on intuition!). Altogether wrong is setting good loans, which return profit, to be equal to refusing bad loans; indeed, making good loans should be -1 times the opportunity cost of refusing good loans, as in Table 7.2(b).

The cost-sensitive classification story, as it turns out, is impervious to these possible refinements of the utilities. Whereas almost any decent classifier can perform at least slightly better than the default classifier in predictive accuracy (e.g., Baesens et al., 2003, report a 72.2% accuracy for their NB model), Zonneveldt et al. (2010) found *no* classifier which performed better than the default when assessed in terms of its expected value, including NB, TAN and augmented TANs, whether using the original cost table or the alternative mentioned above. All of these learning techniques are worthless with the German credit data, despite years of contrary claims in the literature, which were based upon simple assessments of accuracy!

TABLE 7.2

Disutilities (costs) for German credit data: (a) original; (b) alternative.

	(a)		(b)	
	repay	default	repay	default
give loan	0	5	-1	10
refuse loan	1	0	1	0

This cost-sensitive approach to evaluating predictive models is, in fact, quite paradigmatically Bayesian. The typical Bayesian approach to prediction is not to nominate a highest probability value for a query variable given the evidence, but instead to report the posterior probability distribution over the query node. Indeed, not only do Bayesians commonly hesitate about nominating a predicted class, they even hesitate about nominating a specific model for classifying. That is, the ideal Bayesian approach is to compute a posterior distribution over the entire space of models, and then to combine the individual predictions of the models, weighting them according to their posteriors, which is an ensembling technique known as **Bayesian model averaging**. As we have seen, ensemble predictions will typically provide better predictions than those of any of the individual models, including that one with the highest posterior probability. This is the hard-core Bayesian method, as advocated, for example, by Richard Jeffrey (1983). But in most situations this ideal method is unavailable, because it is computationally intractable. More typical of scientific practice, for example, is that we come to believe a particular scientific theory (model), which is then applied *individually* to make predictions or to guide interventions in related domains.

The expected value measurement is the best Bayesian guide to evaluating predictive models; it is the Bayesian gold standard. As everything, this method comes at a price: in particular, we must be able to estimate the utilities of Table 7.1. For many applications this will be possible, by the various elicitation discussed in Chapter 10. But again, in many others the utilities may be difficult or impossible to estimate. This will especially be true when the model being developed has an unknown, unexplored, or open-ended range of potential applications. It may be, for example, that the industrial environment within which the model will be employed is rapidly evolving, so that the future utilities are unknown. Or, in the case of many scientific models, the areas of application are only vaguely known, so the costs of predictive error are only vaguely to be guessed at — pure research is notorious for having unpredictable consequences. Finally, as for other methods, some assessment of the varying performance of classifiers in the expected value of their classifications is important. Even if the utilities are precisely known and invariant, classifiers will estimate varying probabilities for similar instances and this variability needs to be assessed.

Since relevant utilities may be unavailable, we now turn to evaluative methods which share with predictive accuracy and ROC curves the characteristic of being independent of utility structure. They differ, however, by being sensitive to over- and underconfidence, rewarding the well-calibrated predictor.

7.5.6 Proper scoring rules

When computing predictive accuracy or ROC curves the naive Bayes classification approach is to select the class that has maximal probability given the attribute values. I.e.,

$$\text{argmax}_{c_i \in C} P(c_i | a_1, \ldots, a_n)$$

This is not a particularly *Bayesian* approach: it throws away all the probabilistic information we have about the class variable. Of course, most people doing machine learning classification aren't Bayesian and didn't start out working with Bayesian tools, but rather with neural networks, classification trees and other models that have, at best, a dubious probabilistic interpretation. So, it's natural for them to take a Bayesian tool and drop its probabilistic semantics. However, for many reasons, it is both fair and even obligatory for matters to be turned around, and, in particular, for us to insist on probabilistic semantics being applied to the evaluation of classifiers, even when those classifiers have no natural probabilistic semantics, such as neural networks.

The very term "classification" is misleading in this regard: it suggests that the learning goal is one of dropping instances into some fixed set of buckets. Naturally, one cannot drop an instance into two buckets, even if one remains undecided as to which of the two is best. Learning involves reducing uncertainty, but it very rarely involves reducing uncertainty to nothing. *Prediction is inherently and essentially probabilistic.*

When the Bayesian gold standard of expected value cannot be applied to evaluation, we can nevertheless apply a silver standard, that of the **propriety** of **scoring rules** (Savage, 1971).

Definition 7.2 (Scoring rule) *A scoring rule S assigns a score (reward) to a classifier's prediction of the value of a target variable for a partially observed instance based upon the predicted (posterior) probability distribution over possible variable values and the actual value of the target variable. I.e.,*

$$S : \mathscr{P}(C) \times C \to [-\infty, \infty]$$

where $\mathscr{P}(C)$ is the learned posterior distribution over the target variable.

Definition 7.3 (Proper scoring rule) *A proper scoring rule is one which maximally rewards the true (objective) probability distribution over C, if there is one. A* **strictly proper** *scoring rule is one which* uniquely *maximizes the reward for the true probability distribution.*

Predictive accuracy provides a proper scoring rule, but not a strictly proper scoring rule. That is, using the true probability distribution will classify correctly as often as any distribution can, however other, strictly incorrect, distributions can do the same. It is this slackness that allows miscalibration to go unpunished.

Strict propriety implies the requirement of perfectly calibrated probabilities for maximal reward, for only such probabilities will be rewarded maximally. The Bayesian idea of rational gambling, maximizing expected value, as the exemplar for rational decision making thus also supports propriety: it is well known that perfectly calibrated probabilities maximize betting reward (Cover and Thomas, 2006). It is also clear, from its characteristics described above, that minimizing Kullback-Leibler divergence likewise provides a strictly proper scoring rule (for a proof, see §7.8). Now we turn to some other, perhaps less demanding and more accessible, proper scoring rules.

7.5.7 Information reward

Information reward, due to I.J. Good (1952), is just such a measure. Good invented it as a cost neutral assessment of gambling expertise for binomial prediction tasks. Good's definition is:

$$IR_G = \sum_i [1 + \log_2 P(x_i = v)] \qquad (7.6)$$

where v is whichever truth value is correct for the case of x_i. Good introduced the constant 1 in order to fix the reward in case of prior ignorance to zero, which he assumed would be represented by predicting $P(x_i = v) = 0.5$. This reward has a clear and useful information-theoretic interpretation (see Figure 7.7): it is one minus the length of a message conveying the actual event in a language efficient for agents with the model's beliefs (i.e., a language that maximizes the Shannon information content of communications between them). Note that such a message is infinitely long if it is attempting to communicate an event that has been deemed to be impossible, that is, to have probability zero. Alternatively, message length is minimized, and information reward maximized, when the true class is predicted with maximum probability, namely 1. This gives us Property 1.

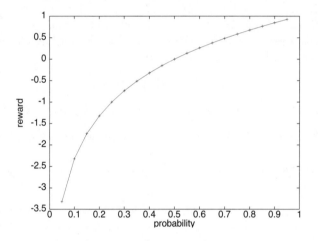

FIGURE 7.7: Good's information reward.

IR_G also has Property 2. Let c be the chance (physical probability, or its stand-in, frequency in the test set) that some test case is in the target class and p the probability estimated by the machine learner. Then the expected IR_G is:

$$[1 + \log_2 p]c + [1 + \log_2(1 - p)](1 - c)$$

To find the maximum we take the derivative with respect to p and set it to zero:

$$
\begin{aligned}
0 &= \frac{d}{dp}\left([1+\log_2 p]c + [1+\log_2(1-p)](1-c)\right) \\
&= \frac{c}{p} + \frac{1-c}{1-p}\frac{d}{dp}(1-p) \\
&= \frac{c}{p} - \frac{1-c}{1-p}
\end{aligned}
$$

So,

$$
c = p
$$

which last equation reports that perfect calibration maximizes reward.

All of this has some interesting general implications for machine learning research and evaluation methodology. For one thing, machine learning algorithms might be usefully modified to monitor their own miscalibration and use that negative feedback to improve their calibration performance while also building up their domain knowledge.

In experimental work, we have shown that standard machine learning problems issue differing verdicts as to which algorithm or model is best, when those verdicts are based on information reward instead of predictive accuracy (Korb et al., 2001). In other words, reliance upon the predictive accuracy measure is demonstrably causing errors in the judgments of performance showing up in the machine learning literature.

7.5.8 Bayesian information reward

There are two respects in which Good's information reward is deficient, leading to potential errors in evaluating models. The first is that IR_G assumes a uniform prior probability over the target classes. If the base rate for a disease, for example, is far lower than 50% in a population, say 1%, then a model which reflects no understanding of the domain at all — beyond that base rate — can accumulate quite a high information reward simply by predicting a 1% chance of disease for all future cases. Rather than Good's intended zero reward for ignorance, we have quite a high reward for ignorance. As a *Bayesian* reward function, we propose one which takes the prior into account and, in particular, which rewards with zero any prediction at the prior probability for the target class.

The second deficiency is simply that IR_G is limited to binomial predictions. Bayesian networks are very commonly applied to more complex prediction problems, so we propose the **Bayesian information reward** (BIR) which generalizes Good's information reward to multinomial prediction (Hope and Korb, 2004):

$$
IR_B = \sum_i \frac{I_i}{n} \tag{7.7}
$$

where n is the number possible target values (arity), $I_i = I_i^+$ for the true class and

$I_i = I_i^-$ otherwise, and

$$I_i^+ = \log \frac{\hat{p}_i}{p_i} \quad \text{(true value)}$$

$$I_i^- = \log \frac{1 - \hat{p}_i}{1 - p_i} \quad \text{(false value)}$$

where $\hat{p}_i$ is the model's probability for the variable taking the particular value at issue and p_i is the prior probability of the variable taking that value. Note that this version of information reward is directly related to Kullback-Leibler divergence, where the prior probability takes the role of the reference probability, but with two differences. First, the weighting is done implicitly by the frequency with which the different values arise in the test set, presumably corresponding to the prior probability of the different values. Second, the ratio between the model probability and the prior is inverted, to reflect the idea that this is a *reward* for the model to *diverge* from the reference prior, rather than to converge on it — so long as that divergence is approaching the truth more closely than does the prior. This latter distinction, rewarding divergence toward the truth and penalizing divergence away from the truth, is enforced by the distinction between IR_B for true value and for false values.

This generalizes Good's information reward to cover multinomial prediction by introducing a reward value for values which the variable in the test case fails to take (i.e., I_i^-). In the case of the binomial IR_G this was handled implicitly, since the probability of the alternative to the true value is constrained to be $1 - \hat{p}_i$. However, for multinomials there is more than one alternative to the true value and they must be treated individually, as we do above. IR_B has the following meritorious properties:

- If a model predicts any value with that value's prior probability, then the IR_B is zero. That is, ignorance is never rewarded. This raises the question of where the priors for the test cases should come from. The simplest, and usually satisfactory, answer is to use the frequency of the value for the variable within the training cases.

- Property 1 is satisfied: as predictions become more extreme, close to one or zero, for values which are, respectively, true or false, the reward increases asymptotically to one. As incorrect predictions become more extreme, the negative reward increases without bound. Both of these are strictly in accord with the information-theoretic interpretation of the reward.

- Property 2 is satisfied: IR_B is maximized under perfect calibration. Furthermore, IR_B is uniquely maximized under calibration, i.e., it is strictly proper (for a proof see Hope, 2008, §3.4.1).

7.6 Summary

Supervised classification is an important part of machine learning, and one in which Bayesian networks, especially simplified, non-causal Bayesian networks, have been shown to perform well. The methods for assessing classifiers range from simple, and often misleading, accuracy counts and ROC curves through to information-theoretic techniques, such as Kullback-Leibler divergence (relative entropy) and Bayesian information reward. These evaluative methods are not limited only to assessing classification performance in classical machine learning, but may also be usefully applied to assessing causal Bayesian networks, whether learned by automated methods or through expert elicitation, as we will see in Chapters 9 and 10.

7.7 Bibliographic notes

One of the best introductions to machine learning, including classification tree learning and a review of orthodox statistical evaluation, is (remains) Tom Mitchell's *Machine Learning* (1997). Other useful introductions include Witten and Frank (2005), which also serves to introduce the popular machine learning testing platform weka, and Bishop (2006). There are about 3,800 others, according to Google Books.

For a good review of naive Bayes models and their predictive performance examined both theoretically and empirically see Hand and Yu (2001). Naive Bayes models have a fairly long history within applied computer science. For some of their early uses see Duda and Hart (1973).

Many machine learning texts introduce or survey ensembling methods such as boosting; for an example, see Bishop (2006, Chap 14).

For a clear introduction to Kullback-Leibler divergence, and information theory in general, see Cover and Thomas (2006). For examples of work on cost-sensitive classification, see Turney (1995) and Provost et al. (1998); there was also some early work by Pearl (1978). Kononenko and Bratko also criticize predictive accuracy for evaluating algorithms and models (Kononenko and Bratko, 1991); furthermore, they too offer a scoring function which takes the prior probabilities of target classes into account. Unfortunately, their measure, while beginning with an information-theoretic intuition, does not have a proper information-theoretic interpretation. See Hope and Korb (2002) for a criticism of the Kononenko-Bratko measure (note however that that paper's measure itself contains an error; IR_B in the text is a further improvement).

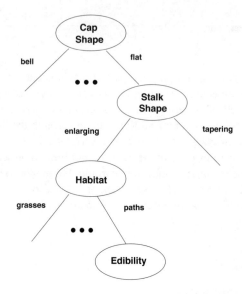

FIGURE 7.8: A classification tree for mushrooms.

7.8   Technical notes

Classification trees

Classification trees (also known as decision trees, a name which unfortunately infringes on the prior vocabulary of decision analysis) invert the ordering of nodes of naive Bayes models and its relatives, with the root node being some attribute, and branches hanging off the root node splitting on more attributes, until the final node (the "leaf" node) represents the target variable, providing a distribution over its possible values, as in Figure 7.8 (showing a single branch).

There are many different methods employed to learn classification trees. Quinlan's original method (in ID3) was to employ an information-theoretic measure ("information gain") to find attributes which result in the most homogeneous leaf nodes (Quinlan, 1983). C4.5 is a popular successor to ID3 (Quinlan, 1993), And is freely available as J48 in `weka`. C5.0 is Quinlan's current version, one which incorporates ensembling techniques.

The "decision stumps" of Holte (1993) are especially simple classification trees, which split on no more than a single attribute. Holte (1993) found that these perform surprisingly well in comparison with the more complicated classification trees discovered by ID3 and its successors, which is a result similar in spirit to the successes of naive Bayes models. In parallel to the success with ensembling naive Bayes models, decision stumps have been ensembled to improve their performance; see Oliver and Hand (1994).

Kullback-Leibler divergence

We prove that KL divergence (restricted to strictly positive distributions) is minimized at zero if and only if the model under evaluation is perfectly calibrated — i.e., it is strictly proper.

Theorem 7.1 *Let p and q of Definition 7.1 be strictly positive probability distributions. Then $KL(p,q) \geq 0$ and $KL(p,q) = 0$ if and only if $p = q$.*

Proof. (See Borgelt and Kruse, 2002, A.15.1.)

First, a lemma:

$$\sum_{x \in X} p(x) \log \frac{q(x)}{p(x)} \leq 0 \tag{7.8}$$

This is true because (since $\forall x [\log x \leq x - 1]$)

$$
\begin{aligned}
\sum_{x \in X} p(x) \log \frac{q(x)}{p(x)} &\leq \sum_{x \in X} p(x) \left(\frac{q(x)}{p(x)} - 1 \right) \\
&= \sum_{x \in X} p(x) - \sum_{x \in X} q(x) \\
&= 1 - 1 \\
&= 0
\end{aligned}
$$

Since,

$$\sum_{x \in X} p(x) \log \frac{p(x)}{q(x)} = - \sum_{x \in X} p(x) \log \frac{q(x)}{p(x)}$$

it follows that

$$\sum_{x \in X} p(x) \log \frac{p(x)}{q(x)} \geq 0 \tag{7.9}$$

And if $\sum_{x \in X} p(x) \log \frac{p(x)}{q(x)} = 0$, then by positivity of p, $\log \frac{p(x)}{q(x)}$ must everywhere be zero; hence $p(x) = q(x)$ everywhere.

7.9 Problems

Problems 2–4 use the German credit data, available from `http://www.stat.auckland.ac.nz/~reilly/credit-g.arff`. For those problems it doesn't matter what tools you use, or if you write your own software to do them. You may use weka (`http://www.cs.waikato.ac.nz/ml/weka/`)) for some of the work, such as AODE (converting numeric attributes to nominal with the filter "NumericToNominal").

Problem 1

Prove that IR_B is a proper generalization of IR_G — in other words, prove that IR_B in the binomial case is equivalent to Good's information reward (assuming that the prior probability is uniform).

Problem 2

Create and parameterize the naive Bayes model for the German credit scoring problem. Compute the following evaluation measures for both the default classifier (i.e., predicting all loans will be repaid, based on the prior probability of a 70% chance of repayment) and your NB model:

1. Predictive accuracy
2. AUC
3. Expected cost, using either (or both) of the cost tables discussed in the text
4. IR_B

Problem 3

Extend your work for Problem 2 to include AODE.

Problem 4

Write an interpretive discussion of your work in Problem 2 (and Problem 3, if you've done that). Assess your own evaluation procedure, including your treatment of the variance in your measures.

8

Learning Linear Causal Models

8.1 Introduction

Thus far, we have seen that Bayesian networks are a powerful method for representing and reasoning with uncertainty, one which demonstrably supports normatively correct Bayesian reasoning and which is sufficiently flexible to support user modeling equally well when users are less than normative. BNs have been applied to a very large variety of problems; many of these applications have been academic exercises — that is to say, prototypes intended to demonstrate the potential of the technology, rather than applications that real businesses or government agencies would be relying upon. A major reason why BNs have not yet been deployed more widely in significant industrial-strength applications is the same reason earlier rule-based expert systems were not widely successful: the **knowledge bottleneck**.

Whether expert systems encode domain knowledge in rules or in conditional probability relations, that domain knowledge must come from somewhere. The main plausible source until recently has been human domain experts. When building knowledge representations from human experts, AI practitioners (called knowledge engineers in this role) must elicit the knowledge from the human experts, interviewing or testing them so as to discover compact representations of their understanding. This encounters a number of difficulties, which collectively make up the "knowledge bottleneck" (cf. Feigenbaum, 1977). For example, in many cases there simply are no human experts to interview; many tasks to which we would like to put robots and computers concern domains in which humans have had no opportunity to develop expertise — most obviously in exploration tasks, such as exploring volcanoes or exploring sea bottoms. In other cases it is difficult to articulate the humans' expertise; for example, every serious computer chess project has had human advisors, but human chess expertise is notoriously inarticulable, so no good chess program relies upon rules derived from human experts as its primary means of play — they all rely upon brute-force search instead. In all substantial applications the elicitation of knowledge from human experts is time consuming, error prone and expensive. It may nevertheless be an important means of developing some applications, but if it is the *only* means, then the spread of the underlying technology through any large range of serious applications will be bound to be slow and arduous.

The obvious alternative to relying entirely upon knowledge elicitation is to employ machine learning algorithms to automate the process of constructing knowledge representations for different domains. With the demonstrated potential of Bayesian networks, interest in methods for automating their construction has grown enormously in recent years. In this and the next chapter we look again at the relation between conditional independence and causality, as this provides the key to causal learning from observational data.

This key relates to Hans Reichenbach's work on causality in *The Direction of Time* (Reichenbach, 1956). Recall from §2.4.4 the types of causal structures available for three variables that are in an undirected chain; these are causal chains, common causes and common effects, as in Figure 8.1. Reichenbach proposed the following principle:

Conjecture 8.1 Principle of the Common Cause *If two variables are marginally probabilistically dependent, then either one causes the other (directly or indirectly) or they have a common ancestor.*

Reichenbach also attempted to analyze time and causality in terms of the different dependency structures exemplified in Figure 8.1; in particular, he attempted to account for time asymmetry by reference to the dependency asymmetry between common causal structures and common effect structures. In this way he anticipated d-separation, since that concept depends directly upon the dependency asymmetries Reichenbach studied. But it is the Principle of the Common Cause which underlies causal discovery in general. That principle, in essence, simply asserts that behind every probabilistic dependency is an explanatory causal dependency. And that is something which all of science assumes. Indeed, we should likely abandon the search for an explanatory cause for a dependency only after an exhaustive and exhausting search for such a cause had failed — or, perhaps, if there is an impossibility proof for the existence of such a cause (as is arguably the case for the entangled systems of quantum mechanics). The causal discovery algorithms we present here explicitly search for causal structures to explain probabilistic dependencies and, so, implicitly pay homage to Reichenbach.

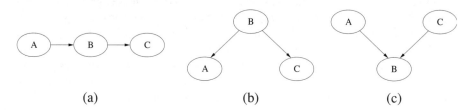

FIGURE 8.1: (a) Causal chain; (b) common cause; (c) common effect.

8.2 Path models

The use of graphical models to represent and reason about uncertain causal relationships began with the work of the early twentieth-century biologist and statistician Sewall Wright (1921, 1934). Wright developed graphs for portraying linear causal relationships and parameterized them based upon sample correlations. His approach came to be called **path modeling** and has been extensively employed in the social sciences. Related techniques include structural equation modeling, widely employed in econometrics, and causal analysis. Behind the widespread adoption of these methods is, in part, just the restriction to linearity, since linearity allows for simpler mathematical and statistical analysis. Some of the restrictiveness of this assumption can be relaxed by considering transformations of non-linear to linear functions. Nevertheless, it is certain that many non-linear causal relations have been simplistically understood in linear terms, merely because of the nature of the tools available for analyzing them.

In using Bayesian networks to represent causal models we impose no such restrictions on our subject. Nevertheless, we shall first examine Wright's "method of path coefficients." Path modeling is preferable to other linear methods precisely because it directly employs a graphical approach which relates closely to that of Bayesian networks and which illustrates in a simpler context many of the causal features of Bayesian networks.

A linear model relates the effect variable Y to parent variables X_i via an additive, linear function, as in:

$$Y = a_1 X_1 + a_2 X_2 + U \qquad (8.1)$$

In this case a_1 is a constant coefficient that reflects the extent to which parent variable X_1 influences, accounts for or explains the value of Y; similarly for a_2 and X_2. U represents the mean value of Y and the variation in its values which cannot be explained by the model. If Y is in fact a linear function of all of its parent variables, then U represents the cumulative linear effect of all the parents that are unknown or unrepresented in this simple model. If U is distributed as a Gaussian $N(\mu, \sigma^2)$ (normal), the model is known as **linear Gaussian**. This simple model is equally well represented by Figure 8.2. Usually, we will not bother to represent the factor U explicitly, especially in our graphical models. When dealing with standardized linear models, the impact of U can always be computed from the remainder of the model in any case.

A linear model may come from any source, for example, from someone's imagination. It may, of course, come from the statistical method of linear regression, which finds the linear function which minimizes unpredicted (residual) variation in the value of the dependent variable Y given values for the independent variables X_i. What we are concerned with in this chapter is the discovery of a joint system of linear equations by means distinct from linear regression. These means may not necessarily strictly minimize residual variation, but they will have other virtues, such as discovering better explanatory models than regression methods can, as we shall see.

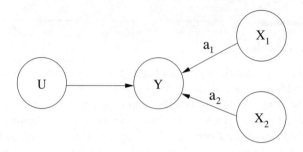

FIGURE 8.2: A linear model.

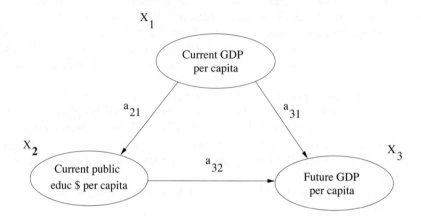

FIGURE 8.3: OECD public education spending model.

Let's consider a concrete example of a path model[1]. Figure 8.3 reports the relations between public educational spending in OECD countries[2] and per capita economic activity at the time of the spending and, again, 15 years later (measured in "GDP per capita," which means gross domestic product per person in the economy; this example is taken, with minor modifications, from Korb et al., 1997). This model indicates that a typical OECD nation will receive some amount of future economic benefit from an extra \$1 added to the current economy in general (for example, via a tax cut) and some different future economic benefit from the dollar being added to public education instead. The graphical model can be transformed into a system of

---

[1]More precisely, we consider here a **recursive path model**. A path model is called recursive just in case it is a directed acyclic graph, as are all those we will consider here. It is possible also to deal with non-recursive path models, which replace some subnetwork relating two variables with a bidirectional arc parameterized with a correlation. Such models are essentially incomplete recursive models: if the subnetwork were known, it would be more informative than the simple correlation and so would be used instead.

[2]OECD stands for *Organization for Economic Co-operation and Development* and includes all of the major developed countries.

linear equations by writing down one equation per dependent variable in the model.

$$X_2 = a_{21}X_1$$
$$X_3 = a_{32}X_2 + a_{31}X_1$$

Focusing on X_3 (Future GDP per capita), there are three kinds of linear causal influence reported by the model: direct, $X_1 \rightarrow X_3$ (as well as $X_2 \rightarrow X_3$ and $X_1 \rightarrow X_2$); indirect, $X_1 \rightarrow X_2 \rightarrow X_3$; and the "spurious" effect of X_1 inducing a correlation between X_2 and X_3. Assuming the Markov property, there can be no other causal influences between variables in the model. This has the further implication, applying Reichenbach's Principle of the Common Cause, that there can also be no other sources of correlation between the variables in the model. This suggests that correlation between the variables and the linear causal influences should be interrelatable: if correlations can only exist where causal influences induce them, and if all causal influences are accounted for in a model (as required by the Markov property), then we should be able to transform correlations into causal weights and vice versa. And so we can.

The interrelation of correlation and linear causal influence is a key result. The causal influences are theoretical: agreeing with David Hume (1962), we cannot simply observe causal forces as such. However, we can observe and measure correlations. Since the two can be precisely interrelated, as we will see immediately below, we can use the observations of correlation to discover, and specify the strength of, linear causal relationships.

8.2.1   Wright's first decomposition rule

Sewall Wright's first Decomposition Rule gives the exact relation between correlations and linear causal coefficients (Wright, 1934). His rule assumes that all variables have been standardized; that is, the scale for each variable has been transformed into standard deviation units. This is easily done for any variable. For example, Future GDP per capita (X_3) values in dollars can be replaced by deviations from the OECD average (Z_3) via the transformation:

$$Z_3 = \frac{X_3 - \mu_3}{\sigma_3} \tag{8.2}$$

where μ_3 is the mean Future GDP per capita and σ_3 is its standard deviation. Standardized variables all have means of zero and standard deviations of 1. The path coefficient p_{ij} is the analog in the standardized model of the linear coefficient in the non-standardized model (the two are related below in Equation 8.5).

Rule 8.1 Wright's Decomposition Rule. *The correlation r_{ij} between variables X_i and X_j, where X_i is an ancestor of X_j, can be rewritten according to the equation:*

$$r_{ij} = \sum_k p_{jk} r_{ki} \tag{8.3}$$

where p_{jk} are path coefficients relating X_j with each of its direct parents X_k.

Application of this rule replaces a correlation, r_{ij}, with products of a path coefficient to a parent X_k and another correlation relating the ancestor X_i to this parent; since this last correlation must relate two nodes X_k and X_i more closely connected than the original pair X_j and X_i, it is clear that repeated application of the rule will eventually eliminate reference to correlations. In other words, using Wright's rule we can rewrite any correlation in terms of sums of products of path coefficients. This will give us a system of equations which we can then use to solve for the path coefficients.

For example, taking each variable in turn from X_1 to X_2 to X_3 in Figure 8.3 we can use Rule 8.1 to generate the following equations for modeling the dependent variables (bearing in mind that $r_{11} = r_{22} = 1$):

$$
\begin{aligned}
\text{Take } X_1: \quad & \text{nothing} \\
\text{Take } X_2 : r_{12} \quad = \quad & \sum_k p_{2k} r_{k1} \\
= \quad & p_{21} r_{11} \\
= \quad & p_{21} \\
\text{Take } X_3 : r_{13} \quad = \quad & \sum_k p_{3k} r_{k1} \\
= \quad & p_{31} r_{11} + p_{32} r_{12} \\
= \quad & p_{31} + p_{32} r_{12} \\
r_{23} \quad = \quad & \sum_k p_{3k} r_{k2} \\
= \quad & p_{31} r_{12} + p_{32} r_{22} \\
= \quad & p_{31} r_{12} + p_{32}
\end{aligned}
$$

We have three equations in three unknowns, so we can solve for the path coefficients:

$$
\begin{aligned}
p_{21} \quad &= \quad r_{12} \\
p_{31} \quad &= \quad \frac{r_{13} - r_{23} r_{12}}{1 - r_{12}^2} \\
p_{32} \quad &= \quad \frac{r_{23} - r_{13} r_{12}}{1 - r_{12}^2}
\end{aligned}
$$

In the public education example, the observed correlations were $r_{12} = 0.8212$, $r_{13} = 0.5816$, and $r_{23} = 0.6697$. Plugging these into the above equations gives us the path coefficients

$$
\begin{aligned}
p_{21} \quad &= \quad 0.8212 \\
p_{32} \quad &= \quad 0.5899 \\
p_{31} \quad &= \quad 0.0972
\end{aligned}
$$

These are standardized coefficients, reporting the impact of causal interventions on any parent variables upon their children.

There is an equivalent rule for decomposing correlations into path coefficients which is even simpler to apply, and which also relates matters back to Reichenbach's discussion of causality and the concept of d-separation.

Rule 8.2 Wright's Second Decomposition Rule. *The correlation r_{ij} between variables X_i and X_j, where X_i is an ancestor of X_j, can be rewritten according to the equation:*

$$r_{ij} = \sum_k v(\Phi_k) \tag{8.4}$$

where Φ_k is an active path between X_i and X_j and $v(\cdot)$ is a valuation of that path.

Intuitively, each active path represents a distinct line of causal influence, while its valuation measures the degree of causal influence. Note that these paths are not simply undirected paths, instead:

Definition 8.1 Active Path. *Φ_k is an active path between X_i and X_j if and only if it is an undirected path (see Definition 2.1) connecting X_i and X_j such that it does not go against the direction of an arc **after** having gone forward.*

The valuation of a path is

$$v(\Phi_k) = \begin{cases} p_{ij} \text{ if } \Phi_k = X_j \rightarrow X_i \\ \prod_{lm} p_{lm} \text{ for all } X_m \rightarrow X_l \in \Phi_k \end{cases}$$

This decomposition of correlation into causal influences corresponds directly to Reichenbach's treatment of causal asymmetry, and therefore also prefigures d-separation. A path traveling always forward from cause to (ultimate) effect identifies a causal chain, of course; a path traveling first backwards and then forwards (but never again backwards) identifies a common ancestry between each pair of variables along the two branches; and the prohibition of first traveling forwards and then backwards accounts for the conditional dependencies induced by common effects or their successors.

This version of Wright's rule also has the property of d-separation that has made that concept so successful: the causal influences represented by a path model can easily be read off of the graph representing that model. Applied to Figure 8.3, for example, we readily find:

$$\begin{aligned} r_{12} &= \sum_k v(\Phi_k) \\ &= v(X_1 \rightarrow X_2) \\ &= p_{21} \end{aligned}$$

$$\begin{aligned} r_{13} &= v(X_1 \rightarrow X_3) + v(X_1 \rightarrow X_2 \rightarrow X_3) \\ &= p_{31} + p_{21}p_{32} \\ r_{23} &= v(X_2 \rightarrow X_3) + v(X_2 \leftarrow X_1 \rightarrow X_3) \\ &= p_{32} + p_{21}p_{31} \end{aligned}$$

which is identical, of course, with what we obtained with Rule 8.1 before.

The path coefficients of Wright's models are directly related to the linear coefficients of regression models: just as the variables are arrived at by standardizing, the path coefficient p_{ij} may be arrived at by standardizing the regression coefficient a_{ij}:

$$p_{ij} = a_{ij} \left(\frac{\sigma_j}{\sigma_i} \right) \tag{8.5}$$

As a consequence of the standardization process, the sum of squared path coefficients convergent on any variable X_i is constrained to equal one:

$$\sum_j p_{ij}^2 = 1$$

This is true assuming that one of the parent variables represents the variance unexplained by the regression model over the known parents — i.e., the U variable is included. Since these sum to one, the square of the path coefficient, p_{ij}^2, can be understood as the proportion of the variation in the child variable X_i attributable to parent X_j, and if p_{iu} is the coefficient associated with the residual variable U, then p_{iu}^2 is the amount of variance in X_i which is left unexplained by the linear model.

Returning again to the education model of Figure 8.3, the path coefficients were

$$
\begin{aligned}
p_{21} &= 0.8212 \\
p_{32} &= 0.5899 \\
p_{31} &= 0.0972
\end{aligned}
$$

Since these are standardized coefficients, they report the square root of the amount of variation in the child variable that is explained by the parent. In other words, for example, variations in public GDP (X_1) account for the fraction $(0.8212)^2 = 0.674 = 67.4\%$ of the variation in public spending on education across OECD countries (X_2).

Standardization makes it very easy to understand the relation between correlations and coefficients, but it isn't everything. By reversing the process of standardization we may well find a more human-readable model. Doing so for public education results in Figure 8.4. From this model we can read off numbers that make more intuitive sense to us than the path coefficients. In particular, it asserts that for a normal OECD country in the period studied, an additional \$1 added to the public education budget will expand the future economy by about \$2, whereas an additional \$1 added at random to the economy (via a tax cut, for example) will expand the future economy by about \$1.

8.2.2   Parameterizing linear models

All of this work allows us to parameterize our path models in a straightforward way: apply Wright's Decomposition Rule (either one) to obtain a system of equations relating correlation coefficients with the path model's coefficients; solve the system of equations for the path coefficients; compute the solution given sample correlations.

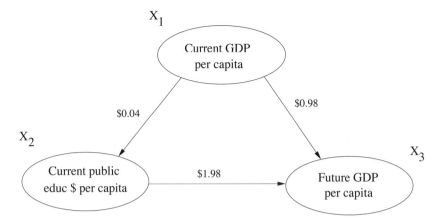

FIGURE 8.4: Non-standardized OECD public education spending model.

This is exactly what was done with the OECD public education model, in fact. It has been proven that recursive path models will always have a system of equations which are solvable, so the method is always available (Boudon, 1968). There are equivalent alternative methods, such as one due to Simon and Blalock (Simon, 1954, Blalock, 1964) and ordinary least squares regression. As we mentioned above, however, we prefer Wright's method because it is more intuitively connected to causal graph representations.

8.2.3 Learning linear models is complex

The ability to parameterize linear models in turn ushers in a very plausible methodology of learning linear models, which indeed is quite widely employed in the social sciences:

1. Find some correlations that are puzzling.
2. Invent a possible causal model that might, if true, explain those correlations (a step which is variously called **abduction** or **inference to the best explanation**).
3. Estimate ("identify") the model parameters, whether via Wright's Rule, least squares or the Simon-Blalock method, in order to complete the model.
4. Test the complete model by predicting new data and comparing reality with the prediction. If reality diverges from the model's expectation, GOTO step 2.

Numerous problems in the social sciences have been investigated in this way, and arguably many of them have been solved. For example, notwithstanding the obfuscation of some politicians, there is no serious doubt amongst economists who investigate such matters that investment in public education is a key ingredient in future economic well-being, partly because of such studies as ours. In other words, we know for a fact that it is possible to learn causal structure from correlation structure, because we humans can and do. Implementing such a process in a computer is another matter, to be sure: the above procedure, in particular, does not immediately commend itself, since no one knows how we perform step 2 — inventing explanatory models.

In AI the standard response to the need to invent or create is to employ search. If we have a clear criterion for evaluating an artifact, but no obvious preferred means for constructing it, then the standard AI move is to search the space of possible artifacts, stopping when we have found one that is satisfactory. Such a search might start with the simplest possible artifact — perhaps a null artifact — and, by testing out all possible single additions at each step, perform a breadth-first search for a satisfactory artifact. Such a search is perfectly fine when the size of some satisfactory artifact is small: since small artifacts are examined first, an acceptable one will be found quickly. But the time complexity of such a search is of the order b^d, where b is the **branching factor** (the number of possible single additions to an artifact) and d is the **solution depth** (the size of the simplest acceptable artifact). So, if we are looking for a sizable artifact, or working with a very large number of possible additions to an artifact at any step, then such a simple search will not work. The exponential time complexity of such brute-force search is what has made world-class computer chess difficult to achieve and world-class computer go (thus far) unattainable. AI as a field has investigated many other types of search, but given a highly complex search space, and no very helpful guide in how to search the space, they will all succumb to exponential search complexity — that is, they will not finish in reasonable time. We will now show that part of these conditions for unsolvability exist in the learning of causal models: the space of causal models is exponentially complex.

How many possible causal models (i.e., dags) are there? The number depends upon the number of variables you are prepared to entertain. A recursive expression for the number of dags $f(N)$ given N variables is (Robinson, 1977):

$$f(N) = \sum_{i=1}^{N} (-1)^{i+1} C_i^N 2^{i(N-i)} f(N-i) \tag{8.6}$$

with $f(0) = f(1) = 1$. With two variables, there are three dags; with three variables there are 25 dags; with five variables there are 25,000 dags; and with ten variables there are about 4.2×10^{18} possible models. As can be seen in (8.6), this grows exponentially in N.

This problem of exponential growth of the model space does not go away nicely, and we will return to it in Chapter 9 when we address metric learners of causal structure. In the meantime, we will introduce a heuristic search method for learning causal structure which is effective and useful.

8.3   Constraint-based learners

We can imagine a variety of different heuristic devices that might be brought to bear upon the search problem, and in particular that might be used to reduce the size of the space. Thus, if we had partial prior knowledge of some of the causal relations between variables, or prior knowledge of temporal relations between variables, that could rule out a great many possible models. We will consider the introduction of specific prior information later, in the section on adaptation (§10.9).

But there must also be methods of learning causal structure which do not depend on any special background knowledge: humans (and other animals), after all, learn about causality from an early age, and in the first instance without much background. Evolution may have built some understanding into us from the start, but it is also clear that our individual learning ability is highly flexible, allowing us severally and communally to adapt ourselves to a very wide range of environments. We should like to endow our machine learning systems with such abilities, for we should like our systems to be capable of supporting autonomous agency, as we argued in Chapter 1.

One approach to learning causal structure directly is to employ experimentation in addition to observation: whereas observing a joint correlation between A and B guarantees that there is *some* causal relation between them (via the Common Cause Principle), a large variety of causal relations will suffice. If, however, we intervene — changing the state of A — and subsequently see a correlated change in the state of B, then we can rule out both B being a cause of A and some common cause being the sole explanation. So, experimental learning is clearly a more powerful instrument for learning causal structure.

Our augmented model for causal reasoning (§3.8) suggests that learning from experimental data is a special variety of learning from observational data; it is, namely, learning from observational samples taken over the augmented model. Note that adding the intervention variable I_A to the common causal structure Figure 8.1 (b) yields $I_A \rightarrow A \leftarrow B \rightarrow C$. Observations of I_A without observations of A can be interpreted as causal manipulations of A. And, clearly, in such cases experimental data interpreted as an observation in the augmented model will find no dependency between the causal intervention and C, since the intervening v-structure blocks the path.

Thus, the restriction to observational learning is apparently not a restriction at all. In any case, there is a surprising amount of useful work that can be done with observational data without augmented models. In particular, there is a powerful heuristic search method based upon the conditional independencies implied by a dag. This was introduced by Verma and Pearl (1990), in their IC algorithm (for "inductive causation"), and first used in a practical algorithm, the PC algorithm, in TETRAD II (Spirtes et al., 1993), presented in §8.3.2.

Algorithm 8.1 *IC Algorithm*

1. **Principle I** *This principle recovers all the direct causal dependencies using undirected arcs. Let $X, Y \in \mathbf{V}$. Then X and Y are directly causally connected if and only if for every $\mathbf{S} \subset \mathbf{V}$ such that $X, Y \notin \mathbf{S}$ $X \not\!\perp\!\!\!\perp Y | \mathbf{S}$.*

2. **Principle II** *This principle recovers some of the arc directions, namely those which are discoverable from the dependencies induced by common effect variables.*

   If $X - Y$ and $Y - Z$ but not $X - Z$ (i.e., we have an undirected chain $X - Y - Z$), then replace the chain by $X \rightarrow Y \leftarrow Z$ if and only if for every $\mathbf{S} \subset \mathbf{V}$ such that $X, Z \notin \mathbf{S}$ and $Y \in \mathbf{S}$ $(X \not\!\perp\!\!\!\perp Z | Y)$. The structure $X \rightarrow Y \leftarrow Z$ is called a v-structure (or: a collider).

3. *Iterate through all undirected arcs $Y - Z$ in the graph. Orient $Y \rightarrow Z$ if and only if either*

   (a) *at a previous step Y appeared as the middle node in an undirected chain with X and Z (so, Principle II failed to indicate Y should be in v-structure between X and Z) and now the arc between X and Y is directed as $X \rightarrow Y$;*

   (b) *if we were to direct $Y \leftarrow Z$, then a cycle would be introduced.*

   Continue iterating through all the undirected arcs until one such pass fails to direct any arcs.

We can illustrate this algorithm with the simple example of Figure 8.5, a causal model intended to reflect influences on the college plans of high school students (Sewell and Shah, 1968). The variables are: *Sex* (male or female), *IQ* (intelligence quotient), *CP* (college plans), *PE* (parental encouragement), *SES* (socioeconomic status).

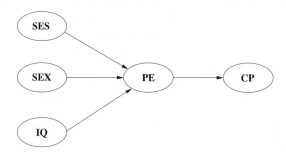

FIGURE 8.5: Causal model for college plans.

We can suppose that we have a causal system in these variables which is governed by this causal model and that we have sampled the variables jointly some reasonably large number of times (Sewell and Shah, 1968 had a sample of 10,318 students) and that we wish to learn the causal model which generated the sample. In order to operate the IC algorithm, let us imagine that we have an oracle who has access to the

Truth — who can peek at the true causal model of Figure 8.5 — and who is forthcoming enough to answer any questions we pose about conditional dependencies and independencies — but no more than that. In that case, the IC algorithm allows us to learn the entire structure of Figure 8.5.

First, Principle I of the IC algorithm will tell us all of the pairs of variables which are directly connected by some causal arc, without telling us the causal direction. That is, Principle I gives us Figure 8.6. Principle II of the algorithm might then examine the chain SES–PE–SEX, leading to SES $\rightarrow$ PE $\leftarrow$ SEX. Subsequently, Principle II will examine IQ in combination with PE and either one of SES or SEX, leading to IQ $\rightarrow$ PE. Principle II will lead to no orientation for the link PE–CP, since this step can only introduce v-structures. Step 3(a), however, will subsequently orient the arc PE $\rightarrow$ CP, recovering Figure 8.5 exactly.

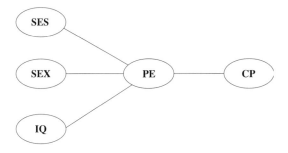

FIGURE 8.6: Causal model for college plans learned after IC Principle I.

The IC algorithm will not always recover the true model. If the true model were the alternative model Figure 8.7, for example, no arc directions would be discoverable: the only two chains available to Principle II are SES–PE–CP and SES–IQ–CP. So, IC will discover the skeleton of Figure 8.7 only.

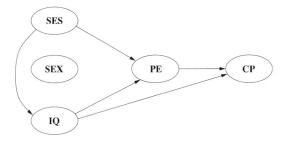

FIGURE 8.7: Alternative causal model for college plans.

8.3.1 Markov equivalence

What Verma and Pearl's IC algorithm is discovering in these cases is the set of Markov equivalent causal models (these are also known as **statistically equivalent** models). Two causal models H_1 and H_2 are **Markov equivalent** if and only if they contain the same variables and any probability distribution that can be represented by one can be represented by the other. In other words, given some parameterization for the one model, we can find a parameterization for the other which yields the very same probability distribution.

That the IC algorithm has this much power and no more follows from Verma and Pearl's proof of the following theorem (Verma and Pearl, 1990). (Note that the theorems of this section are known to hold only for linear Gaussian and discrete probability distributions. In general, we will not consider other probability distributions, and so can use them.)

Theorem 8.1 *(Verma and Pearl, 1990) Any two causal models over the same variables that have the same skeleton and the same v-structures are Markov equivalent.*

Given an infallible oracle (and, of course, the Markov property), Principle I will recover the skeleton. Principle II will recover all the v-structures. Step 3 merely enforces consistency with dag structure. Theorem 8.1, therefore, ensures that the partially oriented graph which IC discovers, called a **pattern** by Verma and Pearl, can only be completed by directing unoriented arcs in Markov equivalent ways. In addition to showing us this, the theorem also provides a simple, easily applied graphical criterion of Markov equivalence. Thus, some examples of Markov equivalence:

- All fully connected models (in the same variables) are Markov equivalent. Fully connected models contain no v-structures, since the end variables in any candidate chain must themselves be directly connected. Therefore, on the graphical criterion, they fall under the same pattern.
- $A \rightarrow B \rightarrow C$ and $A \leftarrow B \leftarrow C$ are Markov equivalent. There is no v-structure here.
- $A \rightarrow B \rightarrow D \leftarrow C$ and $A \leftarrow B \rightarrow D \leftarrow C$ are Markov equivalent. The only v-structure, centered on D, is retained.

For illustration, all patterns in three variables are displayed in Figure 8.8.

Following on Verma and Pearl's ground-breaking work, D. Max Chickering investigated Markov equivalent models and published some important results in the 1995 *Uncertainty in AI* conference (Chickering, 1995).

Theorem 8.2 *(Chickering, 1995) If H_1 and H_2 are Markov equivalent, then they have the same maximum likelihoods relative to any joint sample e; i.e.,*

$$\max_{\theta_1} P(e|H_1, \theta_1) = \max_{\theta_2} P(e|H_2, \theta_2)$$

where θ_i is a parameterization of H_i

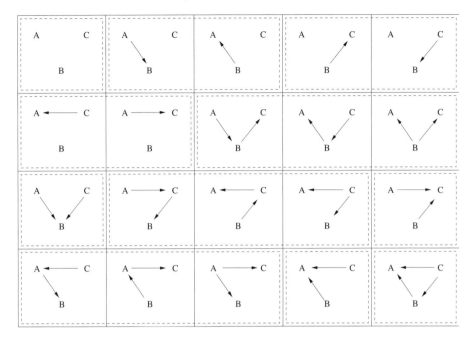

FIGURE 8.8: All patterns in three variables. Patterns are indicated with dotted lines. (Note that the fully connected last pattern shows only one of 6 dags.)

Since standard statistical inference procedures are largely based upon maximum likelihood measures, it is unlikely that they will have the ability to distinguish between Markov equivalent hypotheses. This may sound like a truism — that statistically indistinguishable (equivalent) hypotheses shouldn't be distinguishable statistically. And, indeed, many researchers, including some calling themselves Bayesians, advocate metrics and methods which assume that causal discovery should be aimed at patterns (Markov equivalence classes) rather than at their constituent causal hypotheses. There are at least two reasons to disagree. First, Bayesian methods properly incorporate prior probabilities and do not merely respond to maximum likelihoods. Second, what is statistically indistinguishable using observational data alone is not so using experimental data, which are far more discerning in revealing causal structure. In consequence of this received view, far less attention has been paid to experimental inference and to the causal semantics of Bayesian networks than is warranted. We shall have somewhat more to say about this in Chapter 9.

8.3.1.1   Arc reversal

A second result from Chickering's work is very helpful in thinking graphically about causal models.

Theorem 8.3 *(Chickering, 1995) Any two Markov equivalent dags H_i and H_j are connected in a chain of dags $< H_i, \ldots, H_j >$ where any two adjacent dags differ at most in one covered arc reversal.*

This is related to the arc reversal rule:

Rule 8.3 Arc Reversal. *H_2 can represent any probability distribution represented by H_1[3] if they contain the same variables and arcs except that:*

   (a) for some pair of nodes A and B, $A \to B \in H_1$ and $A \leftarrow B \in H_2$, and

   (b) if this arc is in an uncovered v-structure $A \to B \leftarrow C$ in H_1, then it is covered in H_2, i.e., $C \to A \in H_2$, and

   (c) if the reversal would introduce an uncovered v-structure $C \to A \leftarrow B$ into H_2, then either $C \to B$ or $C \leftarrow B$ must be added to H_2.

The arc reversal rule allows us to change the temporal ordering between any two variables. This will either keep us within a Markov equivalence class, because no v-structures are thereby introduced or destroyed, or else send us into a new equivalence class by requiring us to introduce a covering arc. The covering arc is necessary either to reestablish conditional independence between A and C given B (clause (b)) or else to reestablish conditional dependence (clause (c)). In either case, the arc reversal forces us to move from a simpler model H_1 to a denser model H_2, and, hence, to move from one equivalence class to another.

Many have thought that the ability to reverse arcs arbitrarily while continuing to represent the same probability distributions demonstrates that Bayesian networks are intrinsically *not* about causality at all, that all that matters for understanding them is that they can represent some class of probability distributions. However, since the arc reversal rule can only introduce and never eliminate arcs, it clearly suggests that among all the models which can represent an observed probability distribution, that model which is the sparsest (has the fewest arcs) is the true causal model. Since the true model by definition has no spurious arcs, it will be the sparsest representation of the observed probability distribution[4]; any alternative arrived at by reversing the temporal ordering of nodes (equivalently, reversing the arc directions relating those nodes) must be at least as dense as the true model. In any case, a little reflection will reveal that acceptance of Reichenbach's Principle of the Common Cause implies that non-causal distributions are strictly derivative from and explained by some causal model. Of course, it is possible to reject Reichenbach's principle (as, for example, Williamson does in Williamson, 2001), but it is not clear that we can reject the principle while still making sense of scientific method.

As we shall see below, many causal discovery algorithms evade, or presuppose some solution to, the problem of identifying the correct variable order. In principle, these algorithms can be made complete by iterating through all possible orderings

---

[3]This rule holds if we restrict the probability distributions to linear Gaussian and multinomial distributions.

[4]To be sure, there are cases of measure zero where a non-causal linear model can represent a linear causal distribution more compactly.

to find the sparsest model that the algorithm can identify for each given ordering. Unfortunately, all possible orderings for N variables number $N!$, so this completion is exponential. Possibly, the problem can be resolved by sampling orderings and imposing constraints when subsets of nodes have been ordered. So far as we know, such an approach to causal discovery has yet to be attempted.

8.3.1.2   Markov equivalence summary

In summary, Markov equivalence classes of causal models, or patterns, are related to each other graphically by Verma and Pearl's Theorem 8.1: they share skeletons and v-structures. They are related statistically by having identical maximum likelihoods, and so, by orthodox statistical criteria, they are not distinguishable. Despite that limitation, learning the patterns from observational data is an important, and large, first step in causal learning. We do not yet know, however, how close we can get in practice towards that goal, since the IC algorithm is itself a highly idealized one. So: in reality, how good can we get at learning causal patterns?

8.3.2   PC algorithm

Verma and Pearl's IC algorithm appears to depend upon a number of unrealistic features. First, it depends upon knowledge of the actual conditional independencies between variables. How is such knowledge to be gained? Of course, if one has access to the true causal structure through an actual oracle, then the independencies and dependencies can be read off that structure using the d-separation criterion. But lacking such an oracle, one must somehow infer conditional independencies from observational data. The second difficulty with the algorithm is that it depends upon examining independencies between all pairs of variables given every subset of variables not containing the pair in question. But the number of such alternative subsets is exponential in the number of variables in the problem, making any direct implementation of this algorithm unworkable for large problems.

The causal discovery program TETRAD II copes with the first problem by applying a statistical significance test for conditional independence. For linear models, conditional independence $X \perp\!\!\!\perp Y | \mathbf{S}$ is represented by the zero partial correlation $\rho_{XY \cdot \mathbf{S}} = 0$ (also described as a **vanishing partial correlation**), that is, the correlation remaining between X and Y when the set $\mathbf{S}$ is held constant. The standard significance test on sample correlations is used to decide whether or not the partial correlation is equal to zero. In what Spirtes et al. (1993) call the **PC algorithm**, they combine this significance testing with a small trick to reduce the complexity of the search. The PC algorithm, because it is easy to understand and implement, has recently been taken up by two of the leading Bayesian network software tools, Hugin and Netica (see Appendix B).

Algorithm 8.2 *PC Algorithm*

1. *Begin with the fully connected skeleton model; i.e., every node is adjacent to every other node.*

2. *Set $k \leftarrow 0$; k identifies the order of the set of variables to be held fixed. For all pairs of nodes X and Y set $\mathbf{DSep}(\mathbf{X}, \mathbf{Y}) = \emptyset$; this will keep track of nodes which ought to d-separate the pair in the final graph.*

3. *For every adjacent pair of nodes X and Y, remove the arc between them if and only if for all subsets $\mathbf{S}$ of order k containing nodes adjacent to X (but not containing Y) the sample partial correlation $r_{XY \cdot S}$ is not significantly different from zero. (This corresponds to Principle I of the IC Algorithm.) Add the nodes in $\mathbf{S}$ to $\mathbf{DSep}(\mathbf{X}, \mathbf{Y})$.*

4. *If any arcs were removed, increment k and* goto *Step 3.*

5. *For each triple X,Y,Z in an undirected chain (such that X and Y are connected and Y and Z are connected, but not X and Z), replace the chain with $X \rightarrow Y \leftarrow Z$ if and only if $Y \notin \mathbf{DSep}(\mathbf{X}, \mathbf{Z})$. (This corresponds to Principle II of the IC Algorithm.)*

6. *Apply Step 3 of the IC Algorithm.*

The PC algorithm thus begins with a fully connected model, removing arcs whenever the removal can be justified on grounds of conditional independence. One computational trick is to keep track of d-separating nodes (and, therefore, non-d-separating nodes implicitly) during this removal process, for use in Step 5, avoiding the need to search for them a second time. A second trick for reducing complexity of the algorithm is just that, since the partial correlation tests are applied by fixing the values of sets $\mathbf{S}$ of adjacent nodes only and since arcs are removed early on during low-order tests, by the time the algorithm reaches larger orders there should be relatively few such sets to examine. Of course, this depends upon the connectivity of the true model giving rise to the data: dense models will support relatively fewer removals of arcs in Step 3, so the algorithm will be relatively complex. If in practice most models you work with are sparse, then you can reasonably hope for computational gains from the trick. This is a reasonable hope in general, since highly dense networks, being hard to interpret and use in principle, are less likely to be of interest in any case. Certainly the vast majority of networks actually published in the scientific research literature are very much less dense than, for example, the maximal density divided by two.

Compared with the metric learning algorithms, discussed in Chapter 9, the IC and PC Algorithms are very straightforward. Of course, the IC Algorithm is not really an algorithm, since oracles are not generally available. But statistical significance tests could be substituted directly. The PC Algorithm improves upon that option by finding some computational simplifications that speed things up in ordinary cases. But there remain computational difficulties. For one thing, if an arc is removed in error early in the search, the error is likely to cascade when looking at partial correlations of higher order. And as the order of the partial correlations increases, the number of sample correlations that need to be estimated increases dramatically, since every pair of variables involved has a correlation that must be estimated for the significance test. This will inevitably introduce errors in the discovery algorithm for moderately large networks. We should expect such an algorithm to work well on small models with large samples, but not so well on large models with moderately sized samples; and, indeed, we have found this to be the case empirically in Dai et al. (1997).

8.3.3 Causal discovery versus regression

Regression models aim to reduce the unexplained variation in dependent variables; ordinary least squares regression specifically parameterizes models by computing regression coefficients which minimize unexplained variance. Of course, since the application of Wright's Rule is numerically equivalent, path modeling does the same thing. What causal discovery does, on the other hand, is quite a different matter. Almost no statistician using regression models believes that every independent variable that helps reduce unexplained variation in the dependent variable is actually a relevant causal factor: it is a truism that whenever there is random variation in one variable its sample correlation with another variable subject to random variation will not be identically zero. Hence, the first can be used to reduce, however marginally, "unexplained" variation in the second. So, for a random example, almost certainly variations in seismic activity on Io are apparently correlated with variations in muggings in New York City. But, since it is useless to run around pointing out tiny correlations between evidently causally unrelated events, statisticians wish to rule out such variables.

Regression methods provide no principled way for ruling out such variables. Orthodox statistics claims that any correlation that survives a significance test is as good as any other. Of course, the correlation between Io's seismic events and muggings may not survive a significance test, so we would then be relieved of having to consider it further. However, the probability of a Type I error — getting a significant result when there is no true correlation — is typically set at five percent of cases; so examining any large number of variables will result in introducing spurious causal structure into regressions. This is called the problem of variable selection in statistics. Various heuristics have been invented for identifying spurious causal variables and throwing them out, including some looking for vanishing partial correlations. These variable selection methods, however, have been ad hoc rather than principled; thus, for example, they have not considered the possibility of accidentally inducing correlations by holding fixed common effect variables.

It is only with the concept of causal discovery, via conditional independence learning in particular, that the use of vanishing partial correlations has received any clear justification. Because of the relation between d-separation in causal graphs and conditional independence in probability distributions (or, vanishing partial correlations in linear Gaussian models), we can justify variable selection in causal discovery. In particular, the arc deletion/addition rules of the IC and PC algorithms are so justified. The metric causal learners addressed in Chapter 9 have even better claim to providing principled variable selection, as we shall see.

8.4 Summary

Reichenbach's Principle of the Common Cause suggests that conditional dependencies and independencies arise from causal structure and therefore that an inverse inference from observations of dependency to causality should be possible. In this chapter we have considered specifically the question of whether and how linear causal models can be inferred from observational data. Using the concept of conditional independence and its relation to causal structure, Sewall Wright was able to develop exact relations between parameters representing causal forces (path coefficients) and conditional independence (zero partial correlation, in the case of linear Gaussian models). This allows linear models to be parameterized from observational data. Verma and Pearl have further applied the relation between causality and conditional independence to develop the IC algorithm for discovering causal structure, which Spirtes et al. then implemented in a practical way in the PC algorithm. Thus, the skepticism that has sometimes greeted the idea of inferring causal from correlational structure is defeated pragmatically by an existence proof: a working algorithm exists. Although these methods are here presented in the context of linear models only, they are readily extensible to discrete models, as we shall see in Chapter 9.

8.5 Bibliographic notes

Wright's 1931 paper (Wright, 1934) remains rewarding, especially as an example of lucid applied mathematics. A simple and good introduction to path modeling can be found in Asher's *Causal modeling* (Asher, 1983). For a clear and simple introduction to ordinary least squares regression, correlation and identifying linear models see Edwards (Edwards, 1984). The PC algorithm, causal discovery program TETRAD II and their theoretical underpinnings are described in Spirtes et al. (2000). The current version of this program is TETRAD IV; details may be found in Appendix B.

8.6 Technical notes

Correlation

The correlation between two variables X and Y is written ρ_{XY} and describes the degree of linear relation between them. That is, the correlation identifies the degree to which values of one can be predicted by a linear function of changes in value of the other. For those interested, the relation between linear correlation and linear algebra is nicely set out by Michael Jordan (1986).

Sample correlation is a measurement on a sample providing an estimate of population correlation. Pearson's **product moment correlation coefficient** is the most commonly used statistic:

$$r_{XY} = \frac{S_{XY}}{S_X S_Y}$$

where

$$S_{XY} = \sum_{i=1,\dots n} \frac{(x_i - \overline{X})(y_i - \overline{Y})}{n}$$

is the estimate of covariance between X and Y in a sample of size n and

$$S_X = \sum_i \frac{(x_i - \overline{X})^2}{n-1}$$

$$S_Y = \sum_i \frac{(y_i - \overline{Y})^2}{n-1}$$

are estimates of standard deviation.

Partial correlation

Partial correlation measures the degree of linear association between two variables when the values of another variable (or set of variables) are held fixed. The partial correlation between X and Y when Z is held constant is written $\rho_{XY.Z}$. The "holding fixed" of Z here is *not* manipulating the value of Z to any value; the association between X and Y is measured while *observed* values of Z are not allowed to vary.

The sample partial correlation is computed from the sample correlations relating each pair of variables:

$$r_{XY.Z} = \frac{r_{XY} - r_{XZ}r_{YZ}}{\sqrt{1 - r_{XZ}^2}\sqrt{1 - r_{YZ}^2}}$$

This can be extended recursively to accommodate any number of variables being partialed out.

Significance tests for correlation

The PC algorithm (or IC algorithm, for that matter) can be implemented with significance tests on sample data reporting whether correlations and partial correlations are vanishing. The standard t test for the product moment correlation coefficient is:

$$t = \frac{r_{XY}\sqrt{n-2}}{\sqrt{1 - r_{XY}^2}}$$

with $n-2$ degrees of freedom. For partial correlation the t test is:

$$t = \frac{r_{XY.Z}\sqrt{n-3}}{\sqrt{1 - r_{XY.Z}^2}}$$

with $n-3$ degrees of freedom. This can be extended to larger sets of partialed variables in the obvious way.

Given a t value and the degrees of freedom, you can look up the result in a "t Table," which will tell you how probable the result is on the assumption that the correlation is vanishing. If that probability is less than some pre-selected value (called α), say 0.05, then orthodox testing theory says to reject the hypothesis that there is no correlation. The test is said to be "significant at the .05 level." In this case, the PC algorithm will accept that there is a real positive (or negative) correlation.

8.7   Problems

Problem 1

Consider the model:

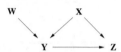

1. Identify all of the active paths between pairs of variables.
2. Use Wright's decomposition rule (in either form) to generate a system of equations for the correlations in terms of the path coefficients.
3. Solve the system of equations — i.e., convert it into a set of equations for the path coefficients in terms of the correlations.
4. Given the following correlation table, compute the path coefficients.

	W	X	Y	Z
W	1			
X	0	1		
Y	.4	.5	1	
Z	.12	.95	.7	1

5. How much of the variation of Z is unexplained? How much for Y?

Problem 2

Consider the model:

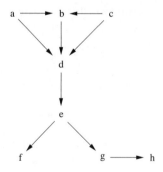

Find all the dags which are Markov equivalent. (You can use the IC algorithm to do this.)

Problem 3

If you have not already done so, complete the oracle from the problems in Chapter 2. Run your d-separation oracle on the additional test network for the IC learning example, `LearningEg.dne`, which can be found in

`http://www.csse.monash.edu.au/bai`.

Problem 4

Implement the IC algorithm using your d-separation oracle. That is, your IC learning algorithm should take as input a list of variables and then, using the oracle supplied with some Bayesian network, discover as much of the Bayesian network structure as possible. It should output the network structure in some convenient ascii format (**not** a graph layout, unless you happen to find that easy!). For example, you can generate an alphabetical list of nodes, with their adjacencies listed and the arc types indicated (parent, child or undirected).

Note that this algorithm is exponential in the number of variables. You will have to deal with this in some way to complete the assignment. You may want to implement some kind of heuristic for which subsets Z are worth looking at relative to a given pair X and Y. Or, you might simply implement a cutoff to prevent your program from examining any more than k subsets, looking at lower order subsets first. Whatever you decide to do, document it.

Run your algorithm on a set of test networks, including at least the IC Learning example, `LearningEg.dne`, `Cancer_Neapolitan.dne`, and `ALARM.dne`. Summarize the results of these experiments.

Problem 5

Instead of, or in addition to, the IC algorithm above, implement the PC algorithm, again using the oracle from Chapter 2 (rather than significance tests). Your PC learning algorithm should take as input a list of variables and then, using the oracle supplied with some Bayesian network, discover as much of the Bayesian network structure as possible. It should output the network structure in some convenient ascii format (**not** a graph layout, unless you happen to find that easy!). For example, you can generate an alphabetical list of nodes, with their adjacencies listed and the arc types indicated (parent, child or undirected).

Run your algorithm on a set of test networks, including at least the IC Learning example, `LearningEg.dne`, `Cancer_Neapolitan.dne`, and `ALARM.dne`. Summarize the results of these experiments. If you have also implemented the IC algorithm, compare the results of the IC with the PC Algorithm.

Problem 6

Reimplement your oracle-based algorithm, whether IC or PC, using significance testing of partial correlations instead of the oracle, with the input being joint samples across the variables rather than a known Bayesian net. Using a variety of α levels (say, 0.01, 0.05, 0.1), compare the results you get with your oracle-based version.

9

Learning Discrete Causal Structure

It is the mark of an educated mind to rest satisfied with the degree of precision which the nature of the subject admits and not to seek exactness where only an approximation is possible.

Aristotle

9.1 Introduction

In Chapter 6 we saw how to parameterize a discrete (multinomial) causal structure with conditional probability tables, regardless of how the structure might have been found. In the last chapter we saw how linear causal structure can be learned and how such structures can be parameterized. Here we will extend the picture of the machine learning of Bayesian networks by considering how to automate the learning of discrete causal structures. We will also look at how to apply expert prior knowledge to the learning process and at the learning of hybrid models of "local structure" — the relation between parents and their child node.

To be sure, we already have in hand a clear and plausible method for structure learning with discrete variables: the PC algorithm (Algorithm 8.2) of TETRAD is easily extended to cope with the discrete case. That algorithm implements the Verma-Pearl constraint-based approach to causal learning by discovering vanishing partial correlations in the data and using them to find direct dependencies (Step 1 of Verma-Pearl) and v-structure (Step 2). But instead of employing a test for partial correlations between continuous variables, we can substitute a χ^2 significance test (see §9.11) comparing $P(Y = y | X = x, Z = z)$ with the expected value if $X \perp\!\!\!\perp Y | Z$, namely $P(Y = y | Z = z)$ — where the $P(\cdot)$ values are actually estimates based on the sample. Thus, we can test directly for conditional independencies with discrete data. This is just what Spirtes et al. in fact did with TETRAD (Spirtes et al., 2000, p. 129).

However, what we will look at in this chapter is the learning of causal structure using **Bayesian metrics**, rather than orthodox statistical tests. In other words, the algorithms here will search the causal model space $\{h_i\}$, with some metric like $P(\cdot | e)$ in hand — or, some approximation to that conditional probability function — aiming to select an h_i that maximizes the function. So, there are two computationally difficult tasks these learners need to perform. First, they need to compute their metric, scoring individual hypotheses. This scoring procedure may itself be computationally intractable. Second, they need to search the space of causal structures, which, as we saw in §8.2.3, is exponential. Most of the search methods applied have been variants of greedy search, and we will spend little time discussing them. We describe a more interesting stochastic search used with the MML metric in more detail, partly because the search process itself results in an important simplification to the MML metric. The resulting MML metric is sufficiently flexible to incorporate prior expertise expressed in a wide variety of forms, as well as a range of representations of local structure.

We will next look at some of the foundational problems that have been raised by various skeptics and some of the means available for avoiding, evading or ammeliorating the difficulties. We will also look at Markov blanket discovery, which is becoming popular as a way of applying causal discovery to problems of high dimensionality (i.e., with large numbers of variables). We will then conclude with a consideration of the problems specific to evaluating causal discovery algorithms and options for improving common evaluation practice.

9.2 Cooper and Herskovits's K2

The first significant attempt at a Bayesian approach to learning discrete Bayesian networks without topological restrictions was made by Cooper and Herskovits in 1991 (Cooper and Herskovits, 1991). Their approach is to compute the metric for individual hypotheses, $P(h_i|e)$, by brute force, by turning its computation into a combinatorial counting problem. This led to their causal discovery program, K2. Because the counting required is largely the counting of possible instantiations of variables and parent sets, the technique is intrinsically restricted to discrete networks. Other restrictions will become apparent as we develop the method.

Since our goal is to find that h_i which maximizes $P(h_i|e)$, we can satisfy this by maximizing $P(h_i, e)$, as we can see from Bayes' theorem:

$$
\begin{aligned}
P(h_i|e) &= \frac{P(e|h_i)P(h_i)}{P(e)} \\
&= \frac{P(h_i, e)}{P(e)} \\
&= \beta P(h_i, e)
\end{aligned}
$$

where β is a (positive) normalizing constant.

To get the combinatorics (i.e., counting) to work we need some simplifying assumptions. Cooper and Herkovits start with these fairly unsurprising assumptions:

1. The data are joint samples and all variables are discrete. So:

$$
P(h_i, e) = \int_{\theta} P(e|h_i, \theta) f(\theta|h_i) P(h_i) d\theta \tag{9.1}
$$

where θ is the parameter vector (e.g., conditional probabilities) and $f(\cdot|h_i)$ is a prior density over parameters conditional upon the causal structure.

2. Samples are independently and identically distributed (i.i.d.). That is, for K sample cases, and breaking down the evidence e into its K components e_k,

$$P(e|h_i, \theta) = \prod_{k=1}^{K} P(e_k|h_i, \theta)$$

Hence, by substitution into (9.1)

$$P(h_i, e) = P(h_i) \int_{\theta} \prod_{k=1}^{K} P(e_k|h_i, \theta) f(\theta|h_i) d\theta \qquad (9.2)$$

Cooper and Herkovits next make somewhat more problematic simplifications, which are nevertheless needed for the counting process to work.

3. The data contain no missing values. If, in fact, they do contain missing values, then they need to be filled in, perhaps by the Gibbs sampling procedure from Chapter 6.

4. For each variable X_k in h_i and for each instantiation of its parents $\pi(X_k)$, $P(X_k = x|h_i, \theta, \pi(X_k))$ is uniformly distributed over possible values $X_k = x$.[1]

5. Assume the uniform prior over the causal model space; i.e., $P(h_i) = \frac{1}{|\{h_i\}|}$.

These last two assumptions are certainly disputable. Assumption 4 does not allow causal factors to be *additive*, let alone interactive! Despite that, it might be reasonable to hope that even this false assumption may not throw the causal structure search too far in the wrong direction: the qualitative fact of dependency between parent and child variables is likely insensitive to the precise quantitative relation between them. And if the qualitative causal structure can be discovered, the quantitative details representing such interactions — the parameters — can be learned as in Chapter 6. Of course, if relevant prior knowledge is available, this uniformity assumption can be readily dismissed by employing non-uniform Dirchlet priors over the parameter space, as we also discussed there.

Something similar might be said of the fifth assumption, that the prior over causal models is uniform: that crude though it may be, it is not likely to throw the search so far off that the best causal models are ultimately missed. However, we shall below introduce a specific reason to be skeptical of this last assumption. In any case, it is an assumption easily dismissed should one have a more useful prior over the hypothesis space.

We now have the necessary ingredients for Cooper and Herskovits' main result.

---

[1]Note that $\pi(X)$ in this chapter refers to *Parents(X)*. We use it here to shorten equations; it has nothing to do with the message passing of Part I.

Theorem 9.1 *(Cooper and Herkovits, 1991)*

Under the assumptions above, the joint probability is:

$$P_{CH}(h_i, e) = P(h_i) \prod_{k=1}^{N} \prod_{j=1}^{|\Phi_k|} \frac{(s_k - 1)!}{(S_{kj} + s_k - 1)!} \prod_{l=1}^{s_k} \alpha_{kjl}! \tag{9.3}$$

Where

- N is the number of variables.
- $|\Phi_k|$ is the number of assignments possible to $\pi(X_k)$.
- s_k is the number of assignments possible to X_k.
- α_{kjl} is the number of cases in sample where X_k takes its l-th value and $\pi(X_k)$ takes its j-th value.
- S_{kj} is the number of cases in the sample where $\pi(X_k)$ takes its j-th value (i.e., $\sum_{l=1}^{s_k} \alpha_{kjl}$).

Each of these values is obtained as the result of some counting process. The point is that, with this theorem, computing $P(h_i, e)$ has become a straightforward counting problem, and is equal to $P(h_i)$ times a simple function of the number of assignments to parent and child variables and the number of matching cases in the sample. Furthermore, Cooper and Herskovits showed that this computation of $P_{CH}(h_i, e)$ is polynomial, i.e., computing $P_{CH}(h_i, e)$ given a particular h_i is tractable under the assumptions so far.

Unfortunately, while the metric may be tractable, we still have to *search* the space $\{h_i\}$, which we know is exponentially large. At this point, Cooper and Herskovits go for a final simplifying assumption:

6. Assume we know the temporal ordering of the variables.

If we rely on this assumption, the search space is greatly reduced. In fact, for any pair of variables either they are connected by an arc or they are not. Given prior knowledge of the ordering, we need no longer worry about arc orientation, as that is fixed. Hence, the model space is determined by the number of pairs of variables: two raised to that power being the number of possible skeleton models. That is, the new hypothesis space has size only $2^{C_2^N} = 2^{(N^2-N)/2}$. The K2 algorithm simply performs a greedy search through this reduced space. This reduced space remains, of course, exponential.

In any case, so far we have in hand two algorithms for discovering discrete causal structure: TETRAD (i.e., PC with a χ^2 test) and K2.

9.2.1 Learning variable order

Our view is that reliance upon the variable order being provided is a significant drawback to K2, as it is to many other algorithms we will not have time to examine in detail (e.g., Buntine, 1991, Bouckaert, 1994, Suzuki, 1996, Madigan and Raftery,

1994).[2] Why should we care? It is certainly the case that in many problems we have either a partial or a total ordering of variables in pre-existing background knowledge, and it would be foolish not to use all available information to aid causal discovery. Both TETRAD II and CaMML, for example, allow such prior information to be used to boost the discovery process (see §9.6.3). But it is one thing to *allow* such information to be used and quite another to *depend upon* that information. This is a particularly odd restriction in the domain of causal discovery, where it is plain from Chapter 8 that a fair amount of information about causal ordering can be learned directly from the data, using the Verma-Pearl IC algorithm.

In principle, what artificial intelligence is after is the development of an agent which has some hope of overcoming problems on its own, rather than requiring engineers and domain experts to hold its hand constantly. If *intelligence* is going to be engineered, this is simply a requirement. One of the serious impediments to the success of first-generation expert systems in the 1970s and 80s was that they were brittle: when the domain changed, or the problem focus changed to include anything new, the systems would break. They had little or no ability to adapt to changing circumstances, that is, to learn. The same is likely to be true of any Bayesian expert system which requires human experts continually to assist it by informing it of what total ordering it should be considering. It is probably fair to say that learning the variable ordering is half the problem of causal discovery: if we already know A comes before B, the only remaining issue is whether there is a direct dependency between the two — so, half of the Verma-Pearl algorithm in constraint-based approach becomes otiose, namely Principle II.

Despite the objection in principle, in practice the requirement for a fixed total ordering prior to operating K2, or other such limited algorithms, can be met by using the simple, polynomial time maximum weighted spanning tree algorithm of Chow and Liu (1968) to generate a total ordering based upon the mutual information between pairs of variables. Since ordering information is embedded within the probabilistic dependency information, however, it is a reasonable suspicion that separating the two tasks — learning variable order and learning causal structure — is inherently suboptimal.

9.3 MDL causal discovery

Minimum Description Length (MDL) inference was invented by Jorma Rissanen (Rissanen, 1978), based upon the Minimum Message Length (MML) approach invented by Wallace and Boulton (1968). Both techniques are inspired by information theory and were anticipated by Ray Solomonoff in interesting early work on information-theoretic induction (Solomonoff, 1964). All of this work is closely re-

---

[2]We should point out that there are again many others in addition to our CaMML which do not depend upon a prior variable ordering, such as TETRAD, MDL (§9.3), GES (Meek and Chickering, 2003).

lated to foundational work on complexity theory, randomness and the interpretation of probability (see von Mises, 1957, Kolmogorov, 1965, Chaitin, 1966).

The basic idea behind both MDL and MML is to play a tradeoff between model simplicity and fit to the data by minimizing the length of a *joint* description of the model and the data assuming the model is correct. Thus, if a model is allowed to grow arbitrarily complex, and if it has sufficient representational power (e.g., sufficiently many parameters), then eventually it will be able to record directly all the evidence that has been gathered. In that case, the part of the message communicating the data given the model will be of length zero, but the first part communicating the model itself will be quite long. Similarly, one can communicate the simplest possible model in a very short first part, but then the equation will be balanced by the necessity of detailing *every* aspect of the data in the second part of the message, since none of that will be implied by the model itself. Minimum encoding inference seeks a golden mean between these two extremes, where any extra complexity in the optimal model is justified by savings in inferring the data from the model.

In principle, minimum encoding inference is inspired by Claude Shannon's measure of information (see Figure 9.1).

Definition 9.1 Shannon information measure

$$I(m) = -\log P(m)$$

Applied to joint messages of hypothesis and evidence:

$$I(h,e) \quad = \quad -\log P(h,e) \tag{9.4}$$

Shannon's concept was inspired by the hunt for an efficient code for telecommunications; his goal, that is, was to find a code which maximized use of a telecommunications channel by minimizing expected message length. If we have a coding scheme which satisfies Shannon's definition 9.1, then we have what is called an **efficient code**. Since efficient codes yield probability distributions (multiply by -1 and exponentiate), efficiency requires observance of the probability axioms. Indeed, in consequence we can derive the optimality of minimizing the two-part message length from Bayes' Theorem:

$$
\begin{aligned}
P(h,e) &= P(h)P(e|h) \\
-\log P(h,e) &= -\log[P(h)P(e|h)] \\
-\log P(h,e) &= -\log P(h) - \log P(e|h) \\
I(h,e) &= I(h) + I(e|h)
\end{aligned}
$$

A further consequence is that an efficient code cannot encode the same hypothesis in two different lengths, since that would imply two distinct probabilities for the very same hypothesis.

Minimum encoding inference metrics thus can provide an estimate of the joint probability $P(h,e)$. Since at least one plausible goal of causal discovery is to find that hypothesis which maximizes the conditional probability $P(h|e)$, such a metric

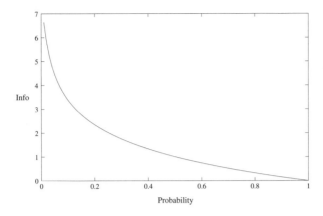

FIGURE 9.1: Shannon's information measure.

suffices, since maximizing $P(h,e)$ is equivalent to maximizing $P(h|e)$. It is worth noting that in order to compute such a metric we need to compute how long the joint message of h together with $e|h$ *would be* were we to build it. It is not actually necessary to build the message itself, so long as we can determine how long it would be without building it.

9.3.1   Lam and Bacchus's MDL code for causal models

The differences between MDL and MML are largely ideological: MDL is offered specifically as a non-Bayesian inference method, which eschews the probabilistic interpretation of its code, whereas MML specifically is offered as a Bayesian technique. As MDL suffers by the lack of foundational support, its justification lies entirely in its ability to produce the goods, that is, in any empirical support its methods may find.

An MDL encoding of causal models was put forward by (Lam and Bacchus, 1993). Their code length for the causal model (structure plus parameters) is:

$$I_{LB}(h) = \sum_{i=1}^{N} \left(k_i \log N + d(s_i - 1) \prod_{j \in \pi(i)} s_j \right) \quad (9.5)$$

where

- N is the number of nodes,
- k_i is the number of parents of the i-th node,
- d is a fixed length precision for all parameters,[3]
- s_i is the number of states of the i-th node.

---

[3] Presumably, this was the word size of the computer being used in bits.

The explanation of this code length is that to communicate the causal model we must specify in the first instance each node's parents. Since there are N nodes this will take $\log N$ bits (using base 2 logs) for each such parent. For each distinct instantiation of the parents (there are $\prod_{j \in \pi(i)} s_j$ such instantiations, which is equal to $|\Phi_i|$ in Cooper and Herskovits' notation), we need to specify a probability distribution over the child node's states. This requires $s_i - 1$ parameters (since, by Total Probability, the last such value can be deduced). Hence, the CPT for each parent instantiation requires $d(s_i - 1)$ bits to specify.

Before going any further, it is worth noting that this falls well short of being an efficient code. It presumes that for each node both sides of the communication knows how many parents the node has. Further, in identifying those parents, it is entirely ignored that the model must be acyclic, so that, for example, the node cannot be a parent of itself. Again, once one parent has been identified, a second parent cannot be identical with it; the code length ignores that as well. Thus, the Lam and Bacchus code length computed for the causal structure is only a loose upper bound, rather than the basis for a determinate probability function.

The code length for parameters is also inefficient, and the problem there touches on another difference between MDL and MML. It is a principle of the MML procedure that the precision with which parameters are estimated be a function of the amount of information available within the data to estimate them. Where data relevant to a parameter are extensive, it will repay encoding the parameter to great precision, since then those data will themselves be encoded in the second part of the message more compactly. Where the data are weak, it rewards us to encode the parameter only vaguely, reserving our efforts for encoding the few data required for the second part. The MDL code of Lam and Bacchus eschews such considerations, making an arbitrary decision on precision. This practice, and the shortcuts taken above on causal structure encoding, might be defended on the grounds of practicality: it is certainly easier to develop codes which are not precisely efficient, and it may well be that the empirical results do not suffer greatly in consequence.

Now let us consider the length of the second part of the message, encoding the data, is given as

$$I_{LB}(e|h) = K \left(\sum_{i=1}^{N} H(X_i) - \sum_{i=1}^{N} H(X_i, \pi(i)) \right) \tag{9.6}$$

where
- K is the number of joint observations in the data,
- $H(X)$ is the entropy of variable X,
- $H(X_i, \pi(i))$ is the mutual information between X_i and its parent set.

Definition 9.2 Entropy

$$H(X) = - \sum_x P(X = x) \log P(X = x)$$

Definition 9.3 Mutual information

$$H(X_i, \pi(i)) = H(X_i) - H(X_i | \pi(i)) = \sum_{x_i, \phi(i)} P(x_i, \phi(i)) \log \frac{P(x_i, \phi(i))}{P(x_i)P(\phi(i))}$$

where $\phi(i)$ ranges over the distinct instantiations of the parent set $\pi(i)$. Intuitively, this mutual information reports the expected degree to which the joint probability of child and parent values diverges from what it would be were the child independent of its parents. I_{LB}, therefore, measures how much entropy in each variable is expected *not* to be accounted for by its parents and multiplies this by the number of data items that need to be reported. Clearly, this term by itself maximizes likelihood, which is as we would expect, being counterbalanced by the complexity penalizing term (9.5) for the final metric:

$$I_{LB}(h,e) = I_{LB}(h) + I_{LB}(e|h) \tag{9.7}$$

Lam and Bacchus need to apply this metric in a search algorithm, of course:

Algorithm 9.1 *MDL causal discovery*

1. *Initial constraints are taken from a domain expert. This includes a partial variable order and whatever direct connections might be known.*

2. *Greedy search is then applied: every possible arc addition is tested; the one with the best MDL measure is accepted. If none results in an improvement in the MDL measure, then the algorithm stops with the current model. Note that no arcs are deleted — the search is always from less to more complex networks.*

3. *Arcs are then checked individually for an improved MDL measure via arc reversal.*

4. *Loop at Step 2.*

The results achieved by this algorithm were similar to those of K2, but without requiring a full variable ordering. Results without significant initial information from humans (i.e., starting from an empty partial ordering and no arcs) were not good, however.

9.3.2 Suzuki's MDL code for causal discovery

Joe Suzuki, in an alternative MDL implementation (Suzuki, 1996), also points out that Lam and Bacchus's parameter encoding is not a function of the size of the data set. This implies that the code length cannot be efficient.[4] Suzuki proposes as a different MDL code for causal structures:

$$I_S(h) = \frac{\log K}{2} \sum_{i=1}^{N} \left((s_i - 1) \prod_{j \in \pi(i)} s_j \right) \tag{9.8}$$

which can be added to $I_{LB}(e|h)$ to get a total MDL score for the causal model. This gives a better account of the complexity of the parameter encoding, but entirely does away with the structural complexity of the Bayesian network! To be sure, structural complexity is reflected in the number of parameters. Complexity in Bayesian networks corresponds to the density of connectivity and the arity of parent sets, both

---

[4] Suzuki further points out that this means the Lam and Bacchus code violates the Kraft inequality.

of which increase the number of parameters required for CPTs. Nevertheless, that is by no means the whole story to network complexity, as there is complexity already in some topologies over others entirely independent of parameterization, as we shall see below in §9.5.1.1.

9.4   Metric pattern discovery

From Chapter 8 we know that DAGs within a single pattern are Markov equivalent. That is, two models sharing the same skeleton and v-structures can be parameterized to identical maximum likelihoods and so apparently cannot be distinguished given only observational data. This has suggested to many that the real object of metric-based causal discovery should be patterns rather than the DAGs being coded for by Lam and Bacchus. This is a possible interpretation of Suzuki's approach of ignoring the DAG structure entirely. And some pursuing Bayesian metrics have also adopted this idea (e.g., Anderson et al., 1997, Chickering, 1996). The result is a form of metric discovery that explicitly mimics the constraint-based approach, in that both aim strictly at causal patterns.

The most common approach to finding an **uninformed prior** probability over the model space — that is, a general-purpose probability reflecting no prior domain knowledge — is to use a uniform distribution over the models. This is exactly what Cooper and Herskovits did, in fact, when assigning the prior $1/|\{h_i\}|$ to all DAG structures. Pattern discovery assigns a uniform prior over the Markov equivalence classes of DAGs (patterns), rather than the DAGs themselves. The reason adopted is that if there is no empirical basis for distinguishing between the constituent DAGs, then empirical considerations begin with the patterns themselves. So, considerations *before* the empirical — prior considerations — should not bias matters; we should let the empirical data decide what pattern is best. And the only way to do that is to assign a uniform prior over patterns.

We believe that such reasoning is in error. The goal of causal discovery is to find the best causal model, not the best pattern. It is a general and well-accepted principle in Bayesian inference that our prior expectations about the world should not be limited by what our measuring apparatus is capable of recording (see, e.g., Lindley, 1972). For example, if we thought that all four possible outcomes of a double toss of a coin were equally likely, we would hardly revise our opinion simply because we were told in advance that the results would be reported to us only as "Two heads" or "Not two heads." But something like that is being proposed here: since our measuring apparatus (observational data) cannot distinguish between DAGs within a pattern, we should consider only the patterns themselves.

The relevant point, however, is that the distinct DAGs within a pattern have quite distinct implications about causal structure. Thus, all of

- $A \leftarrow B \rightarrow C$
- $A \rightarrow B \rightarrow C$
- $A \leftarrow B \leftarrow C$

are in the same pattern, but they have entirely different implications about the effect of causal interventions. Even if we are restricted to observational data, we are not *in principle* restricted to observational data. The limitation, if it exists, is very like the restriction to a two-state description of the outcome of a double coin toss: it is one that can be removed. To put it another way: the pattern of three DAGs above can be realized in at least those three different ways; the alternative pattern of $A \rightarrow B \leftarrow C$ can be realized only by that single DAG. Indeed, the number of DAGs making up a pattern over N variables ranges from a pattern with one DAG (patterns with no arcs) to a pattern with $N!$ DAGs (fully connected patterns). If we have no prior reason to prefer one DAG over another in representing causal structure, then *by logical necessity* we do have a prior reason to prefer patterns with more DAGs over those with fewer.

Those advocating pattern discovery need to explain why one DAG is superior to another and so should receive such a hugely greater prior probability. We, in fact, shall find prior reasons to prefer some DAGs over others (§9.5.1.1), but not in a way supporting a uniform distribution over patterns.

9.5 CaMML: Causal discovery via MML

We shall now present our own method for automated causal discovery, implemented in the program CaMML, which was developed in large measure by Chris Wallace, the inventor of MML (Wallace and Boulton, 1968).[5] First we present the MML metric for linear models. Although we dealt with linear models in Chapter 8, motivating constraint-based learning with them, we find it easier to introduce the causal MML codes with linear models initially. In the next section we present a stochastic sampling approach to search for MML, in much more detail than any other search algorithm. We do so because the search is interestingly different from other causal search algorithms, but also because the search affords an important simplification to the MML causal structure code. Finally, only at the end of that section, we present the MML code for discrete models.

Most of what was said in introducing MDL applies directly to MML. In particular, both methods perform model selection by trading off model complexity with fit to the data. The main conceptual difference is that MML is explicitly Bayesian and it takes the prior distribution implied by the MML code seriously, as a genuine probability distribution.

---

[5]In this and the next section we provide an introductory, simplified account of CaMML. A more complete description of its design and operation can be found in O'Donnell (2010).

One implication of this is that if there is a given prior probability distribution for the problem at hand, then the correct MML procedure is to encode that prior probability distribution directly, for example, with a Huffman code (see Cover and Thomas, 2006, Chapter 5, for an introduction to coding theory). The standard practice to be seen in the literature is for an MML (or, MDL) paper to start off with a code that makes sense for communicating messages between two parties who are both quite ignorant of what messages are likely. The simplest messages (the simplest models) are given short codes; messages communicating more complex models are typically composites of such shorter messages. That standard practice makes perfect sense (and, we shall follow it here) — *unless* a specific prior distribution is available, which implies something less than total prior ignorance about the problem. In such a case the relation between code length and probability in (9.4) requires the prior information to be used in the code, fixing a prior distribution over the model space. In consequence, MML is a genuine Bayesian method of inference.

9.5.1   An MML code for causal structures

We shall begin by presenting an MML code for causal DAG structures. We assume in developing this code no more than was assumed for the MDL code above (indeed, somewhat less). In particular, we assume that the number of variables is known. The MML metric for the causal structure h has to be an efficient code for a DAG with N variables. h can be communicated by:

- First, specify a total ordering. This takes $\log N!$ bits.
- Next, specify all arcs between pairs of variables. If we assume an arc density = 0.5 (i.e., for a random pair of variables a 50:50 chance of an arc),[6] we need one bit per pair, for the total number of bits:

$$\binom{N}{2} = \frac{N(N-1)}{2}$$

- However, this allows multiple ways to specify the DAG if there is more than one total ordering consistent with it (which are known as **linear extensions** of the DAG). Hence, this code is inefficient. Rather than correct the code, we can simply compensate by reducing the estimated code length by $\log M$, where M is the number of linear extensions of the DAG. (Remember: we are not actually in the business of *communication*, but only of computing how long efficient communications would be!)

Hence,

$$I_{MML}(h) = \log N! + \frac{N(N-1)}{2} - \log M \tag{9.9}$$

---

[6]We shall remove this assumption below in §9.6.3.

9.5.1.1   Totally ordered models (TOMs)

Before continuing with the MML code, we consider somewhat further the question of linear extensions. MML is, on principle, constrained by the Shannon efficiency requirement to count linear extensions and reduce the code length by the redundant amount. This is equivalent to increasing the prior probability for DAGs with a greater number of linear extensions over an otherwise similar DAG with fewer linear extensions. Since these will often be DAGs within the same Markov equivalence class (pattern), this is a point of some interest. Consider the two DAGs of Figure 9.2: the chain has only one linear extension — the total ordering $< A, B, C >$ — while the common cause structure has two — namely, $< B, A, C >$ and $< B, C, A >$. And both DAGs are Markov equivalent.

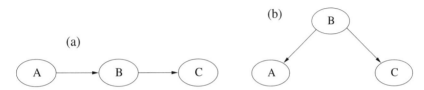

FIGURE 9.2: A Markov equivalent: (a) chain; (b) common cause.

The most common approach by those pursuing Bayesian metrics for causal discovery thus far has been to assume that DAGs within Markov equivalence classes are inherently indistinguishable, and to apply a uniform prior probability over all of them; see, for example, Madigan et al. (1996). Again, according to Heckerman and Geiger (1995), equal scores for Markov equivalent causal structures will often be appropriate, even though the different DAG structures *can* be distinguished under the causal interpretation.

Note that we are not here referring to a uniform prior over patterns, which we considered in §9.4, but a uniform prior *within* patterns. Nonetheless, the considerations turn out to be analogous. The kind of indistinguishability that has been justified for causal structures within a single Markov equivalence class is *observational* indistinguishability. The tendency to interpret this as *in-principle* indistinguishability is not justified. After all, the distinguishability under the causal interpretation is clear: a causal intervention on *A* in Figure 9.2 will influence *C* in the chain but not in the common causal structure. Even if we are limited to observational data, the differences between the chain and the common cause structure will become manifest if we expand the scope of our observations. For example, if we include a new parent of *A*, a new v-structure will be introduced only if *A* is participating in the common causal structure, resulting in the augmented DAGs falling into distinct Markov equivalence classes.

We can develop our reasoning to treat linear extensions. A **totally ordered model**, or **TOM**, is a DAG together with one of its linear extensions. It is a plausible view of causal structures that they are, at bottom, TOMs. The DAGs represent causal processes (chains) linking together events which take place, in any given instance, at particular times, or during particular time intervals. All of the events are ordered by time. When we adopt a DAG, without a total ordering, to represent a causal process, we are representing our ignorance about the underlying causal story by allowing multiple, consistent TOMs to be entertained. Our ignorance is not in-principle ignorance: as our causal understanding grows, new variables will be identified and placed within it. It is entirely possible that in the end we shall be left with only one possible linear extension for our original problem. Hence, the more TOMs (linear extensions) that are compatible with the original DAG we consider, the more possible ways there are for the DAG to be realized and, thus, the greater the prior probability of its being true.

In short, not only is it correct MML coding practice to adjust for the number of linear extensions in estimating a causal model's code length, it is also the correct Bayesian interpretation of causal inference.

This subsection might well appear to be an unimportant aside to the reader, especially in view of the fact that counting linear extensions is exponential in practice (Brightwell and Winkler, 1990). In consequence, the MML code presented so far does not directly translate into a tractable algorithm for scoring causal models. Indeed, its direct implementation in a greedy search was never applied to problems with more than ten variables for that reason (Wallace et al., 1996). However, TOMs figure directly in the sampling solution of the search and MML scoring problem, in section §9.6 below.[7]

9.5.2   An MML metric for linear models

So, we return to developing the MML metric. We have seen the MML metric for causal structure; now we need to extend it to parameters given structure and to data given both parameters and structure. We do this now for linear models; we develop the MML metric for discrete models only after describing CaMML's stochastic search in §9.6, since the discrete metric takes advantage of an aspect of the search.

The parameter code. Following Wallace and Freeman (1987), the message length for encoding the linear parameters is

$$I_{MML}(\theta|h) = \sum_{X_j} -\log \frac{f(\theta_j|h)}{\sqrt{F(\theta_j)}} \tag{9.10}$$

$f(\theta_j|h)$ is the prior density function over the parameters for X_j given h. $F(\theta_j)$ is the Fisher information for the parameters to X_j. It is the Fisher information which controls the precision of the parameter estimation. In effect, it is being used to discretize the parameter space for θ_j, with cells being smaller (parameters being more precisely estimated) when the information afforded by the data is larger. (For details see Wallace and Freeman, 1987.)

---

[7]For an interesting further defence of the use of a uniform prior over TOMs rather than DAGs, see O'Donnell (2010, §4.2.3). Across 29 Bayesian networks in widespread use in the empirical literature (and selected for their availability), he shows that the uniform TOM prior biases the search in favor of the true model relative to the uniform DAG prior.

Wallace et al. (1996) describe an approximation to (9.10), using the standard assumptions of parameter independence and the prior density $f(\theta_j|h)$ being the normal $N(0, \sigma_j)$, with the prior for σ_j being proportional to $1/\sigma_j$.

The data code. The code for the sample values at variable X_j given both the parameters and DAG is, for i.i.d. samples,

$$I_{MML}(x_j|h, \theta_j) = -\log P(x_j|h, \theta_j) = -\log \prod_{k=1}^{K} \frac{1}{\sigma_j \sqrt{2\pi}} e^{-\delta_{jk}^2/2\sigma_j^2} \qquad (9.11)$$

where K is the number of joint observations in the sample and δ_{jk} is the difference between the observed value of X_j and its linear prediction. Note that this is just a coding version of the Normal density function, which is the standard modeling assumption for representing unpredicted variation in a linear dependent variable.

The MML linear causal model code. Combining all of the pieces, we have the total MML score for a linear causal model being:

$$I_{MML}(h, e) = I_{MML}(h) + I_{MML}(\theta|h) + \sum_{j=1}^{N} I_{MML}(x_j|h, \theta_j) \qquad (9.12)$$

9.6 CaMML stochastic search

Having the MML metric for linear causal models in hand, we can reconsider the question of how to search the space of causal models. Rather than apply a greedy search, or some variant such as beam search, we have implemented two stochastic searches: a genetic algorithm search (Neil and Korb, 1999) and a Metropolis sampling process (Wallace and Korb, 1999).

9.6.1 Genetic algorithm (GA) search

Genetic algorithms have been applied successfully to causal discovery search using metrics from K2 (Larrañaga et al., 1996), MDL (Wong et al., 1999) and MML, in CaMML. The CaMML genetic algorithm searches the DAG space; that is, the "chromosomes" in the population are the causal model DAGs themselves. The algorithm simulates some of the aspects of evolution, tending to produce better DAGs over time — i.e., those with a lower MML score. The best DAGs, as reported by the MML metric, are allowed to reproduce. Offspring for the next generation are produced by swapping a subgraph from one parent into the graph of the other parent and then applying mutations in the form of arc deletions, additions and reversals. A minimal repair is done to correct any introduced cycles or reconnect any dangling arcs. Genetic algorithms frequently suffer from getting trapped in local minima, and ours was no exception. In order to avoid this, we introduced a temperature parameter (as in simulated annealing) to encourage a more wide-ranging search early in the process. None of this addresses the problem of counting linear extensions for the

MML metric itself. We handled that by an approximate estimation technique based upon that of Karzanov and Khachiyan (1990). For more details about our genetic algorithm see Neil and Korb (1999). Although the GA search performs well, here we shall concentrate upon the Metropolis sampling approach, as it offers a more telling solution to the problem of counting linear extensions.

9.6.2 Metropolis search

Metropolis sampling allows us to approximate the posterior probability distribution over the space of TOMs by a sampling process.[8] From that process we can estimate probabilities for DAGs. The process works by the probability measure providing a kind of pressure on the sampling process, ensuring that over the long run models are visited with a frequency approximating their probabilities (Metropolis et al., 1953, Mackay, 1998). This is a type of **Markov Chain Monte Carlo** (MCMC) search, meaning that the probabilistically selected sequence of sampled models forms a Markov Chain (where the probability of reaching any particular model depends only upon what model is currently being sampled and not on the sampling history). The probabilities for our CaMML implementation are, of course, derived from the MML metric. Since the models being sampled are TOMs, rather than DAGs, the MML metric at each step is computed for TOMs only. Hence there is no (direct) problem with counting linear extensions.[9]

Since we are searching in the space of TOMs, we need an adjustment of $I_{MML}(h)$ (9.9) to reflect the change. This is, very conveniently, simply to drop the term requiring us to count linear extensions. Hence,

$$I_{MML}(h) = \log N! + \frac{N(N-1)}{2} \tag{9.13}$$

with h now ranging over TOMs rather than DAGs.

Algorithm 9.2 *CaMML Metropolis Algorithm*

1. *First, an initial TOM is selected at random. Call this M.[10]*

2. *Randomly choose between the following possible changes to the current model M producing M′:*

   (a) **Temporal order change:** *Swap the order of two neighboring nodes. (If there is an arc between them, it reverses direction.)*

---

[8]Incidentally, a number of other causal discovery algorithms also search the TOM space, although they derive no special benefit from that in computing a metric. E.g., Larrañaga et al. (1996), de Campos et al. (2000), Teyssier and Koller (2005).

[9]Note that this implies that the MML metric derived below in (9.16) does *not* apply to DAGs. A direct application of it to DAGs would once again require counting the number of linear extensions, so as to adjust the code length in the manner above.

[10]To make the search more effective, in practice this step is replaced by a simulated annealing search for a good initial TOM.

(b) **Skeletal change:** *Add (or delete) an arc between two randomly selected nodes.*

(c) **Double skeletal change:** *Randomly choose three nodes. Add (or delete) arcs between the first two nodes and the last (in the given temporal order). (Double changes are not necessary for the Metropolis algorithm to work, but are included to accelerate the sampling process.)*

3. *Accept M' as the new M if* $\frac{P_{MML}(M')}{P_{MML}(M)} > U[0,1]$*; otherwise retain M*

   where

   $$P_{MML}(x) = e^{-I_{MML}(x)}$$

   and $U[0,1]$ is a uniform random variate in the interval $[0,1]$.

4. *Update counts associated with the current visit to M.*

5. *Loop at 2 (until a set number of steps has been completed).*

This Monte Carlo process meets the sufficient conditions for the Metropolis algorithm to apply, viz., the number of possible transitions from each TOM is constant, and every transition is reversible (Metropolis et al., 1953). The process will therefore visit every TOM with a frequency proportional to its joint probability with the data, and hence proportional to its MML posterior.

We are not directly interested in the posterior probabilities of individual TOMs. Sampled TOMs are counted as visits to their corresponding DAGs, instead. DAGs, in turn, are members of patterns, and visits to these patterns are counted as well. These counts are maintained in two hash tables. DAG hash keys are constructed using N^2 64-bit random integers, with a view to keeping the probability of two distinct TOMs getting the same key extremely low.

The counts for DAGs are actually accumulated for "clean" representative DAGs, rather than all DAGs. What this means is that an arc whose presence is not sufficiently supported by the data is removed, so that visits to a neighborhood of a clean DAG consisting of DAGs with spurious arcs will increase the sample weight for their clean DAG representative. The choice of whether to remove an arc is based upon a greedy search for arcs to delete: if an arc deletion leads to an improved MML score, it is deleted. This cleaning process only affects counting and not the sampling procedure; that is, the mutation process in the Metropolis search continues with the unclean TOM. Of course, since counts are kept only for (clean) DAGs, and since these counts are based upon a search through the TOM space, there is never any need to count the number of TOMs contributing to any DAG count; counting linear extensions is completely unnecessary. It's also worth noting that since counts are accumulated only for clean DAGs, pattern counts are likewise limited to those patterns which contain those clean DAGs — i.e., unclean patterns are simply ignored.

After the sampling process is complete, the highest posterior (clean) DAGs are reported to the user. A further MML grouping over patterns is conducted and the highest posterior MML pattern groups are also reported the user. The grouping of patterns is performed in a way similar to the cleaning of DAGs. That is, when separate patterns are merged there is some savings in the MML code representing the patterns, as well as some loss in the MML code representing the data (since some data better encoded with a pattern being merged must now be encoded with a new, simpler "representative" pattern). A greedy search is conducted over the pattern space to find all groupings which provide overall savings on code length.

The grouping of DAGs and patterns is intrinsic to MML and based on the same idea as the use of Fisher information to determine the precision of parameters: the theory being used to explain (encode) the data should only be as detailed or complex ("precise") as the data being encoded can reasonably justify. It is particularly important in grouping DAGs, which form the basis for the stochastic counting over the model space. Since the TOM space is extremely large, it is possible without this kind of grouping to sample huge numbers of models without ever visiting the same model twice, leading to a useless estimate of the posterior distribution. In other words, without MML cleaning of DAGs, the sampling process would take unreasonable (exponential) amounts of time to yield useful estimates of the posterior distribution (Wallace and Korb, 1999).

9.6.3　Expert priors

Automated causal discovery is hampered by many difficulties. These include being overwhelmed by high dimensional data sets and problems which violate foundational assumptions, such as faithfulness, both of which will be discussed below in §9.7. One common difficulty is a massive failure of common sense, such as "discovering" that a patient's age is caused by her medical condition. Such results lead some to consider causal discovery programs to be "stupid", but that strikes us as unfair: humans interpret causal processes against a rich background of prior knowledge, whereas the causal discovery programs are commonly run with no background understanding at all. To rectify matters — that is, to speed up the process of automatically finding sensible models — it is plausible to boost automated discovery by feeding our programs with at least a modest slice of our own prior understanding of the problem. Many causal discovery programs allow that.

TETRAD, for example, allows prior variable orderings to be specified, by grouping variables in *tiers*. Variables in earlier tiers are constrained during search not to be effects of variables in later tiers. Variables within a tier can be discovered to exist in causal chains in any (non-cyclic) order. Of course, by having as many tiers as variables, one can impose a total ordering upon the variables. Tiers provide hard constraints for the search algorithm.

A better method of specifying prior probability is that adopted by Heckerman and Geiger (1995) for their Bayesian BDe and BGe metrics. They have the user generate a "best guess" causal model in the beginning. Their prior probability metric then rewards models which are closer to this best guess model (in edit distance terms of arc deletions, additions and reversals). Since the initial model provides a prior distribution over the model space (which is only natural for a Bayesian search process!), the prior information acts as a soft (probabilistic) constraint, which can be overcome by enough data suggesting a different structure. How much data is needed can be controlled by specifying an **equivalent sample size (ESS)** for the prior model (i.e., sample data in the same size will be equal weighted with the prior information).

CaMML has a still more flexible system of specifying prior constraints. In the first place, for each pair of variables the user can specify a prior probability for the existence of a particular directed arc between them. In the structure code, instead of coding each possible arc, or absence of an arc, as one bit, as in Equations (9.9) and (9.13), those equations are modified to reflect the prior probability p_i for each possible directed arc i. In particular, for each arc in a model CaMML adds $-\log p$ bits and for each missing possible arc it adds $-\log(1 - p)$. Thus we get the following MML structural metric:

$$I_{MML}(h) = \log N! - \sum_i \log p_i - \sum_j \log(1 - p_j) \qquad (9.14)$$

where i indexes the arcs in a TOM and j indexes the possible arcs that are absent in the TOM.[11]

CaMML goes far beyond arc probabilities, allowing all of the following kinds of soft constraints (O'Donnell et al., 2006):

DAG structure. A complete (or incomplete) network can be specified, generating a prior distribution over the model space via edit distance, as in Heckerman's scheme.

Direct causal connections between variables may be indicated (e.g., $A \rightarrow B$).

Direct relation $(A - B)$. It may be known that two variables are related directly, but the direction of causality is unknown.

Causal dependency $(A \Rightarrow B)$. This allows the user to indicate that one variable is an ancestor of the other, while the mechanism between them remains unknown.

Temporal order $(A \prec B)$, i.e., A precedes B. In many domains it is clear that some variables come before others irrespective of any causal chain linking them. (This can be used to implement tiers of variables, but is more general than tiers.)

Dependency $(A \sim B)$. The most general sort of prior information provided for is a generic dependency. This implies that there is some prior d-connection between the nodes, whether via a chain or a common ancestor.

These kinds of prior information may be applied in any combination, so long as the combination is coherent. Each of these except the DAG structure is specified as a probabilistic constraint, which implies that the absence of connections and relations of each type may be specified as well. A prior DAG structure implies a prior probability for any given TOM via: (1) the bubble-sort distance from the closest linear

---

[11] During sampling these arc probabilities are treated as priors; i.e., the probabilities are updated during the sampling process according to the estimated posterior distribution over arc probabilities.

extension for that DAG to the target TOM; (2) combined with the undirected edit
distance between these two TOMs. Any other constraints provided contribute to the
completed prior distribution over the TOM space in ways fully described only in
Rodney O'Donnell's PhD (O'Donnell, 2010). The use of priors to combine expert
elicitation with causal discovery is discussed further in Part III.

9.6.4 An MML metric for discrete models

With the stochastic search through TOM space and counts maintained per DAG and
pattern we no longer need to represent the linear extension count in the MML structure code, as (9.13) records. We can now readily extend the linear MML code to the
discrete case. The MML structural score for TOMs is unaffected by the nature of
the variables, as is the Metropolis search process. What we need is a new metric for
parameters and data, given the TOM, which here amounts to encoding a multinomial distribution and its data for each instantiation of a variable's parents, a problem
which MML has long since been applied to (Boulton and Wallace, 1969). With iteration over the different parent states and, further, over all the variables in the network
this solution yields:

$$I_{MML}(d, \theta | h) = \sum_{j=1}^{N} \left[s_j \left(\frac{1}{2} \log_2 \frac{\pi e}{6} \right) - \log_2 \left(\prod_{j=1}^{|\Phi_k|} \frac{(s_k - 1)!}{(S_{kj} + s_k - 1)!} \prod_{l=1}^{s_k} \alpha_{kjl}! \right) \right] \qquad (9.15)$$

where we use the notation of (9.3), except replacing e (for evidence) with d (for
data), so we can refer to the natural log base with e. The right-hand log term (with the
product over parent states) is equivalent to K2's metric, while the left-hand summand
is a standard MML bit penalty per parameter estimate (presented in bits, and equal
to $\frac{1}{2} \ln \frac{\pi e}{6}$ **nits** — meaning units to the base of the natural log). The penalty reflects
the difference between the Bayesian way of integrating the parameters out of the
computation (as in §9.1) and the MML method, which *estimates* the parameters by
partitioning the parameter space and selecting the optimal cell for communicating
that estimate. The specifics of the derivation of this penalty are beyond our scope
here — see Wallace and Freeman (1987) for details.

The direct relation to the probability distribution of Cooper and Herkovits'
$P_{CH}(h,d)$, is not hard to understand intuitively, bearing in mind that the most significant distinction between their approach and the MML approach is their assumption
of a uniform prior over hypotheses (assumption 5), implicitly denying the relevance
to a discovery metric for DAGs of multiple linear extensions (as well as that of any
structural complexity not directly reflected in the number of parameters). Since the
MML search is being conducted over the TOM space, the MML metric is being applied to TOMs, not DAGs, so this difference falls by the wayside — so long as we
are applying the metric to TOMs and not DAGs!

The final MML metric for TOMs must add in the measure for the structural complexity of the TOM:

$$I_{MML}(h, d) = I_{MML}(d, \theta | h) + I_{MML}(h) \qquad (9.16)$$

where the right summand is from (9.14).

Given the search mechanics described above, this metric is never applied to DAGs; DAG probabilities are estimated by accumulating the estimates for their associated TOMs instead. Any direct application of this metric to DAGs will be misleading.

9.6.5 Learning hybrid models

The probabilistic relation between a variable and its parents is normally encoded in the child variable's CPT. For discrete networks CPTs can encode any kind of relationship, with each parameter (the probability of some child state given some joint state of its parents) encoded independently of the values for any other parameters (i.e., under an assumption of parameter independence). While CPTs are maximally flexible, having the greatest representational power for discrete local structure, they are also maximally demanding, requiring that each state probability be estimated independently of all the rest using whatever data or expertise is available. The number of parameters required is, in fact, exponential in the number of parent variables. This is highly burdensome when probabilities are being elicited from experts, since for complex local structures they are unlikely to have well-founded opinions about some large number of parameters (or the patience required to generate so many opinions). But it is also burdensome when parameterizing from data. When any particular instantiation of parent values is uncommon, then the data available to estimate the relevant conditional probabilities may be very sparse; in many cases, there are no relevant samples. As we saw in §6.4, there has been interest in using alternative, and at least potentially simpler, representations of local structure, ones which allow dependencies between parameters and so broaden the set of sample points relevant to interdependent parameters. Here we look briefly at combining the learning of local structure with causal discovery.

Friedman and Goldszmidt (1999) first applied this idea of integrating the learning of local structure into a global causal discovery algorithm, learning classification trees instead of CPTs. They used MDL and BDe scores of both global and local structures, applying greedy search to find optimal global causal network structure, by adding, deleting and reversing arcs, and simultaneously to find optimal local classification trees, by adding best-first split nodes to reach a maximal tree and then greedily trimming to find the best tree. Applying full CPT learning within the global structure search, they showed that learning classification trees instead achieved results at least as good, but with the added value of learning with fewer parameters. In somewhat similar work, Neil et al. (1999) investigated the use of logit models (cf. §6.4.4) to represent local structures during global causal discovery with MML.

Currently, CaMML takes a very different approach. Instead of presenting the user with a global choice amongst CPTs, classification trees and logit models, meaning that any learned model will *only* contain local structure representations of the selected type, in hybrid model discovery the discovered models contain a mixture of all these types of local structure, depending upon which one performs best for the local situation (O'Donnell et al., 2006). This produces something like the best of all possible worlds: when data are scarce, or when parameter interdependencies are strong, then representations such as logit and classification trees produce low variance representations of the data that are robust; when data are plentiful, and when there are no simplifying interdependencies between parameters, then hybrid learning allows either CPTs or fully extended classification trees to represent all parameters independently and as precisely as the data will allow (see §7.5.2). The scoring of CPTs, classification trees and logit functions has in each case been integrated into the overall MML score of the causal model, allowing the normal MCMC search process to be used. Empirical results show hybrid CaMML outperforming CaMML restricted to CPTs across a wide range of cases. The BDe metric implementation of Friedman and Goldszmidt (1999) performs similarly, while that of the MDL metric is generally outclassed (O'Donnell et al., 2006). In short, hybrid CaMML matches the best performing causal discovery program with any particular local structure representation, selecting simple trees or logit models when relationships are simple or data scarce and selecting CPTs or complex trees when needed.

9.7   Problems with causal discovery

The technology of causal discovery is problematic. Some of its problems have been solved, or are being solved; others remain. Here we address three which have received widespread attention. First we consider the twin problems of the assumptions of the causal Markov condition and of faithfulness — they are twins in the sense that they are converses of each other, and also in the sense that both have aroused a fair degree of philosophical indignation and opposition. The third problem is more practical and is an active area of research, namely how to adapt causal discovery to very large problems.

9.7.1   The causal Markov condition

What has come to be known as the **causal Markov condition (CMC)** is essentially a more formal version of Reichenbach's Principle of the Common Cause (Conjecture 8.1; Reichenbach, 1956):

> If two variables are marginally probabilistically dependent, then either one causes the other (directly or indirectly) or they have a common ancestor.[12]

---

[12]To be sure, this is no quotation from Reichenbach, who instead wrote: "If an improbable coincidence has occurred, there must exist a common cause." On the face of it, Reichenbach's literal formulation is far stronger than our version above. Indeed, taken literally it is so strong as to provide an unbelievable interpretation of Reichenbach. For one thing, Reichenbach was interested in explanations of enduring dependencies, so the improbability referred to must have been an improbability conditioned on a *lack* of a common cause. For another, his work is throughout concerned with ancestral causal relations, so the alternative explanation via direct or indirect causation was not being ignored. In short, our paraphrase is a better rendition of Reichenbach's intent than that of some literalist interpreters.

The Markov property (aka "Markov condition"; see §2.2.4) asserts

> For any X_i and $X_j \notin Descendant(X_i)$ $X_i \perp\!\!\!\perp X_j | \pi_{X_i}$. I.e., any two variables not related by direct descent are independent of each other conditioned upon the parent set of either one of the variables.

The causal Markov condition is the very same thing, with the additional proviso that all the arcs in a network satisfying it are understood as expressing direct causal connections.

The problem with CMC is just that, as we noted in §8.1, it appears to be an assumption of all causal discovery programs, and it is known to be false.

CMC, in the form of Reichenbach's Principle, has been the center of a long-standing debate in the philosophy of science, with a fairly large number of proposed counterexamples having been produced by now. Here is a selection:

Sober's Bread and Water. Perhaps the best known counterexample is that of Sober (1988). The water level in Venice has been rising relatively steadily for many decades (at least, relative to Venice's buildings!). But also the price of bread in Britain has been rising relatively steadily over the same period. We can suppose that there is an enduring dependency between the two. However, it is far-fetched to suppose that this dependency must be explained in terms of one causing the other, or again that they are jointly the result of some prior cause. There is nothing in particular that causally relates Venetian water levels and bread prices, so CMC is false.

Williamson's Mean and Variance. Williamson (2005) reports the case of a Bayesian network containing a node for the mean of a binomial and another for its variance. Since for binomials $\sigma_x^2 = \mu_x(1 - \mu_x)$ these two nodes are always maximally informative of one another: they have an unshakable dependency, regardless of what other nodes may be conditioned upon. However, means do not cause variance, nor vice versa.

Cartwright's Constraints. Nancy Cartwright (1989) pointed out that if we go into a shop looking to buy, say, a drink and a sandwich, then a limited budget of $10 necessarily imposes a functional constraint between the two purchases; i.e., the price of the second item cannot be greater than $10 minus the price of the other. Again, this constraint induces a dependency without implying a causal relationship.

Salmon's Interactive Fork. Salmon (1984) aimed to further Reichenbach's program by developing a process theory of token causation (individual instances of causes and their effects) while rejecting probabilistic analyses of type causation (e.g., smoking causing cancer). As a part of the latter, he offered a number of cases of "interactive forks", where two effects of a common cause remain probabilistically dependent even when the state of the cause is known. For example, he supposed it is possible to arrange two pool balls so that one is pocketed if and only if the other one is; furthermore, this (supposedly) could remain true regardless of any knowledge we might have about the cue ball and how it is struck.

Salmon's Lawful Constraints. Salmon (1984) also pointed out that physical conservation laws entail functional relationships between variables which cannot be screened off (rendered conditionally independent) by any common cause.

Williamson sums up his discussion with: "Hence, the Principle of the Common Cause fails" (Williamson, 2005, p. 55). The skeptics of causal discovery urge us to give up in despair in the face of such examples. As we've agreed, causal discovery depends upon CMC; and, from these examples and more, we can see that CMC is false; hence, the only intellectually legitimate course is to abandon causal discovery. The two premises to this argument are true. Nevertheless, the conclusion is false and the argument is invalid. There are a great many procedures in this world which have false assumptions and yet remain extremely useful and used. For example, almost all construction engineering depends upon the truth of Newtonian mechanics; however, 20th century physics has well and truly established the falsity of Newtonian mechanics. Adhering strictly and utterly to the truth would proscribe, amongst other things, all uses of computer simulation, whether using Bayesian networks, cellular automata or individual-based modeling. The proposed proscription is nuts!

Rather than write off a very successful and potentially very important technology, we prefer to draw some lessons about its limits from these kinds of counterexamples. Certain kinds of variables, for example, are off-limits for causal networks. In particular, those which are strict functions of one another, such as the mean and variance of a single binomial, are properly replaced by a single variable. In semantic Bayesian networks, and other *non-causal* networks, duplicating or subsetting variables may be useful and important; but when trying to talk a strictly causal language, duplication is simply wrong. The same applies when the functional relations between variables are imposed not by semantics but by boundary conditions such as conservation laws or Cartwright's bank balance. The point is not that variables representing relata subject to these constraints are illegitimate in general, but we know in advance that they cannot be causes or effects of each other, nor joint effects of a common variable. I.e., we know in advance that by putting them all into a single Bayesian network we will *not* have a causal network. There's nothing wrong with that, but there is something wrong with then also pretending we *might* have a causal network after all, which is the only way in which the *causal* Markov condition can reenter the story.

In Elliot Sober's example a first impulse to invoke a *Time* variable. If we fix time, any dependency between water levels and bread prices must vanish, salvaging CMC. We may have doubts, however, that *Time* is actually a legitimate causal variable — for example, *Time* cannot be an effect variable, since there is no known physical process with which to manipulate it (Handfield et al., 2008). We can also have doubts, however, about the example itself. A curiosity about it is that it is so readily extended: it's not just water levels and bread prices that have regularly increased over the last few decades, but also the human population, global GDP, CO_2 in the ocean and air, light pollution, the size of the universe, your age, our ages, and indefinitely many other things, together with the inverses of everything in decline, such as biodiversity and total untapped reserves of oil. Every animal that matures (or declines) has systematic dependencies in the levels of maturation (decline) of its body parts. Likewise every solar system and galaxy. If Reichenbach's Principle has the absurd consequence that all of these things are either causing each other or being caused by the hidden hand of Time, then surely the time has come... either to reject or revise the Principle. Reichenbach was interested in principles which would help science to discriminate between substantive alternative hypotheses, not ones which would assert the world is made up of an undifferentiated mess, so revision rather than rejection would seem to be more in keeping with the principle of charity. We accept that nonstationary time series (e.g., inflating or deflating time series) should not be analysed with statistical measures of dependency that assume the data are stationary (Hoover, 2003). Indeed, more generally, the assumption that sample data are independently and identically distributed, an assumption we have relied upon throughout, implies that the system being sampled is a *single* system, rather than one that is busy transforming itself.[13] Objectors might hold that that itself is the problem and one that limits the causal discovery project. However, causal discovery is not limited to the discovery of static Bayesian networks; dynamic Bayesian networks readily represent time series, without the spurious correlations, and they can be learned as well. Learning DBNs is beyond our scope here, but there is a small and growing literature on the subject (see, e.g., Van Berlo et al., 2003, Ross and Zuviria, 2007).

The skeptical questions raised about the causal Markov condition serve to remind us of the limits of causal discovery. They also should encourage us to go further to develop more exact ideas about what kinds of variables are legitimate in causal models, what not, and why. If data sets record variables that are best merged, or excluded, we need to know that in advance and deal with it by preprocessing the data. If data sets record nonstationary time series, then we may need to preprocess them to remove spurious dependencies or else to apply alternative discovery methods. Inevitably, some problems will emerge only after applying causal discovery to data, which may require another iteration, either revising the data or the models discovered, or else supplying prior constraints to the discovery process. In short, without a definitive solution to all objections to CMC (or other assumptions of causal discovery), the learning algorithms will continue to need care in handling.

9.7.2 A lack of faith

Another important assumption of causal discovery, and another source of controversy about it, is the assumption of faithfulness between the causal model and the probabilistic dependencies being explained.

Definition 9.4 *A causal model is* **faithful** *to a probability distribution if and only if for any* $\mathbf{X}, \mathbf{Y}, \mathbf{Z}$ *in the model, if* $\mathbf{X} \not\perp \mathbf{Y} | \mathbf{Z}$, *then* $\mathbf{X} \not\!\perp\!\!\!\perp \mathbf{Y} | \mathbf{Z}$.

In other words, if a causal model *suggests* a causal dependency between variables in some circumstance, via a d-connection, then there must be a probabilistic de-

---

[13]Spirtes et al. (1993) formulation of CMC (p. 29) explicitly rules out sampling from changing, or mixed, systems and so was never subject to Sober's kind of counterexample.

pendency between them in that circumstance. This condition is the opposite of the (causal) Markov condition, which asserts that if a causal model suggests a lack of causal dependency between variables in some circumstance, then there is no probabilistic dependency between them in that circumstance. If both faithfulness and the Markov property hold, then we have what Judea Pearl called a **perfect map**: the Bayesian network has d-connections where, and only where, the probability distribution has conditional dependencies.

The problem for causal discovery is that learning algorithms assume that the true model for a problem is faithful, and that assumption is often false. That fact is precisely what is wrong with causal discovery, according to Nancy Cartwright (2001) and other skeptics.

Again, in the end we will not accept the implicit conclusion, that we should abandon causal discovery, but it's actually quite easy to see that the premise is right: not all true causal models are faithful to the probability distributions that they generate. Perhaps the most famous example of this is the neutral Hesslow model of the contraceptive pill (Hesslow, 1976). The contraceptive pill reduces the probability of pregnancy (of course!); and pregnancy is a fairly strong cause of thrombosis. We might hope, then, that the pill is a net preventative of thrombosis. However, the pill *also* directly causes thrombosis; Hesslow asked us to consider the possibility that the causal efficacy of the pill as a direct cause and its efficacy as an indirect preventative exactly cancel out, so that its net effect is nil. In that case the true model (the one that generates the data) is given in Figure 9.3. And that model is unfaithful to its distribution, since $P(Thrombosis) = P(Thrombosis|Pill)$ even though there is a direct arc between them.

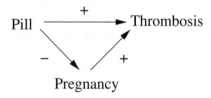

FIGURE 9.3: Neutral Hesslow.

Since most causal discovery algorithms will prefer models which are simpler to those that are complex, so long as they both can represent the same conditional dependencies (in a version of Ockham's razor), when presented with data generated by the neutral Hesslow model, they will typically return the faithful, but false, model of Figure 9.4.

This kind of unfaithfulness is a version of "Simpson's paradox", where two variables are directly related, but also indirectly related through a third variable, and so depends upon there being multiple paths between two variables of interest.

A plausible response to this kind of example, that of Spirtes et al. (2000), is to point out that it depends upon an exact choice of parameters for the model. If the

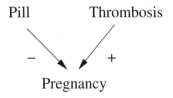

FIGURE 9.4: Faithful Hesslow.

parameters were ever slightly so different, a non-zero correlation would result, and faithfulness would be saved. In measure theory (which provides the set-theoretic foundations for probability theory) such a case is said to have *measure zero* — with the idea that the probability of this circumstance arising at random is zero (see Spirtes, Glymour, and Scheines, 2000, Theorem 3.2). Accepting the possibility of zero-probability unfaithful cases implies merely that our causal discovery algorithms are fallible. But no proponent of causal discovery can reasonably be understood as claiming infallibility.[14] Clearly, faithfulness cannot always be right — otherwise, how could we ever come to admit that it had been violated, as in Figure 9.3? A better interpretation of the faithfulness condition is as a *presumption*: that, until we have good reason to come to doubt that we can find a faithful model, we should presume that we can. In the thrombosis case, with precisely counterbalancing paths, we begin with knowledge that the true model is not faithful, so we are obliged to give up the presumption.

Cartwright objects to this kind of saving maneuver. She claims that the "measure zero" cases are far more common than this argument suggests. Also, many of the systems we wish to understand are artificial, rather than natural, and in many of these we specifically want to cancel out deleterious effects. In such cases we can anticipate that the canceling out will be done by introducing third variables associated with both cause and effect, and so introducing *by design* a "measure-zero" case. For example, an alcohol tax could have no net effect upon tax revenue, precisely because it is tuned to reduce alcohol consumption as much as possible before impacting upon net revenue. Again, Steel (2006) points out that there are many cases of biological redundancy in DNA, where if the allele at one locus is mutated, the genetic character will still be expressed due to a backup allele; in that case the mutation and the genetic expression will fail the faithfulness test. As Steel emphasizes, the point of all these cases is that the measure-zero premise fails to imply the probability zero conclusion: the system parameters have not been generated "at random" but as a result of intelligent or evolutionary design, leading to unfaithfulness.

Although we think the measure-zero defence is pertinent in at least some contexts, it cannot be the final defence of causal discovery against unfaithfulness. For one

---

[14]To be sure, Cartwright thinks otherwise: "Bayes-net methods...will bootstrap from facts about dependencies and independencies to causal hypotheses—and, claim the advocates, *never get it wrong*" (Cartwright, 2001, p. 254; italics ours). But, Cartwright's straw-man is what is wrong.

thing, not all cases of unfaithfulness involve multiple paths. An example of Richard Neapolitan makes this clear: finesteride reduces DHT (a kind of testosterone) levels in rats; and low DHT can cause erectile dysfunction. However, finesteride doesn't reduce DHT levels sufficiently for erectile dysfunction to ensue (in at least one study); in other words, there is a threshold above which variations in DHT have no effect on dysfunction.[15] Graphically, this is simply represented by:

$$Finesteride \;\rightarrow\; DHT \rightarrow\; Dysfunction$$

Since, there is no marginal dependency between finesteride and erectile dysfunction, we have a failure of transitivity in a simple causal chain. In general, threshold effects can cause failures of transitivity without any measure-zero balancing acts! So, what can a causal discovery algorithm do?

A practical heuristic when faced with inductive difficulties is to consider how humans go about responding to like difficulties. However problematic and tortuous human science may be, it has managed to discover the relationships between the contraceptive pill and thrombosis, and that between finesteride and dysfunction. It has managed it by a complex mixture of investigative techniques, including *experimental interventions*. If we allow ourselves access to experimental data, it would be absurd to deny it to our algorithms. And, indeed, we think intervention data is the real key to unlocking the secrets of unfaithful models.

In order to see what we can learn from intervention we can consider *fully* augmented models, meaning those where *every* original variable X gets a new intervention parent I_X, doubling the number of variables (see §3.8.2). In the case of the two Hesslow models, faithless (true) and faithful (false), full augmentation results in Figure 9.5.

By introducing new v-structures over all endogenous variables, fully augmented models can always empirically distinguish the unfaithful original model from the faithful imposter, as in Figure 9.5. Furthermore, full augmentation also eliminates the problem repeatedly alluded to throughout our discussion of causal discovery: the indistinguishability of Markov equivalent models. In the full augmentation space no two distinct models in the original space appear in the same Markov equivalence class, and so no two original models are statistically indistinguishable (Korb and Nyberg, 2006, Theorem 3).[16] Although our theorem is particular to cases isomorphic to the neutral (and linear) Hesslow case, the result is of much wider interest, since the neutral Hesslow structure is the only way in which a true linear causal model can be unfaithful to the probability distribution which it generates, through balancing multiple causal paths of influence.

Perhaps of more interest will be interventions upon discrete causal models, which are more commonly the center of attention in causal discovery and which certainly

---

[15]This example was meant to appear in Neapolitan (2003), however his publisher thought their readership too delicate to be exposed to it!

[16]Spirtes et al. (2000) proved a different theorem, namely that no two distinct causal models that are strongly indistinguishable remain so under intervention (Theorem 4.6). This implies that there is *some* possible world capable of distinguishing between the two models under intervention. Our theorem proves that *this world* is such a possible world.

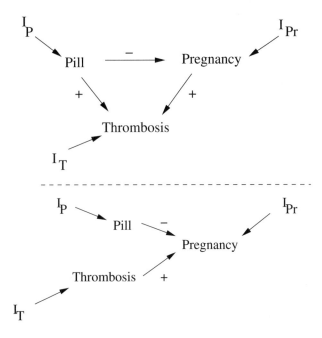

FIGURE 9.5: Augmented Hesslow models: the faithless, but true, causal model (top); the faithful, but false, model (bottom).

introduce more complexity than do linear models. In particular, in non-linear cases there are many more opportunities for intransitivities to arise, both across multiple paths and, unlike linear models, across isolated paths. Nevertheless, similar theoretical results are available for discrete causal models, at least when the interventions themselves can be faithfully represented by the augmented models. The threshold effect cases of unfaithfulness, by the way, do not by themselves cause undo difficulties for causal discovery. The local dependencies between *Finesteride* and *DHT*, on the one hand, and between *DHT* and *Dysfunction*, on the other, are readily discoverable, and they are not interfered with by long-range intransitivity. The detailed treatment of discrete unfaithfulness is beyond our scope here; however, see instead Nyberg and Korb (2006).

Neither the presumption of faithfulness by causal discovery algorithms nor their inability to penetrate beneath the surface of the statistical indistinguishability within Markov equivalence classes offer any reason to dismiss the causal discovery project. The contrary view is plausible only when ignoring the possibility of extending causal modeling and causal discovery by intervening in physical systems, much as human scientists do.

9.7.3   Learning in high-dimensional spaces

The "curse of dimensionality" is to be presented with the problem of learning in the face of very high dimensional data — i.e., samples across a large number of variables. As machine learning is getting applied to larger problems all the time, such as in bioinformatics, commercial data mining, and web mining, the curse is biting ever harder. The algorithms we have discussed so far are all algorithms for global causal discovery which take in all available data as a whole and attempt to find the best Bayesian network(s) to explain the data. They all suffer in performance from too much data, and especially from too many variables.

An alternative local discovery approach, based upon discovering Markov blankets (MBs), has grown in popularity in recent years. One use of MB discovery is just to use the MB in prediction tasks: given a particular variable of interest for prediction, the variables in the MB in the true model provide all information relevant for predicting the target variable, since all other variables are conditionally independent of the target given the MB (via the Markov property). In other words, while assessing what is in MB may be problematic (and is done using normal statistical tests of conditional dependency), should one exactly identify what is in the blanket, then there is nothing further to know for predicting the target, other than the target variable's value itself. Such a reduction of dimensionality is called **feature selection** in classification learning, and once it has been done, any classification learner, such as naive Bayes from Chapter 7 or multiple regression, can be used on the selected features to find a predictor function. This works fine for standard prediction tasks, but is not the same as discovering causal structure and, indeed, is not much use in predicting the consequences of causal interventions. If some of the predictor variables are externally modified rather than observed, naive Bayes and regression models will likely give wrong results, since some of the predictor variables may be, e.g., children of the target variable, when interventions upon them will only incorrectly be supposed to inform us about the target variable.

When the aim is not prediction, however, but indeed discovering local causal relations around some particular variable, then this can be described as local causal discovery.[17] For example, the PC algorithm then can be applied, restricting the data used to the target variable and its discovered blanket. Early work on MB discovery includes Koller and Sahami (1996), Cooper (1997) and the Sparse Candidate (SC) algorithm of Friedman et al. (1999).

Given a method of finding the Markov blanket for any given variable, it's natural to apply that method iteratively over all of the variables in a data set and then gluing the results together, turning local causal discovery into global causal discovery in a way that evades the curse of dimensionality. An alternative approach is to use the information obtained from MB discovery as constraints upon an otherwise normal global causal discovery method. Either approach offers significant computational savings compared with global causal discovery applied to the full data set. MB

---

[17]Note that this is distinct from the local structure learning of §9.6.5. In that case, it is a matter of learning parameters given a local arc structure, whereas here we are concerned with learning the local arc structure.

discovery has become an active area of research in recent years. Tsamardinos et al. (2006) developed the max-min hill climbing (MMHC) algorithm out of Friedman's SC, compounding MB discovery into a global discovery procedure. Nägele et al. (2007) demonstrated the application of MMHC in global causal discovery in very high-dimensional spaces, with thousands of variables. Jia et al. (2007) have applied MMHC to global pattern discovery.

9.8 Evaluating causal discovery

Having seen a number of algorithms for learning causal structure developed, the natural question to ask is: Which one is best? Unfortunately, there is no agreement about exactly what the question means, let alone which algorithm answers to it. There has been no adequate experimental survey of the main contending algorithms. Some kind of story could be pieced together by an extensive review of the literature, since nearly every publication in the field attempts to make some kind of empirical case for the particular algorithm being described in that publication. However, the evaluative standards applied are uneven, to say the least.

Note that the evaluative question here is not identical to the evaluative question being asked by the algorithms themselves. That is, the causal discovery algorithms are aiming to find the best explanatory *model* for whatever data they are given, presumably for use in future predictive and modeling tasks. For this reason they may well apply some kind of complexity penalty to avoid overfitting the training data. Here, however, we are interested in the question of the best model-discoverer, rather than the best model. We are not necessarily interested in the complexities of the model representations being found, because we aren't necessarily interested in the models as such. Analogically, we are here interested in the intelligence of a learner, rather than in the "smartness" (or, elegance) of its solutions.[18]

Here we shall review the most common methods of causal discovery evaluation in the literature and then introduce a new, information-theoretic alternative which employs the idea of causal interventions, causal Kullback-Leibler divergence (CKL).

9.8.1 Qualitative evaluation

Perhaps the most obvious way of assessing the adequacy of a learned model relative to a known original model is the edit distance between the two structural models. TETRAD publications (e.g., Spirtes et al., 2000, Scheines et al., 1994) have used

---

[18]It is for this reason that metrics used in causal discovery, such as MML and MDL, are not directly applicable to our problem in this section — in addition to the obvious point that such metrics would be biased in favor of their respective discovery algorithms! Similarly, "active learning" metrics, such as that of Cooper and Yoo (1999), being applied to causal discovery with experimental data are also not immediately applicable, and they are also not immediately comparable to our CKL evaluation metric discussed below.

stochastic sampling from known causal models to generate artificial data and then reported percentages of errors of four types:

- Arc omission, when the learned model fails to have an arc in the true model
- Arc commission, when the learned model has an arc not in the true model
- Direction omission, when the learned model has not directed an arc required by the pattern of the true model
- Direction commission, when the learned model orients an arc incorrectly according to the pattern of the true model

This is an attempt to quantify qualitative errors in causal discovery. It is, however, quite crude. For example, some arcs will be far more important determiners of the values of variables of interest than others, but these metrics assume all arcs, and all arc directions within a pattern, are of equal importance.

Regardless, this kind of metric is by far the most common in the published literature. Indeed, the most common evaluative report consists of using the ALARM network (Figure 5.2) to generate an artificial sample, applying the causal discovery algorithm of interest, and counting the number of errors of omission and commission. Every algorithm reported in this chapter is capable of recovering the ALARM network to within a few arcs and arc directions, so this "test" is of little interest in differentiating between them. Cooper and Herskovits's K2, for example, recovered the network with one arc missing and one spurious arc added, from a sample size of 10,000 (Cooper and Herskovits, 1992). TETRAD II also recovers the ALARM network to within a few arcs (Spirtes et al., 2000), although this is more impressive than the K2 result, since it needed no prior temporal ordering of the variables. Again, Suzuki's MDL algorithm recovered the original network to within 6 arcs on a sample size of only 1000 (Suzuki, 1996).

Perhaps slightly more interesting is our own empirical study comparing TETRAD II and CaMML on linear models, systematically varying arc strengths and sample sizes (Dai et al., 1997). The result was a nearly uniform superiority in CaMML's ability to recover the original network to within its pattern faster than (i.e., on smaller samples than) TETRAD II.

9.8.2 Quantitative evaluation

Because of the maximum-likelihood equivalence of DAGs within a single pattern, it is clear that two algorithms selecting identical models, or Markov equivalent models, will be scored alike on ordinary evaluation metrics. But it should be equally clear that non-equivalent models may well deserve equal scores as well, which the qualitative scores above do not reflect. Thus, if a link reflects a nearly vanishing magnitude of causal impact, a model containing it and another lacking it but otherwise the same may deserve (nearly) the same score without receiving it. Again, the parameters for an association between parents may lead to a simpler v-structure representing the very same probability distribution as a fully connected network of three variables (as in the neutral Hesslow example above). So, we clearly want a more discriminating and accurate metric than the metrics of omission/commission.

The traditional such metric is **predictive accuracy**: the percentage of correctly predicted values of some target variable. In classification problems this has a clear meaning, but for learning causal models it is generally less than clear which variable(s) might be "targeted" for classification. Predictive accuracy, in any case, suffers from other problems, which we examined in §7.5.1.

Kullback-Leibler divergence (§3.6.5) of the learned model from the model generating the data is preferable and has been used by some. Suzuki, for example, reported the KL metric over the ALARM network, showing a better result for his MDL algorithm than that for K2 (Suzuki, 1996). KL divergence is normally measured over all the variables in a model. Because of the Markov equivalence of all models within a pattern (Theorem 8.1), this implies that what is being evaluated is how close the learned model is to the original *pattern*, rather than to the original causal model. Furthermore, since fully connected models can be parameterized to fit *any* distribution, fully connected models will always be assessed as optimal. So KL is not obviously the best method for evaluating *causal* discovery.

9.8.3   Causal Kullback-Leibler (CKL)

As we have noted, the statistical equivalence (e.g., maximum likelihood identity) of Markov equivalent models has led many to give up on causal discovery, satisfying themselves with the easier goal of pattern discovery. But we've also noted that there is a large difference between what is observationally indistinguishable in current practice and what is observationally indistinguishable in principle. It is only in principle indistinguishability that can justify the metaphysical pessimism implicit in conflating Markov equivalent models. Furthermore, we have also already seen (in §9.7.2) that Markov equivalent models are in fact observationally distinguishable, so long as we can draw upon intervention data. Here we describe a metric which is capable of distinguishing the indistinguishable, **causal Kullback-Leibler divergence (CKL)**, which utilizes hypothetical interventions.

The scenario presumed for this metric is the same as that presumed elsewhere in this section: we have one or more causal models to begin with; we stochastically generate artificial data; we run multiple causal discovery programs on the data; we apply some metric to measure the "distance" from the generating models to the learned models in order to assess which programs have come closest in their learned models to the true models.[19] The first point to notice is that, since we have the true model in hand, we *can* intervene upon it in any way we like to generate any interventional data we like. But all of our experiments, indeed all of our observations, are occurring inside computer simulations, and in that sense they are hypothetical. Issues such as the ethical, economic or political costs of interventions, which often make trouble for experimental scientists, make no trouble for us whatsoever. (This is one of the advantages of computer simulation!)

As we pointed out in §9.7.2, a full augmentation of a model (intervening on ev-

---

[19]For the very different problem of evaluation "in the wild", i.e., where the data are real and the true model unknown, see Chapter 10.

ery original variable), will put every model in any given Markov equivalence class in the original model space into distinct equivalence classes in the augmented model space. In consequence, the KL measure from the augmented true model to augmented learned models will appropriately discriminate between (originally) equivalent models. It is in this fully augmented space where Kullback-Leibler divergence becomes an effective evaluation measure for causal discovery.[20]

Formally, we can define CKL as:

Definition 9.5 (CKL)

$$CKL(P_1, P_2) = \sum_{\vec{x}'} \sum_{\vec{x}} P_1'(\vec{x}', \vec{x}) \log \frac{P_1'(\vec{x}', \vec{x})}{P_2'(\vec{x}', \vec{x})}$$

where P_1 is the original model's distribution, P_2 is the learned model's distribution, while $\vec{x}'$ and $\vec{x}$ range over instantiations of intervention and original variables, and P_1' and P_2' extend the probability distributions P_1 and P_2, respectively, to the fully augmented space.

Examination of this definition raises a difficult question: when we extend a probability distribution over N variables to one over $2N$ variables there are infinitely (uncountably) many options open to us; which one do we use? What we should like to use is whatever extension provides us with the greatest power of discriminating learned models that are similar to one another, while retaining much of any rank order produced by simple KL or edit distance measures. In other words, we want a tool that works like a microscope rather than a kaleidoscope! Unfortunately, we do not have any pat answer for how to generate these extensions. Together with a uniform distribution over interventions (i.e., each subset of original variables having an equal chance of being intervened upon), we have tried: (a) uniform distributions over the intervened upon variables (i.e., the variable takes any of its possible values with equal probability); (b) distributions over the intervened upon variables equal to their prior distributions (i.e., when intervened upon, a variable's distribution is set to its prior distribution in the original model). A third form, CKL_3, we have used is an "all-but-one" intervention set; that is, each variable has an equal chance of not being intervened upon, and, when that happens, *all* other variables *are* intervened upon. This corresponds to a kind of ideal experiment in which all possible factors are controlled except for a target variable of interest. In this last version of CKL, we again force all intervened upon variables to take their prior distributions.

O'Donnell et al. (2007) (and O'Donnell, 2010) report a variety of experiments demonstrating the ability of CKL, and especially CKL_3, to better assess learned models, in comparison with edit distance and KL, when those learned models are either within or without the original model's equivalence class and where the divergences

---

[20]To be sure, the problem of augmented models embedding the augmented true model (e.g., some fully connected model in the augmentation space), and so receiving the same KL score, in principle remains. However, the interventions involved in the augmentation space are under our control, so if we apply the natural restriction that all interventions target one and only one of the original variables, no fully augmented model will embed any other.

are either extreme or the result of omissions or additions of arcs reflecting minor dependencies. In short, for testing learning programs in the lab — using artificial data generated from known models — CKL uniquely combines the structural sensitivity of edit distance with the parametric sensitivity of KL to provide the best known evaluation metric for causal discovery.

9.9 Summary

There are two distinct approaches to learning causal structure from data. Constraint-based learning attempts to identify conditional independencies in isolation and to construct causal models, or patterns, from them. Given perfect access to the conditional independencies, and Reichenbach's Principle of the Common Cause, we can reasonably infer that the true causal model lies in one and only one pattern. However, it is generally not optimal to judge conditional independencies in isolation: the presence (or absence) of one dependency frequently will confer support or undermine the presence of another. Metric learners can take advantage of such evidential relevance, since they score causal models (or patterns) as a whole. Constraint-based learners are the more popular, because they are simple to understand and implement, and they can now be found in the leading Bayesian network tools, as well as in TETRAD. The different metric learners appear to have more promise for the long haul, however. Such experimental literature as exists favors them, although the methodology of evaluation for Bayesian network learners remains unclear. We expanded on the question of evaluation, and especially the evaluation of the Bayesian networks themselves, in §7.5.

9.10 Bibliographic notes

In addition to what we have presented here, there are a host of other learning techniques that attempt to ply the tradeoff between model simplicity and data fit by applying "penalties" to some measure of model complexity, including BIC (Bayesian information criterion) (Schwarz, 1978) and AIC (Akaike information criterion) (Sakamoto et al., 1986).

A good early review of the causal discovery literature is Buntine's guide (Buntine, 1996). Jordan's anthology contains a number of useful articles on various aspects of causal discovery (Jordan, 1999). Neapolitan's *Learning Bayesian Networks* (Neapolitan, 2003) treats some aspects of causal discovery we do not have the time for here. A good recent review of the causal discovery literature is that of Daly et al. (2010).

A detailed examination of Markov Blanket approaches to causal discovery is that of Aliferis et al. (2010a,b). For an interesting collection of papers mostly applying some kind of Markov blanket discovery approach to causal prediction tasks see the collection Guyon et al. (2008).

Glymour and Cooper (1999) is an interesting anthology on causal discovery. Much of the best recent work in the area shows up in the *Uncertainty in AI* conferences and in the online *Journal of Machine Learning Research*.

9.11 Technical notes

χ^2 test.

The χ^2 test is a standard significance test in statistics for deciding whether an observed frequency fails to match an expected frequency. It was proposed in the text as a substitute in the PC algorithm for the partial correlation significance test when applying the algorithm to discrete variables. In particular, it can be used to test whether $P(Y = y | X = x, Z = z) = P(Y = y | Z = z)$ across the different possible instantiations of the three variables (or, sets of variables), in order to decide whether $X \perp\!\!\!\perp Y | Z$.

The assumption that $P(Y = y | X = x, Z = z) = P(Y = y | Z = z)$ can be represented by taking the expected frequencies in the CPT cells (treating X and Z as parents of Y) to be $P(Y = y | X = x_1, Z = z) = \ldots = P(Y = y | X = x_j, Z = z) = f(Y = y | Z = z)$, where this last is the frequency with which $Y = y$ given that $Z = z$. The observed frequencies are then just $f(Y = y | X = x_1, Z = z), \ldots, f(Y = y | X = x_j, Z = z)$. The statistic for running a significance test on the discrepancy between these two measures is:

$$\chi^2 = \sum_{ijk} \frac{n_{jk}(f(y_i | x_j, z_k) - f(y_i | z_k))^2}{f(y_i | z_k)}$$

where i, j, k index the possible instantiations $\Omega_Y, \Omega_X, \Omega_Z$ of Y, X, Z respectively and there are n_{jk} samples where X takes value j and Z takes value k.

For conducting such a significance test, it is generally recommended that $n_{jk} f(y_i | z_k) \geq 5$ for each possible combination. The degrees of freedom for the test are $v = |\Omega_{YXZ}| - |\Omega_{YZ}|$.

9.12 Problems

Programming Problems

Problem 1

Implement the Bayesian metric $P_{CH}(h_i, e)$ of (9.3) for discrete causal models.

Problem 2

Implement one of the alternative metrics for discrete causal models, namely MDL, or MML (or BDe, after reading Heckerman and Geiger, 1995).

Problem 3

Implement a simple greedy search through the DAG or TOM space. Test the result on artificial data using one or more metrics from prior problems.

Problem 4

Implement the PC algorithm using the χ^2 from §9.11. Compare the results experimentally with one of the metric causal discovery programs from prior problems.

Problem 5

Obtain or write your own Markov blanket discovery algorithm, after reading some of the references to the literature above. Generate some artificial data from a Bayesian net (e.g., using Netica) and see how well its Markov blankets are discovered from the data.

Evaluation Problem

For this problem you should get and install one or more of the causal discovery programs: TETRAD IV, WinMine (free from Microsoft) or CaMML (from us). For instructions see Appendix B.

Problem 6

Run the causal discovery program on some of the data sets at our book web site. Try the program with and without giving it prior information (such as variable order). Evaluate how well it has done using two or more of the measures: predictive accuracy, KL distance, information reward, edit distance, or CKL_3.

Part III

KNOWLEDGE ENGINEERING

By now we have seen what Bayesian networks are, how they can represent uncertain processes of a considerable variety, how they can be conditioned to reflect new information, make decisions, perform causal modeling and optimize planning under uncertainty. We have seen how causal structures can be parameterized from data once learned or elicited. And we have also seen how such structures can be learned in the first place from data, either by taking advantage of conditional independencies in the data or by using Bayesian metrics. What is largely missing from the story so far is a *method* for putting all of these ingredients together in a systematic way.

In Chapter 10 we apply some of the more useful ideas of software engineering, and recent experiences of ours and others working with Bayesian networks, toward the development of such a method, which we call **KEBN**: Knowledge Engineering with Bayesian Networks. We describe this in terms of stages, generating arc structure, parameters and dealing with utilities, but in practice KEBN works best in an incremental, prototyping approach, which we explain. Throughout we also record mistakes we have frequently encountered when dealing with Bayesian models and what can be done about them.

In our concluding chapter we return to the presentation of Bayesian network applications, but this time with the idea of illustrating the KEBN processes, evaluation methods and other issues raised in Chapter 10.

10

Knowledge Engineering with Bayesian Networks

10.1 Introduction

Within the Bayesian network research community, the initial work in the 1980s and early 1990s focused on inference algorithms to make the technology computationally feasible. As it became clear that the "knowledge bottleneck" of the early expert systems was back — meaning the difficulties of finding human domain experts, extracting their knowledge and putting it into production systems — the research emphasis shifted to automated learning methods. That is necessary and inevitable. But what practitioners require then is a overarching methodology which combines these diverse techniques into a single "knowledge engineering" process, allowing for the construction Bayesian models under a variety of circumstances, which we call **Knowledge Engineering with Bayesian Networks**, or **KEBN**.

In this chapter we tie together the various techniques and algorithms we have previously introduced for building BNs and supplement them with additional methods, largely draw from the software engineering discipline, in order to propose a general iterative and incremental methodology for the development and deployment of BNs. No one has fully tested such a methodology, so our KEBN model must remain somewhat speculative. In any case, we can illustrate some of its major features with a number of case studies from our experience, which we proceed to do in Chapter 11.

We emphasize the importance of continual evaluation throughout the development process, and present additional methods for evaluating Bayesian networks, to supplement those already presented in Chapter 7. Throughout this chapter, we identify a number of common modeling mistakes (summarized in Table 10.1) and suggest ways to avoid them.

10.1.1 Bayesian network modeling tasks

When constructing a Bayesian network, the major modeling issues that arise are:

1. What are the variables? What are their values/states?
2. What is the graph structure?
3. What are the parameters (probabilities)?

When building decision nets, the additional questions are:

4. What are the available actions/decisions, and what impact do they have?

TABLE 10.1 Some common modeling mistakes.

KEBN aspect	Mistake	Section
The Process	Parameterizing before evaluating structure	§10.3.10
	Trying to build the full model all at once	§10.8
The Problem	Not understanding the problem context	§10.3.1.1
	Complexity without value	
Structure - Nodes	Getting the node values wrong	§10.3.2
	Node values aren't exhaustive	
	Node values aren't mutually exclusive	
	Incorrect modeling of mutually exclusive outcomes	
	Trying to model fuzzy categories	
	Confusing state and probability	
	Confusion about what the node represents	
Structure - Arcs	Getting the arc directions wrong	§10.3.9
	(a) Modeling reasoning rather than causation	
	(b) Inverting cause and effect	
	(c) Missing variables	
	Too many parents	§10.3.6
Parameters	Experts' estimates of probabilities are biased	§10.4.1
	(a) Overconfidence	
	(b) Anchoring	
	(c) Availability	
	Inconsistent "filling in" of large CPTs	
	Incoherent probabilities (not summing to 1)	§10.4.2
	Being dead certain	

5. What are the utility nodes and their dependencies?

6. What are the preferences (utilities)?

Expert elicitation is a major method for all of these tasks. Methods involving automated learning from data, and adapting from data, can be used for tasks 1-3 (if suitable data are available). We have described the main techniques for tasks 2 and 3 in Chapters 6, 7, 8, and 9. Task 1 has been automated as well, although we do not go into these methods in this text. Identifying variables is known in the machine learning literature as unsupervised classification (or, clustering); see, for example, Chris Wallace's work on Snob (Wallace and Boulton, 1968, Wallace and Dowe, 1994). There are many techniques for automated discretization, for example Monti and Cooper (1998). On the other hand, very little has been done to automate methods for tasks associated with building decision networks and we do not cover them in this text. In the remainder of this chapter we shall first focus on how to perform all the tasks using expert elicitation, then we shall consider methods for **adaptation**, combining elicitation with machine learning, in §10.9.

10.2 The KEBN process

10.2.1 KEBN lifecycle model

A simple view of the software engineering process construes it as having a lifecycle: the software is born (design), matures (coding), has a lengthy middle age (maintenance) and dies of old age (obsolescence). One effort at construing KEBN in such a **lifecycle model** (also called a "waterfall" model) is shown in Figure 10.1. Although we prefer a different view of KEBN (presented just below), the lifecycle is a convenient way of introducing many aspects of the problem, partly because it is so widely known and understood.

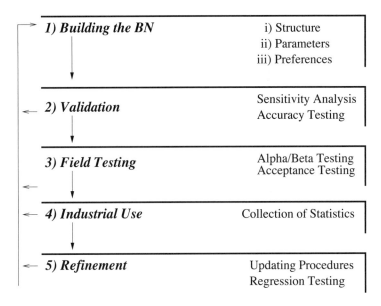

FIGURE 10.1: A KEBN lifecycle model.

Building the Bayesian network is where the vast majority of research effort in KEBN has gone to date. In the construction phase, the major network components of structure, parameters and, if a decision network, utilities (preferences) must be determined through elicitation from experts, or learned with data mining methods, or some combination of the two.

    Evaluation aims to establish that the network is right for the job, answering such questions as: Is the predictive accuracy for a query node satisfactory? Does it respect any known temporal order of the variables? Does it incorporate known causal structure? **Sensitivity analysis** looks at how sensitive the network is to changes in input and parameter values, which can be useful both for validating that the network is correct and for understanding how best to use the network in the field.

Field testing first puts the BN into actual use, allowing its usability and performance to be gauged. **Alpha testing** refers to an intermediate test of the system by inhouse people who were not directly involved in developing it; for example, by other inhouse BN experts. **Beta testing** is testing in an actual application by a "friendly" end-user, who is prepared to accept hitting bugs in early release software. For software that is not being widely marketed, such as most BNs, this idea may be inapplicable — although domain experts may take on this role. Acceptance testing is surely required: it means getting the end users to accept that the BN software meets their criteria for use.

Industrial use sees the BN in regular use in the field and requires that procedures be put in place for this continued use. This may require the establishment of a new regime for collecting statistics on the performance of the BN and statistics monitoring the application domain, in order to further validate and refine the network.

Refinement requires some kind of change management regime to deal with requests for enhancement or fixing bugs. **Regression testing** verifies that any changes do not cause a degradation (regression) in prior performance.

In this chapter we will describe detailed procedures for implementing many of these steps for Bayesian network modeling. Those which we do not address specifically, such as how to do regression testing and acceptance testing, do not seem to have features specific to Bayesian network modeling which are not already addressed here. We refer you to other works on software engineering which treat those matters in the notes at the end of the chapter (§10.11).

10.2.2　Prototyping and spiral KEBN

We prefer the idea of prototyping for the KEBN process to the lifecycle model. Prototyping interprets the analogy of life somewhat differently: as an "organism," the software should grow by stages from childhood to adulthood, but at any given stage it is a self-sufficient, if limited, organism. **Prototypes** are functional implementations of software: they accept real input, such as the final system can be expected to deal with, and produce output of the type end-users will expect to find in the final system. What distinguishes early form prototypes from the final software is that the functions they implement are limited, so that they are fairly easily developed, whereas later ones approximate the full functionality envisioned. Since each prototype in the entire sequence supports a limited form of the targeted functionality, end-users can experiment with them in just the way they are intended to use the final product, and so they can provide feedback and advice from the early stages of development. Much of the testing, then, is done in a setting as close to the target usage environment as possible.

The **initial prototypes** should be used for planning the KEBN process. The subproblem addressed should be self-contained but reasonably representative of the global problem. It should be scoped to minimize development risk, with the prototype employing available capabilities as much as possible and using simplified variables and structure. As a result you avoid one of the main risks in the BN development process, overselling the capabilities of the final system (Laskey and Mahoney, 2000).

Since initial prototypes are both functional and representative, they provide a working product for provisional assessment and planning.

The **incremental prototypes** require relatively simple extensions to the preceding prototype. They should attack a high priority, but small, subset of the remaining difficult issues. The size of the subset of issues tackled, and their difficulty, is used to control the continuing development risk. The incremental development refines both the domain expert's and the knowledge engineer's understanding of the requirements and approach.

Why prototype? We believe that it just is the best software development process overall, agreeing with Fred Brooks (1995) and Barry Boehm (1988). Prototyping allows the organic growth of software, tracking the problem specifications and growing in manageable spurts. The use of prototypes attacks the trade-off between comprehensiveness and intelligibility from the right starting point, namely the small end. Those are general attributes favoring prototyping for software engineering of all kinds. However, because Bayesian network development usually involves building graphical models using visual aids from early on, it is even easier to provide end-users with the kind of graphical user interface the end product is likely to have, easing the prototyping approach.

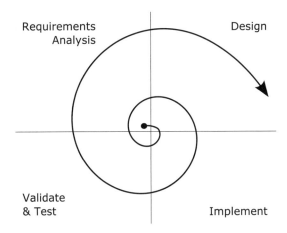

FIGURE 10.2: A spiral model for KEBN.

In order to highlight the differences between prototyping and the usual approach to software development, as typified in lifecycle models, Barry Boehm introduced the **Spiral Model** of software development, which we illustrate in Figure 10.2. The construction of a sequence of prototypes can be viewed as a repeating cycle of analyzing requirements, design, implementation, operation and evaluation. In evaluation the behavior of each prototype model on sample problems is explored, in conjunction with end-users, and the next stage is planned as problems in the current prototype are uncovered.

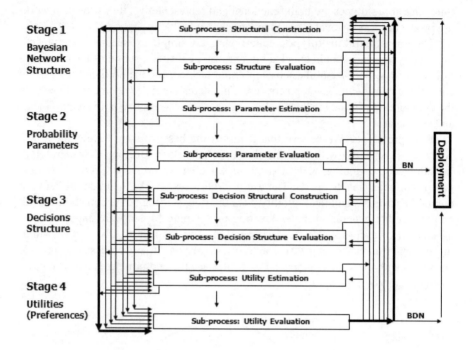

FIGURE 10.3: Boneh's iterative lifecyle model for KEBN (Boneh, 2010). (Reproduced with permission.)

10.2.3   Boneh's KEBN process

Figure 10.3 shows a more detailed KEBN process model (Boneh, 2010). It consists of four main stages, one for each aspect of the BDN: BN structure, probability parameters, BDN structure and utilities. Each stage is broken down into of two common sub-processes: (1) construction/estimation, i.e., building that part of the model, and (2) evaluation of the model to date.

The methodology is iterative and descriptive (rather than prescriptive), allowing complete flexibility. The stages and sub-processes may be undertaken in any order, while each sub-process and stage can be revisited as many times as necessary. This process model reflects our experience that there is no real division between development before and after deployment, Rather it depicts the integration of development, deployment and subsequent ongoing adaptation and refinement, with deployment itself a part of the cycle.

This methodology is clearly based on an underlying natural sequential order for the execution of sub-processes (as represented by the middle links in Figure 10.3). The structure must be built before the parameters can be added, while the BN must exist before it can be extended with decisions and utilities. While this sequential order will be followed in most occasions, variations may be driven by factors such as the intended model size and complexity, project time constraints, the individual style and expertise of the developers, the availability of resources, and so on.

Critical to developing Bayesian networks is evaluative feedback. One of the major advantages of the spiral, prototyping development of software is that from the very beginning a workable (if limited) user interface is available, so that end-users can get involved in experimenting with the software and providing ideas for improvement. This is just as true with the knowledge engineering of Bayesian networks, at least when a GUI front-end is available.

In the following section we consider each stage of the KEBN process in turn: BN structure, parameters, decision structure and preferences (utilities). For each stage, we introduce methods for elicitation and evaluation. We also show how the methods for automated learning of structure and parameters from data, presented in Chapters 7-9, fit into the KEBN process.

10.2.4 Are BNs suitable for the domain problem?

As we have already seen, Bayesian networks can handle a wide variety of domains and types of application. They explicitly model uncertainty, allow for the representation of complex interactions between variables, and they can be extended with utilities and decision nodes for planning and decision making. Furthermore, the representations developed are not just black boxes, but have a clear semantics, available for inspection. Nevertheless, BNs are not suitable to any and every application, so before launching any large-scale KEBN process, it is important to make sure that BN technology is suitable for the particular problem. It is easiest to list features that would suggest BNs are *not* appropriate.

- If the problem is a "one-off," for which there is no data available and any model built won't be used again, then the overhead of the KE process may not be worth it. Bayesian networks might still be used in a one-off modeling process, of course, without going through all of the KEBN knowledge engineering overhead.

- There are no domain experts, nor useful data.

- If the problem is very complex or not obviously decomposable, it may not be worth attempting to analyze into a Bayesian network.

- If the problem is essentially one of learning a function from available data, and a "black box" model is all that is required, an artificial neural network or other standard machine learning technique may be applied.

10.2.5 Process management

It is important that the knowledge engineering process be properly managed. Foremost is the management of human relations. It is unrealistic to expect that putting the domain expert and the knowledge engineer in a room together will result in the smooth production of a BN model for the domain. It is far more likely that they will talk past each other! The knowledge engineer must learn about the problem domain and the domain expert must learn what BN models are and what they can do. This learning aspect of the process can be time-consuming and frustrating, but both sides must expect and tolerate it.

One productive way of training the participants is to start by constructing very simple models of simplified, throw-away domain problems. These are "pre-proto-types": whereas prototypes are intended to feed into the next stage, at least conceptually, the throw-away models are intended to build mutual understanding. As knowledge engineering proceeds, both the domain expert and the knowledge engineer's understanding of the problem domain deepens. Throughout this process building communication between them is vital!

It is particularly important to get commitment from the expert to the project and the knowledge engineering process. It is going to take much time and effort from the expert, so the expert has to be convinced early on that it will be worth it. Certain characteristics are desirable in an expert: their expertise is acknowledged by their peers, they are articulate and they have the interest and the ability to *reason about* the reasoning process. Support, or at least tolerance, from the expert's management is, of course, necessary.

The rationale for modeling decisions should be recorded in a "style guide," which can be used to ensure that there is consistency across different parts of the model, even if they are developed at different times and by different people. Such a style guide should include naming conventions, definitions and any other modeling conventions. Most important is documenting the history of significant design decisions, so that subsequent decision making can be informed by that process. This is an aspect of **change management**. Another aspect is archiving the sequence of models developed. An automated version control tool, such as CVS, git or SVN, can be used to generate such archives, as well as avoid colliding changes when many people are working on the same project.

10.3   Stage 1: BN structure

There are several competing goals when building the graphical structure of a network. First, we would like to minimize the number of parameters, both in order to make the probability elicitation task easier and to simplify belief updating. These goals suggest fewer nodes, fewer arcs and smaller state spaces. On the other hand, we would obviously like to maximize the fidelity of the model, which sometimes requires more nodes, arcs and states (although excess detail can also decrease accuracy). A tradeoff must be made between building a more accurate model and the cost of additional modeling.

10.3.1 Nodes and values

10.3.1.1 Understanding the problem context

> *There are known knowns. These are things we know that we know. There are known unknowns. That is to say, there are things that we now know we don't know. But there are also unknown unknowns. These are things we do not know we don't know.*
>
> Donald Rumsfeld, on an occasion of rare accuracy.

Common Modeling Mistake 1 *Not understanding the problem context*

It is crucial for the knowledge engineer to gain a clear understanding of the problem context. Ideally, this should be available in some form of project description. In practice, the problem being addressed is usually clarified and refined during the knowledge engineering process. Thus, the knowledge engineer must first ask basic questions such as:

Q: *"What do you want to reason about?"*
Q: *"What don't you know?"*
Q: *"What information do you have?"*
Q: *"What do you know?"*

Answers to these questions give a list of **known** and **unknown** variables, which provides a good starting point for structure building.

When attempting to model large, complex domains it is important to limit the number of variables in the model, at least in the beginning, in order to keep the KE task tractable. The key is to determine which are the most important variables/nodes.

Common Modeling Mistake 2 *Complexity without value*

A very common impulse, when something is known about the problem, is to want to put it in the model. But including everything known, just because it *is* known, simply adds complexity to the model without adding any value (and in fact often reduces value). Instead, the knowledge engineer must focus on the question:

Q: *"Which of the known variables are most relevant to the problem?"*

10.3.1.2 Types of node

- One class of variables consists of those whose values the end-user wants to know about, the "output" nodes of the network, and are often referred to as the **target** or **query** or **output** nodes.
- The **evidence** or **observation** or **input** nodes play the role of "inputs" and can be identified by considering what sources of information about the domain are available, in particular, what evidence could be observed that would be useful in inferring the state of another variable.

- **Context** variables can be determined by considering sensing conditions and background causal conditions.
- **Controllable** variables are those whose values can be set by intervention in the domain environment (as opposed to simply observing their value).

It is important to note that the roles of nodes may change, depending how the BN is to be used. It is often useful to work backwards by identifying the query variables and "spreading out" to the related variables.

Let us return to the cancer diagnosis example used throughout Chapter 2 and look at it in terms of variable identification. The main interest of medical diagnosis is identifying the disease from which the patient is suffering; in this example, there are three candidate diagnoses. An initial modeling choice might be to have the query node *Disease*, with the observation nodes being *Dyspnoea* (shortness of breath), which is a possible symptom, and *X-ray*, which will provide another source of information. The context variables in this case are the background information about the patient, such as whether or not he is a *Smoker*, and what sort of exposure to *Pollution* he has had. In this simple diagnosis example, nothing has been described thus far that plainly falls into the category of a controllable variable. However, the doctor may well prefer to treat *Smoker* as a controllable variable, instead of as context, by attempting to get the patient to quit smoking. That may well turn into an example of a not-fully-effective intervention, as many doctors have discovered!

10.3.1.3 Types of values

When considering the variables, we must also decide what states, or values, the variable can take. Some common types of discrete nodes were introduced in §2.2.1: Boolean nodes, integer valued or multinomial categories. For its simplicity, and because people tend to think in terms of propositions (which are true or false), Boolean variables are very commonly employed. Equivalently, two-valued (binary) variables may be used, depending upon what seems most natural to the users. For example, when modeling the weather, the main weather node could be called *Weather*, and take the values {*fine, wet*}, or the node could be made a Boolean called *FineWeather* and take the values {*T, F*}. Other discrete node types will likely be chosen when potential observations are more fine-grained.

10.3.1.4 Discretization

While it is possible to build BNs with continuous variables without discretization, the simplest approach is to **discretize** them, meaning that they are converted into multinomial variables where each value identifies a different subrange of the original range of continuous values. Indeed, many of the current BN software tools available (including Netica) require this. Netica provides a choice between its doing the discretization for you crudely, into even-sized chunks, or allowing the knowledge engineer more control over the process. We recommend exercising this control and discretize manually or else using a more sophisticated algorithm, such as that of Monti and Cooper (1998).

TABLE 10.2

Alternative discretizations of an *Age* node with 4 values.

Whole Population	Pension Population	Students
0-18	50-60	4-12
19-40	61-67	13-17
41-60	68-72	18-22
61-110	73-110	23-90

Consider the situation where you must discretize an *Age* node. One possible discretization might reflect the age distribution in the population. However, the optimal discretization may vary depending on the situation, from the whole population, to people who receive government pensions, or people engaged in full time study — see Table 10.2. Here, each range covers 25% of the target population.

This discretization is still based on the idea of even-sized chunks. It may be better to base the discretization on differences in effect on related variables. Table 10.3 shows a possible discretization when modeling the connection between *Age* and number of children, represented by the node *NumChildren*.

TABLE 10.3

Discretization of an *Age* node based on differences in number of children.

Age	P(NumChildren|Age)				
	0	1	2	3	>4
0-11	1	0	0	0	0
12-16	0.95	0.04	0.01	0	0
17-21	0.90	0.07	0.02	0.01	0
22-25	0.80	0.12	0.05	0.02	0.01
26-30	0.40	0.25	0.18	0.10	0.07
31-34	0.30	0.25	0.25	0.14	0.06
35-42	0.25	0.20	0.30	0.20	0.05
43-110	0.22	0.23	0.25	0.22	0.08

10.3.2 Common modeling mistakes: nodes and values

Discrete variable values must be **exhaustive** and **exclusive**, which means that the variable must take on exactly one of these values at a time. Modeling mistakes relating to each of these factors are common.

Common Modeling Mistake 3 *Node values aren't exhaustive*

Sometimes this happens when a node is discretized, and the discrete states do not span the range of the actual variable. For example, for a node *Frequency* (say of an event), the values $\{< 1/20yrs, \leq 1/5yrs\}$ are not exhaustive; they don't allow the event occurring more often than one year in five. The solution is usually to either

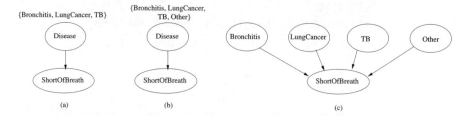

FIGURE 10.4: Node values must be exhaustive and mutually exclusive: disease example (a) not exhaustive (b) not mutually exclusive (c) correct

change the discretization or to modify the states covering the end(s) of the variable range; for the *Frequency* node, the set of values $\{< 1/20yrs, \leq 1/5yrs, > 1/5hrs\}$ is now exhaustive.

Common Modeling Mistake 4 *Node values aren't mutually exclusive*

For example, suppose that a preliminary choice for a *Disease* query node is to give it the values {*lungCancer, bronchitis, tuberculosis*}, as shown in Figure 10.4(a). This modeling choice isn't exhaustive, as it doesn't allow for the possibility of another disease being the cause of the symptoms; adding a fourth alternative *other* alleviates this problem, as shown in Figure 10.4(b). However, this doesn't solve the second problem, since taking these as exclusive would imply that the patient can only suffer from one of these diseases. In reality it is possible (though of course uncommon) for a patient to suffer from more than one of these, for example, *both* lung cancer and bronchitis. The best modeling solution here is to have distinct Boolean variables for each disease of interest, say the nodes *LungCancer*, *Bronchitis* and *Tuberculosis*, as in Figure 10.4(c). This model does not explicitly represent the situation where the patient suffers from another disease, but doesn't exclude it either.

Common Modeling Mistake 5 *Separate nodes for different states of the same variable.*

Another common problem made by naive BN modelers is the creation of separate nodes for different states of the same variable. For example, they might create both a *FineWeather* variable and a *WetWeather* variable (both Boolean). These states ought to be mutually exclusive.[1] Furthermore, once they are created as separate variables, the error is often "solved" by the addition of an additional arc between the nodes and a deterministic CPT that enforces the mutual exclusion. This is not very satisfactory, however, as the resultant structure is more complex than necessary. This is also an example of how the choices made when modeling nodes and states affects structure modeling.

---

[1] This example may strike the reader as trivial and/or unlikely, so we note it becomes much more likely if the nodes are created by different knowledge engineers working on a separate sub-networks, that are later combined.

FIGURE 10.5: Mixed breed dog, half pointer and half greyhound. (From http://en.wikipedia.org/wiki/File/puppymix.jpg.)

Common Modeling Mistake 6 *Trying to model fuzzy categories*

Suppose the modeling task required the representation of two dog breeds, say Pointer and Greyhound. It is straightforward to have a node *Breed* with possible values {*pointer, greyhound*}. But suppose dogs from those two breeds mate and produce a half-breed offspring (see Figure 10.5). A common modeling mistake is to try to represent this using the prior, if *Breed* is a root node: $P(Breed = pointer) = 0.5$, $P(Breed = greyhound) = 0.5$. Or for non-root nodes, a 50-50 likelihood evidence is added for the *Breed* node (see § 3.4).

Both are **incorrect** ways to model the fuzzy category of a half-breed dog, as they both allow later evidence (say from elsewhere in the network) to change the posteriors $Bel(Breed|e)$. Instead, the correct modeling solution is to extend the state space of the *Breed* node to {*pointer, greyhound, pointer greyhound mix*}. If the breed in a particular reasoning situation is then observed to be the mixed breed, this can be be added as evidence *Breed=pointer greyhound mix*.

Common Modeling Mistake 7 *Confusing state and probability*

We have seen BNs containing nodes with names such as "Likelihood" and "Risk", which we consider to be poor BN modeling. For example the LUIM BN framework (McNeill et al., 2006), shown in Figure 10.6 is using the BN for qualitative "meta-level" reasoning only, by directly mapping *Risk = Likelihood × Consequences* into the BN structure. Rather, BN nodes should represent the state of the world being modeled, while the probability (or "likelihood") of those states is the posterior probabilities computed after adding any evidence. Presumably the "Consequences" in LUIM are intended to represent the cost/benefits of outcomes. It is far preferable to use decision networks to model consequences quantitatively and to combine probabilities and utilities through the computation of an expected utility.

Common Modeling Mistake 8 *Confusion about what the node represents*

Consider the problem of modeling pest infestation of an orchard. Suppose the BN contains a node *OrchardInfested* with posteriors *Bel(OrchardInfested=T)* = 0.3, as

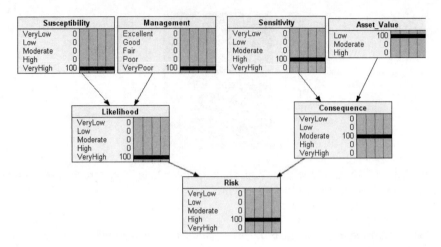

FIGURE 10.6: LUIM BN model is an example of poor BN modeling, with *Likelihood* and *Consequence* nodes. Reproduced with permission from Monash University ePress (McNeill et al., 2006, Fig. 5).

shown in Figure 10.7(a). This should probably mean that there is a 30% chance that the orchard is infested (which might be based on a known frequency of infested orchards in a particular region). But it is a common modeling mistake to take this to mean that 30% of the trees in the orchard, or 30% of the fruit coming from the orchard, are infested. This is better modeled by a *LevelOfOrchardInfestation*, as in Figure 10.7(b). How should these nodes be related? The direction of the arc between them may depend on what data (unconditional or conditional) is available about orchard infestation, but clearly an orchard is pest-free (*OrchardInfested=F*) if and only if it has a zero level of infestation (*LevelOfOrchardInfestation=none*), which must be reflected in the CPTs.

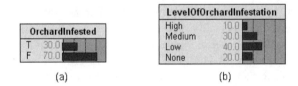

(a)   (b)

FIGURE 10.7: Orchard infestation example: (a) Boolean node (b) Node values represent levels of infestation.

10.3.3 Causal relationships

When deciding on the structure of the network, the key is to focus on the **relation-ships** between variables. There are many types of qualitative understanding that can help determine the appropriate structure for a domain.

The first, and most important, are the **causal** relationships. As we discussed earlier (see §2.4), while orienting the arcs in a causal direction is not required, doing so maximizes the representation of conditional independence, leading to a more compact, simpler model. To establish the causal relationships, the knowledge engineer must identify the variables that could cause a variable to take a particular state, or prevent it from taking a particular state. Once identified, arcs should be added from those causal variables, to the affected variable. In all the following examples, we will see a natural combination of variable elicitation with the identification of relationships between variables. Sometimes it is appropriate to ask direct questions about causes. For example:

Q: *"What can cause lung cancer?"*
A: *"Smoking and pollution."*
Modeling: suggests arcs from those nodes to the *LungCancer* node.

Q: *"Is there anything which prevents TB?"*
A: *"There is a TB immunization available."*
Modeling: suggests an arc from *Immunization* to *TB*.

Alternatively, the same cause-to-effect structure may be identified by asking about effects. For example:
Q: *"What are the effects of lung cancer?"*
A: *"Shortness of breath and a spot on the lungs that may show up on the X-ray."*
Modeling: suggests the arcs from *LungCancer* to *X-ray* and *Dyspnoea*.

Another kind of causal relationship is **prevention**, when an effect will occur *unless* a preventative action is taken first.
Q: *"Is there anything that can prevent HIV causing AIDS?"*
A: *"Anti-viral drugs such as AZT can prevent the development of AIDS."*
Modeling: suggests arcs from both *HIV* and *AZT* to *AIDS*.

Another way of looking at this is to consider the possibility of **interference** to the causal relationship.
Q: *"Is there any factor that might interfere with cholesterol-lowering medication treating heart disease?"*
A: *"Yes, if the patient doesn't modify her diet, the medication won't be effective."*
Modeling: suggests arcs from both *Medication* and *Diet* to *HeartDisease*.

We have already seen other ways of describing this sort of interference relationship in §5.6.4 when we looked at problems with a sensor that affect its measuring capacity, with terms such as **moderates** and **invalidates** being used to describe an **unless** condition. **Enabling relationships** exist where the enabling variable may under certain conditions permit, enhance or inhibit the operation of a cause.

Q: *"Is anything else required to enable the cholesterol-lowering medication to be affective against heart disease?"*

A: *"Yes, the patient must also modify her diet."*

Modeling: again, suggests arcs from both *Medication* and *Diet* to *HeartDisease*.

As we discussed earlier in §2.3.1, a v-structure in Bayesian networks, with two parents sharing a common effect, gives rise to a form of reasoning called "explaining away." To investigate whether this substructure exists, the knowledge engineer should ask a series of questions around **explanations** such as:

Q: *"Are both X and Y possible explanations for Z?"*

Q: *"Would finding out that Z is true increase your belief that both X and Y may be true?"*

Q: *"Would then finding out that X is true undermine your previously increased belief in Y?"*

We look further at causal interactions in §10.4.5.

10.3.4 Dependence and independence relationships

As we know, Bayesian networks encode conditional dependency and independency relationships. Very generally, a dependency can be identified by asking the domain expert:

Q: *"Does knowing something about the value of one variable influence your beliefs as to the value of the other variable, and vice versa?"*

Once we obtain a positive answer to this, we must determine what kind of dependency exists. If two variables are dependent regardless of the values of all other variables, then they should be directly connected.

Alternatively, some pairs of variables will be dependent only through other variables, that is, they are only conditionally dependent. d-separation tests can be used to check that the encoded relationships agree with the domain expert's intuitions. Unfortunately, the concept of d-separation can be difficult to interpret and takes time for domain experts to understand. We have been involved in the development of Matilda (Boneh et al., 2006), a software tool that can help domain experts explore these dependencies. It supports a visual exploration of d-separation in a network, supplemented by a non-technical explanation of possible interactions between variables.

Matilda uses terms like 'direct causes,' 'causes' and 'paths' to explain graph relations. The term 'd-separation' is replaced by the more intuitive notion of 'blocking.' Extending this, a d-separating set found among parent nodes is called a 'simple blocking set.' This is a d-separating set but may contain proper subsets that can also d-separate. A minimal d-separating set found among parent nodes is called an 'essential blocking set.' This is a d-separating set from which no node can be removed without impairing the d-separation. A minimal d-separating set found anywhere in the network is called 'minimal blocking set.' Matilda's focus on the parent nodes reflects the importance of these relationships in the modeling process.

In order to describe the general relation "X is d-separated from Y by Z" (i.e., $X \perp Y | \mathbf{Z}$) Matilda gives the terms X, Y and Z a specific temporary role in the model. The nodes X become the **query** nodes, the nodes we want to reason about. The nodes Y become the **observation** nodes, the nodes we may observe. The nodes Z are **known** nodes, the values of which we already have. The notion of **influence**, in the context of "a change in the value of node X does not influence the value of node Y," is described by the term **change of belief** (*"knowing the value of node X does not change our belief in the value of node Y"*). To be sure, by giving the sets of nodes these roles the symmetry of the relation could be lost. To overcome this, the phrase *"vice versa"* needs to be added. Using these terms, Matilda describes the relation $X \perp Y | \mathbf{Z}$ as:

> If the known nodes are observed, then knowing the value of the observation nodes will not change the belief about the value of the query nodes and vice versa.

Matilda's visualization of the relation "X is d-separated from Y given Z," with the corresponding verbal explanation is shown in Figure 10.8. The blocking relationship is shown using a common symbol for 'stop' or 'not allowed' (in red), for the Z node, with query nodes X indicated by "?" (purple), and observation nodes Y by an upside down "?" (blue). The color scheme ties the verbal and visual explanations together.[2]

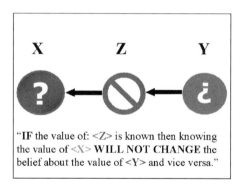

FIGURE 10.8: Matilda's visualization and explanation of "X is d-separated from Y given Z."

Matilda allows the user to ask various types of questions about the relationships between variables. A BN for a fire alarm domain (extending one used in Poole et al., 1998), shown in Figure 10.9, will be used to illustrate this process.

Example: Fire alarm. *We receive a phone call saying that everyone is leaving the building. The call can come from three different sources: a security center gets a*

---

[2]Note that we cannot reproduce the color scheme in this text.

report from a special sensor in the building. If the sensor reports 'Leaving' the security center calls us. We also get calls from kids who are playing practical jokes (mainly during the holidays) as well as from seriously concerned people who notice smoke coming out of the building (mainly after work-hours). The sensor is noisy. It sometimes does not report when everyone is leaving and sometimes reports leaving for no reason. If the fire alarm goes off, that causes people in the building to leave. The fire alarm can go off either because there is a fire or because someone tampers with it. The fire also causes smoke to rise from the building. In winter, fireplaces in the building can also cause the presence of smoke.

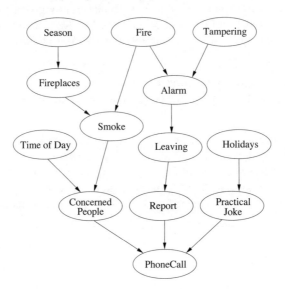

FIGURE 10.9: A BN solution for the fire alarm example.

Matilda's Type 1 question. What is the relationship between two nodes?

This option allows the user to ask "What information d-separates two selected nodes?" by selecting two nodes X and Y and choosing between the three types of d-separating sets: simple, essential and minimal blocking sets (as described above). Matilda then computes the selected type of d-separating set and highlights it using the blocking symbol.

Example: selected X=*Alarm*, Y=*Phone Call* (Figure 10.10).

A simple blocking set is the set of parents of *Phone Call*: {*Concerned People, Report, Practical Joke*}. The verbal explanation is: "IF the value of: <Practical Joke>, <Report>, <Concerned People> is known then knowing the value of <Alarm> WILL NOT CHANGE the belief about the value of <Phone Call> and vice versa."

An essential blocking set is the set {*Concerned People, Report*}. The verbal explanation is: "IF the value of: <Report>, <Concerned People> is known then knowing the value of <Alarm> WILL NOT CHANGE the belief about the value of <Phone Call> and vice versa."

The possible minimal blocking sets are: {*Fire, Report*}, {*Fire, Leaving*}, {*Report, Concerned People*}, {*Report, Smoke*}, {*Concerned People, Leaving*}, {*Leaving, Smoke*}. The verbal explanation of the first one is: "IF the value of: <Fire>, <Report> is known then knowing the value of <Alarm> WILL NOT CHANGE the belief about the value of <Phone Call> and vice versa."

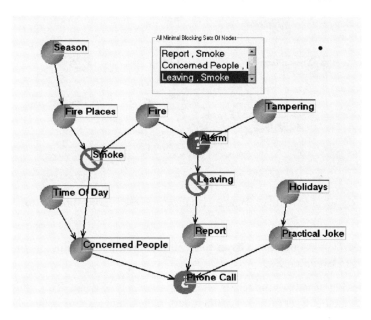

FIGURE 10.10: Matilda's Type 1 visualization of d-separation in the fire alarm example.

Matilda's Type 2 question. When does a node become irrelevant?

Here, Matilda visualizes the relationships between one node and the rest of the network. This option allows the user to select a single node *X* and ask for a set of nodes that d-separates ("blocks" in Matilda terminology) this node from the rest of the structure. The number of such sets is potentially exponential, so Matilda highlights only one set, namely the Markov blanket (see §2.2.2).

Example: selected *X=Report* (Figure 10.11).

The Markov blanket for this node is the set {*Leaving* (parent), *Phone Call* (child), *Practical Joke, Concerned People* (child's parents)}. The verbal explanation is: "IF the value of: <Leaving>, <Phone Call>, <Practical Joke>, <Concerned People> is known then knowing the value of <Report> WILL NOT CHANGE the belief about the value of any other node and vice versa."

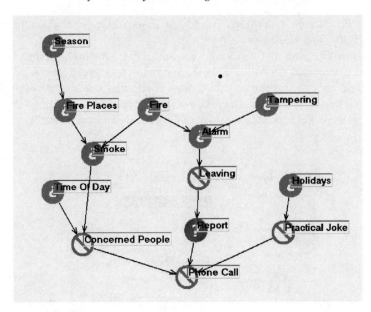

FIGURE 10.11: Matilda's Type 2 visualization of a Markov blanket for *Report*.

Matilda's Type 3 question. Given some information, what happens to the relationships between nodes?

Here, Matilda visualizes the relationships between *sets* of nodes. This option allows the user to select a set of nodes *X* (the query nodes) and a set of nodes *Z* (the prior information) and request the set of all *Y* nodes that are d-separated (blocked) from *X*. Matilda highlights all the nodes *Y* in response.

Example: selected *X*={*Phone Call*}, with prior information for nodes *Z*={*Smoke, Fire*} (Figure 10.12).

The nodes that are d-separated from *Phone Call* by *Smoke* and *Fire* are *Fire Places* and *Seasons*. The verbal explanation is "IF the value of: <Fire>, <Smoke> is known then knowing the value of <Phone Call> WILL NOT CHANGE the belief about the value of <Season>, <Fire Places> and vice versa."

In short, for various types of query, Matilda visualizes the relation "X is d-separated from Y given Z." In the first type, the user chooses *X* and *Y*; in response the tool highlights *Z*. In the second question type, the user chooses *X*; in response the tool highlights *Z*. In the third question type, the user chooses *X* and *Z*; in response the tool highlights *Y*. Note that in the first two types of queries the tool identifies the d-separating sets of nodes, whereas in the latter type of query the user is asking the question with regard to a specific d-separating set of nodes.

Case-studies (Boneh et al., 2006) suggest that Matilda is useful in understanding networks as they are being built and in understanding the consequences of different possible design choices. It can be used not just by BN experts, but also by domain experts to validate the network and identify potential problems with the structure. It

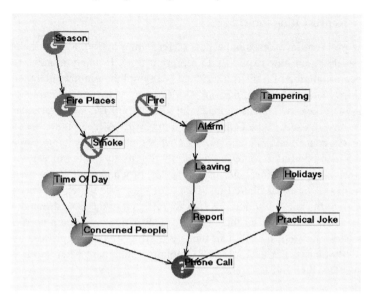

FIGURE 10.12: Matilda's Type 3 visualization.

can be used to investigate an existing network at any stage during development. It is especially helpful in investigating networks built by automated methods, when prior intuitive understanding may be weak.

10.3.5 Other relationships

There are other less explicit indications of the correct network structure.

Association relationships occur when knowing a value of one variable provides information about another variable. By Reichenbach's Principle of the Common Cause, some causal nexus must explain the association; although any active (un-blocked) path will do this, just the information that there is *some* active path may serve as a beginning in building the causal structure. The absence of an unconditional association is also useful to know. Thus, in the fire alarm example of Figure 10.9, there is no marginal dependence between *Time of Day* and *Fire*, which is an implication worth checking. On the other hand, there is an implied association between *Report* and *Smoke* (through their common ancestor *Fire*).

In many domains, there may be a known **temporal ordering** of variables, where one event or value change occurs *before* another. The known ordering may be either total or partial. In either case, the temporal information will restrict the orientation of some of the arcs, assuming you are building a causal network.

10.3.5.1   Representing time

When the problem involves some aspect of reasoning over time, a modeling choice must be made about how explicitly to represent time. In many models, there is no explicit representation of time. Rather there is an implicit temporal element in causal relationships, for example in *Smoking* → *Cancer*. The simplest explicit temporal modeling is to have two nodes representing the state of the same variable at different times; for example, in the Goulburn Catchment ERA (see Figure 5.13) there are nodes for *CurrentAbundance* and *FutureAbundance*. Or, a dynamic Bayesian network (§4.5) can be used to represent explicitly the full dynamic process over time.

  Whether there is a before/after pair of nodes, or whether there are multiple time slices for all the variables, there must be a time step over which the change is happening. Often this is left implicit, i.e., not represented in the BN, although of course affects the change modeled in the CPTs. An alternative is to add a node *delta-T* to represent the time step; the BN can then be used to model the process at different temporal granularity. For example, the Goulburn Catchment ERA had a node *Time* with values *OneYear* and *FiveYear*.

Common Modeling Mistake 9 *Thinking BNs can't represent feedback loops*

Because Bayesian networks are directed *acyclic* graphs, it is sometimes thought they cannot be used to model feedback loops. However it is straightforward to model feedback processes with a dynamic Bayesian network. Figure 10.13 shows two examples. In the first, there is a feedback loop between socio-economic status (SES) and education. The DBN version breaks that down into the generations, that is, a person's education level depends on the SES of their parents, whereas that education level influences the SES they attain in their own life. The second example shows a feedback loop between the amount of grass in a pasture and the size of the herd that pasture can carry; the loop from *Grass* back to itself indicates a dynamic process

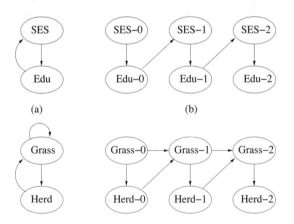

FIGURE 10.13: (a) Feedback loops modeled with (b) DBNs.

for the grass itself. The DBN version shows the grass and herd size interacting over some unspecified time step.[3]

10.3.6 Controlling the number of arcs

Common Modeling Mistake 10 *Too many parents!*

If a node has many parents, then the size of the CPT can become very large. One way to reduce the number of parents is to remove the so-called "weak" arcs, those that do not have much impact on the child node.

Another way to reduce the number of parameters is to alter the graph structure; **divorcing multiple parents** is a useful technique of this type. It is typically applied when a node has many parents (and so a large CPT), and when there are likely groupings of the parents in terms of their effect on the child. Divorcing means introducing an intermediate node that summarizes the effect of a subset of parents on a child.

An example of divorcing is shown in Figure 10.14; the introduction of variable *Z* divorces parent nodes *A* and *B* from the other parents *C* and *D*. The method of divorcing parents was used in MUNIN (Andreassen et al., 1989). Divorcing involves a trade-off between new structural complexity (the introduction of additional nodes and arcs) and parameter simplification. We note that divorcing may also be used for purposes other than reducing the number of parameters, such as dealing with overconfidence (Nikovski, 2000).

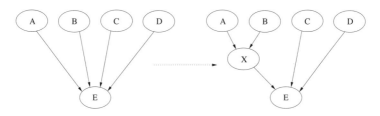

FIGURE 10.14: Divorcing example.

Divorcing example: search and rescue. *Part of the land search and rescue problem is to predict what the health of a missing person will be when found. Factors that the domain expert initially considered were external factors, such as the temperature and weather, and features of the missing person, such as age, physical health and mental health.*

This example arises from an actual case of Bayesian network modeling (Boneh et al., 2006). A BN constructed by the domain expert early in the modeling process for a part of the search and rescue problem is shown in Figure 10.15(a). The number of parameters required for the node *HealthWhenFound* is large. However, there is a natural grouping of the parent nodes of this variable into those relating to

---

[3]Obviously this example is very simplified, with no representation of season, temperature, rain, etc. that would all effect the grass growth rate.

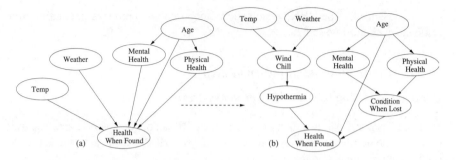

FIGURE 10.15: Search and rescue problem: (a) the original BN fragment; (b) the divorced structure.

weather and those relating to the missing person, which suggests the introduction of intermediate nodes. In fact, it turns out that what really matters about the weather is a combination of temperature and weather conditions producing a dangerous wind chill, which in turn may lead to hypothermia. The extent to which the person's health deteriorates due to hypothermia depends on *Age* and the remaining personal factors, which may be summarized by a *ConditionWhenLost* node, hence, the introduction of three mediating variables, *WindChill*, *Hypothermia* and *ConditionWhenLost*, as shown in Figure 10.15(b). The result is a structure that clearly reduces the number of parameters, making the elicitation process significantly easier.[4]

NB: Note that divorcing comes at a price, namely an inability to reflect any interaction across divorced parents. E.g., in Figure 10.14 if A and C interact in their influence on E, then the proposed variable X does not allow that to be represented. That is the reason why groups of related parents are sought (such as climate in Figure 10.15), since their merger should not generally break important interactions.

10.3.7   Combining discrete and continuous variables

The initial work on BNs with continuous variables (Pearl, 1988, Shachter and Kenley, 1989) only allowed variables with linear Gaussian distributions. A method for combining discrete and continuous variables was proposed (Lauritzen and Wermuth, 1989) and implemented in cHugin (Olesen, 1993); this approach did not allow for discrete children of continuous parents. The usual solution is to discretize all continuous variables at some stage. As suggested above, this is best done manually at the moment. In the future, we expect the best discretization methods will be incorporated in Bayesian network tools, so that the process can be automated, given sufficient statistical data.

---

[4]This example comes from a case study undertaken when evaluating Matilda (Boneh et al., 2006). Examination of the initial structure with Matilda led to the expert identifying two missing variables, *Wind Chill* and *Hypothermia*. (Note that these two could be combined in the network shown without any loss of information; this would not work, however, in the larger network.) Subsequent analysis led to the divorce solution.

There has also been some research on extending the inference algorithms to cope with some special cases of continuous variables and inference, and these no doubt will eventually be applied in application tools (e.g., Lerner et al., 2001).

10.3.8 Using other knowledge representations

We also note that building the BN structure is a form of conceptual modeling, whose aim is to represent 'the meaning of terms and concepts used by domain experts to discuss the problem, and to find the correct relationships between different concepts' (Wikipedia, 2010). There are many different flavours of conceptual models, together with techniques and tools for building them.

For example, concept maps were developed by Joseph D. Novak and his research group at Cornell University in the 1970s (Novak and Cañas, 2008), while alternatives include argument maps, cognitive maps, conceptual graphs, mind maps, mental models, topic maps, semantic network and the semantic web! In the KEBN process, early BN structure prototypes can certainly involve production of knowledge representation structures, such as concept maps or ontologies (e.g., Boneh, 2010, Helsper and van der Gaag, 2007), which are later transformed or extended into the actual BN structure.

Qualitative probabilistic networks (QPN) (Wellman, 1990) provide a qualitative abstraction of Bayesian networks, using the notion of positive and negative influences between variables. Wellman shows that QPNs are often able to make optimal decisions, without requiring the specification of the quantitative part of the BN, the probability tables. QPNs can also be used as a preliminary model in the knowledge engineering process, which can later be fleshed out.

10.3.9 Common modeling mistakes: arcs

Common Modeling Mistake 11 *Getting the arc direction wrong*

It is very common to see BNs that model the human **reasoning** process instead of the **causal** process. For example, typical human diagnostic reasoning would be that if someone has a temperature, it is quite likely she or he has the flu. Thinking about the problem in this way leads quite naturally to a BN containing Temperature $\rightarrow$ Flu. We discussed earlier (e.g., § 2.4) why the BN *should* model Cause $\rightarrow$ Effect relationships. The main advantages are that causal networks represent the dependencies more efficiently (with fewer arcs and so fewer parameters) and that causal networks allow themselves to be used in causal reasoning, and especially reasoning about possible interventions.

It is easier to make this sort of error if relevant causal mechanisms are omitted.

Example: Type 1 Diabetes. *The presence of relevant antibodies is a risk factor for later developing Type 1 diabetes. Infants who are never breastfed have twice the risk of developing Type 1 diabetes as those who are breastfeed for 3 months or longer. Vitamin D deficiency in the first year of life is also thought to increase the risk of* *developing Type 1 diabetes. Data is available to show "that 2% of people that have a relative with type 1 diabetes will have the relevant antibodies present, while people who do not have a relative with type 1 diabetes only a 0.1% chance of having the relevant antibodies present."*

Figure 10.16(a) shows a possible structure that might result if only the known risk factors as described are modeled explicitly. This is an example of the more general modeling mistake of **inverting cause and effect**. A consideration of the causal mechanisms suggests that: a relative with diabetes may indicate a genetic predisposition, antibodies are unlikely to cause the diabetes, but can be the result of an inappropriate autoimmune response related to the genetic disposition, while the presence of antibodies may be a symptom of Type 1 diabetes. This in turn suggests an alternative structure, shown in Figure 10.16(b), reflecting these possible causal mechanisms.[5]

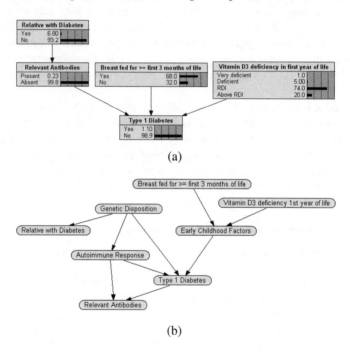

(a)

(b)

FIGURE 10.16: Type 1 diabetes example: (a) confusing cause and effect; (b) reflecting causal mechanisms.

The difficulty with getting the arc direction right can be a problem with **missing variables**. Consider the problem of modeling the relationships between the following variables (this arose in our work on the so-called IOWA medical data set, reported in

---

[5]Disclaimer: this is a teaching example developed from our own superficial reading about Type 1 diabetics, with no input from a medical expert!

Flores et al., 2010).

SHRBRLIE: Shortness of breath when lying flat? (Y/N)
PRSSCHST: Ever had any pressure in chest? (Y/N)
ABLWALK: Able to walk 1/2 mile without help? (Y/N)

The data shows these variables are all dependent on each other, but while the structures shown in Figure 10.17(a) and (b) both model that dependence, neither is correct. Instead, the solution is to add another node, which is a common cause. This sort of modeling errors occurs when the modeling is too focused on the data that is available.

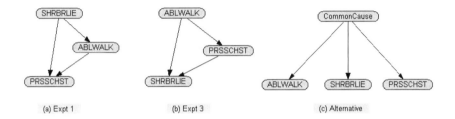

(a) Expt 1   (b) Expt 3   (c) Alternative

FIGURE 10.17: Missing variables lead to incorrect causal structures.

10.3.10 Structure evaluation

Common Modeling Mistake 12 *Parameterizing before evaluating the structure*

It is *crucial* that the structure be evaluated *before* adding the parameters. Structure evaluation may reveal that an important variable has been overlooked, or if irrelevant nodes have been included, that the node values aren't right, that an important arc is missing, or that an arc shouldn't be there. All these require a structural change that will result in different conditional probabilities being required. If the network has already been parameterized, very often the effort expended to acquire some parameters will be completely wasted, as they are just thrown away!

Q: *"Are the nodes in the BN the right ones?"*
Q: *"Are the state space (values) for each node correct?"*
Q: *"Are the arcs rights?"*

If the answer is 'no' to *any* of these, the structure will change and many of the parameters in the CPTs will have to be thrown away; the time spent learning or eliciting them will have been wasted. **Always evaluate the structure before parameterizing!**

10.3.10.1   Elicitation review

An elicitation review is a structured review of the major elements of the elicitation process. It allows the knowledge engineer and domain expert to take a global view of the work done and to check for consistency. This is especially important when working with multiple experts or combining expert elicitation with automated learning methods.

First, the variable and value definitions should be reviewed, encompassing:

1. A clarity test: do all the variables and their values have a clear operational meaning — i.e., are there definite criteria for when a variable takes each of its values? "Temperature is high" is an example that does not pass the clarity test;[6] this might be refined to "Temperature $\geq$ 30 degrees" which does.

2. Agreement on variable definitions:

   - Are all the relevant variables included? Are they named usefully?
   - Are all states (values) appropriate? Exhaustive and exclusive?
   - Are all state values useful, or can some be combined?
   - Are state values appropriately named?
   - Where variables have been discretized, are the ranges appropriate?

3. Consistency checking of states: are state spaces consistent across different variables? For example, it might cause misunderstanding if a parent variable takes one of the values {*veryhigh, high, medium, low*} and its child takes one of {*extremelyhigh, high, medium, low*}.

The graph structure should be reviewed, looking at the implications of the d-separation dependencies and independencies and at whether the structure violates any prior knowledge about time and causality.

It is often useful to do a **model walk-through**, where a completed version of the model is presented to domain experts who have not been involved in the modeling process to date. All components of the model are evaluated. This can be done by preparing a set of scenarios with full coverage of the BN — i.e., sets of assumed observations with various sets of query nodes, when the reasoning required over the scenarios collectively exercises every part of the network. Ideally, there will be scenarios testing the network under both extreme and normal conditions. Scenario walk-throughs perform an analogous function to code reviews in the software engineering process.

10.4   Stage 2: Probability parameters

The parameters for a BN are a set of conditional probability distributions of child values given values of parents. There is one distribution for each possible instantia-

---

[6]We have been guilty of exactly this in §3.2.

tion of parent variables; so the bad news is that the task of probability assessment is exponential in the number of parent variables. If there is local structure (see §10.4.5), of course, the number of parameters to be estimated is reduced.

10.4.1 Parameter sources

There are three possible parameter sources.

1. Data

We have previously described some specific methods for learning parameters from domain data (see Chapter 6). General problems with data include: noise, missing values and small samples. These can be overcome to a certain extent by using robust data mining techniques. It is also useful to instrument the domain for collecting data to use in the future, via the adaptation of §10.9.

2. Domain Experts

Common Modeling Mistake 13 *Experts' estimates of probabilities are biased, including (a) overconfidence, (b) anchoring and (c) availability*

The most basic problem is finding suitable experts who have the time and interest to assist with the modeling process. Another difficulty is that humans, including expert humans, almost always display various kinds of bias in estimating probabilities. These include:

- **Overconfidence**, the tendency to attribute higher than justifiable probabilities to events that have a probability sufficiently greater than 0.5. Thus, an event which objectively has a probability of 0.9 will usually be attributed a probability that is somewhat higher (see Figure 10.18). To be sure, expertise itself has been found to have a moderating effect on this kind of miscalibration (Lichtenstein, Fischhoff, and Phillips, 1982).

- **Anchoring**, the tendency for subsequent estimates to be "weighed down" by an initial estimate. For example, if someone is asked to estimate the average age of coworkers and begins that process by estimating the average age of those in a meeting, a high (or low) age at the meeting will very probably bias the estimate upwards (or downwards) (Kahneman and Tversky, 1973).

- **Availability**, that is, assessing an event as more probable than is justifiable, because it is easily remembered or more salient (Tversky and Kahneman, 1982).

There is a large, and problematic, literature on assessing these biases and proposals to debias human probability estimates. The latter have met with limited success. The best advice we can give, short of an in-depth exploration of the literature (and we do recommend some such exploration, as described in §10.11 below), is to be aware of the existence of such biases, discuss them with the experts who are being asked to make judgments and to take advantage of whatever statistics are available, or can be made available, to test human judgments against a more objective standard.

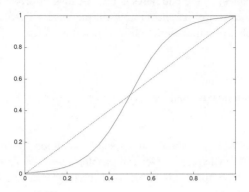

FIGURE 10.18: Overconfidence curve: subjective probability (vertical axis) vs. objective probability (horizontal axis).

Two attributes of a good elicitation process are (adapted from Morgan and Henrion, 1990, pp. 158-159):

1. The expert should be apprised of what is known about the process, especially the nearly universal tendency to overconfidence and other forms of bias. In order to avoid some of the problems, values should be elicited in random order and the expert *not* given feedback on how the different values fit together until a complete set has been elicited.

2. The elicitation process is not simply one of requesting and recording numbers, but also one of refining the definitions of variables and terms to be used in the model. What values are elicited depends directly upon the interpretation of terms and these should be made as explicit as possible and recorded during the elicitation. This is a part of the process management described earlier in §10.2.5.

3. The Literature

There may be a published body of knowledge about the application domain. This can be useful if the original data are not available, by providing other models learned from the data that can be incorporated into the BN. For example, in § 5.3 we described BN models for predicting coronary heart disease (CHD) that were based on (equation-based) epidemiological models from the medical literature. Using expert knowledge from the literature can reduce the elicitation burden on domain experts, or allow a model to be parameterized even when no domain experts are available. Expert information from the literature also has the advantage of having been peer reviewed, compared to information directly elicited from experts for the BN modeling process.

One common problem with published statistics is sparseness. For example, in a medical diagnosis domain, the probability of observing a symptom given a disease,

$P(Symptom|Disease = T)$, may be available in the medical textbooks ("80% of patients with TB will present with a cough"), but not $P(Symptom|Disease = F)$, the information about the frequency of the symptom occurring when the disease is not present. There is also a bias in what information is available in the literature: the fact of its publication reflects interest. These problems can be moderated by using expert opinion to review the proposed parameterization.

There is a risk involved with combining parameters from different sources, with different biases. It is not necessarily the case that combining sources with multiple biases smoothly averages out to no bias, or to a readily measurable and manageable bias. For an informal example, one author shot baskets at the Exploratorium in San Francisco with some biased goggles. At first, this resulted in missing to the left, but fairly soon the brain accommodated the misdirection and the basketball started hitting the hoop again. As soon as the goggles were removed, however, the shots missed to the right! The unbiased new input was being combined with an older bias, leading to error.

10.4.2   Probability elicitation for discrete variables

For discrete variables, one approach is direct elicitation, where an expert provides a number such as "the probability is 0.7." However, given the problems people have with such estimation, noted above, other elicitation techniques might be considered. People are often better at providing frequencies rather than probabilities, such as "1 in 4" or "1 in 10,000" (Gigerenzer and Hoffrage, 1995), especially for situations where the probabilities involved are very large or very small. Assessing extreme probabilities directly is difficult, and orders of magnitude assessments might be tried.

Assessing by odds is often useful. For example, given that $Y = y$, a domain expert may report:

"It is three times more likely that variable X *has the value* x_1 *than* x_2*."*

This information gives the equation $P(X = x_1|Y = y) = 3P(X = x_2|Y = y)$, and only one probability has to be elicited.

Common Modeling Mistake 14 *Inconsistent "filling in" of large CPTs*

When the CPT is large, there is often inconsistency in the parameterization across it. This inconsistency may be at the individual probability level; for example, the expert uses 0.99 for "almost certain" in one part of the CPT, and 0.999 in another. Or it may be inconsistency across the CPT; for example, using different distributions for combinations of parents that in fact are very similar. Such inconsistencies can occur because different parts of the CPT are provided by different experts, or are elicited at different times. Or they may occur simply because the expert can't keep track of enough "anchoring" cases.

An alternative approach is to use qualitative assessment, where the domain expert describes the probability in common language, such as "very high" or "unlikely." Such probability elicitation using a scale with numerical and verbal anchors is described in van der Gaag et al. (1999). The verbal cues in that scale were `certain`, `probable`, `expected`, `fifty-fifty`, `uncertain`, `improbable` and `impossible`. Having been translated from Dutch some of these cues are inappropriate, for example "`uncertain`" is ambiguous and could be replaced with "`unlikely`." Because these qualitative phrases can be ambiguous (in fact, this is the problem of **linguistic uncertainty**), they can cause miscommunication, especially where more than one domain expert is involved, unless they are themselves calibrated. It is advisable to do the mapping of verbal levels to actual probabilities (called the **verbal map**) separately from the probability elicitation exercise. The verbal map should be customized to suit the individual expert. We have developed a software tool called VE (for Verbal Elicitor) in conjunction with the Netica BN software, which supports the qualitative elicitation of probabilities and verbal maps (Hope et al., 2002).

An example of VE's main window for eliciting qualitative probabilities is shown in Figure 10.19, together with the window editing the verbal map for this tool. The probabilities associated with each verbal cue are set using a slider bar, and other verbal cues can be added if desired.

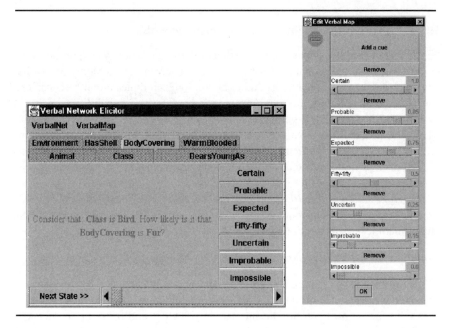

FIGURE 10.19: VE: main elicitation window (left) and the verbal map editing window (right).

Common Modeling Mistake 15 *Incoherent probabilities (not summing to 1)*

A common problem when eliciting probabilities is that an expert may specify probabilities that are **incoherent**, failing to satisfy the probability axioms. For example, suppose the verbal mapping shown in Figure 10.19 is being used and that the ex-

TABLE 10.4
Incoherent qualitative assessments for the *X-ray* CPT.

$P(X\text{-}ray = Pos|Cancer = T)$ = Probable
$P(X\text{-}ray = Neg|Cancer = T)$ = Improbable
$P(X\text{-}ray = Pos|Cancer = F)$ = Uncertain
$P(X\text{-}ray = Neg|Cancer = F)$ = Certain

pert provides the qualitative assessment of the probabilities for X-ray results given in Table 10.4.

These probabilities are incoherent, since they do not sum to one for each conditioning case. VE provides a function to correct incoherent probabilities automatically; it uses an iterative, linear optimization to find verbal mappings that are coherent and as close as possible to the original map.

Common Modeling Mistake 16 *Being dead certain*

> *It ain't what you don't know that gets you into trouble. It's what you know for sure that just ain't so.*
>
> Mark Twain

> *I beseech you, in the bowels of Christ, think it possible that you may be mistaken.*
>
> Oliver Cromwell, writing to the synod of the Church of Scotland on August 5, 1650

The extremity of overconfidence, being dead certain, is almost always a mistake. If the prior probability of a hypothesis is set to either extreme, that is, to 0 or 1, then Bayes' theorem tells us that the posterior probability (the probability given the evidence) must be the same, regardless of how much evidence there is to the contrary. Dennis Lindley (1985) dubbed this Cromwell's Rule, after the quote above, which states that one should avoid using prior probabilities of 0 or 1, except when applied to statements that are logically true or false. Lindley describes the amusing example of a coherent Bayesian who attaches a zero prior to the hypothesis that the Moon is made of green cheese and cannot then change his belief even after whole armies of astronauts coming back bearing tasty slices. Examples closer to home may include the beliefs of some about Creationism and global warming.

Similar care should be taken to avoid setting an entry in the CPT to 0. Such a 0 entry means that the node *cannot* take that value given the particular combination of parents. If in future, that combination of parents' values and the node value *are* observed, it causes problems, it is considered an impossible combinations of evidence that the software won't allow to be entered. So, even if a situation appears "extremely unlikely", it should be given a very small probability, e.g., $\varepsilon = 0.001$.

10.4.3 Probability elicitation for continuous variables

The easiest continuous variables to parameterize are those which are distributed according to a **parametric model**, that is, those which are fully characterized by a limited number of parameters. The most common example is the Normal distribution (also known as the Gaussian or "bell curve"), $N(\mu, \sigma^2)$, which is characterized by just two parameters, the mean μ and the variance σ^2. Normal distributions are used to model: noisy measurement processes (e.g., velocities, the positions of stars); the central tendencies (average values) of almost any process (e.g., average age or income of samples from a population), which is justified by the central limit theorem of probability theory; and, by a kind of metaphorical extension, summative measures of a complex of processes when the individual processes are not well understood (e.g., IQ). Another popular parametric distribution is the exponential, which is often used to model life- and death-processes, such as the life span of an electrical part. Still other parametric continuous distributions are the Gamma, Chi-square and F distributions. These, and the discrete parametric distributions (e.g., binomial and Poisson), form much of the material of probability and statistics classes. They are worth learning about since in every case one need only obtain a good estimate of only a few parameters from an expert in order to obtain a good estimate of the entire probability distribution — assuming, of course, that the variable in question is properly modeled by the chosen family of distributions!

If the problem is not as simple as estimating a couple of parameters, like the mean and variance of a normal distribution, then most common is to elicit estimates of key values for the probability density function. Recall from §1.3.2 that if the continuous density function is $f(x)$, then the cumulative distribution function $F(x)$ is

$$F(x) = P(X \le x) = \int_{x' \le x} f(x')dx' \tag{10.1}$$

A likely route to estimating the density function is by bi-sectioning it. First, elicit the median, the value at which X is equally likely to be found either above or below. Then elicit the 25th percentile by "bisecting" the region below the median into two further equally likely regions, and then the 75th percentile analogously. This process can be continued until the density has been sufficiently well refined for the problem, or until the expert can no longer make meaningful distinctions. This kind of estimation may be usefully accompanied by the expert simply sketching her or his impression of the shape of the density function; otherwise one might overlook something simple and important, such as whether the density is intended to be unimodal or bimodal, skewed or symmetric, etc.

Having estimated a distribution in this fashion, it may be best and simplest to find a parametric model (with parameter values) which reproduces the estimated distribution reasonably well and use that for your Bayesian network model.

10.4.4 Support for probability elicitation

Visual aids are known to be helpful and should be used for probability elicitation (see Figure 10.20). With a **pie chart** the expert aims to size a slice of the "pie" so that a spinner will land in that region with the probability desired. A **histogram** may help the expert to order discrete events by probability. As we mentioned, simple freehand drawings of probability distributions can also be informative.

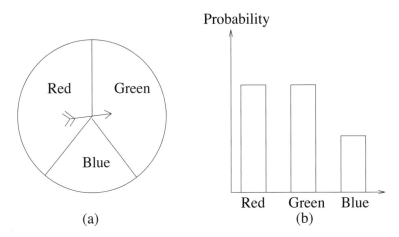

FIGURE 10.20: Examples of visual aids for probability elicitation: (a) pie chart; (b) histogram.

Lotteries can be used to force estimates of either probabilities or utilities, in techniques going back to Ramsey (1931). Given clear utility values, say dollars, you can elicit someone's estimated probability of an uncertain event E by finding at what point the person is indifferent between two gambles: the first one paying, say, $100 if E comes true; the second one paying, say, $1 million if a (free) lottery ticket *Wins*. Since the two gambles are considered equivalued, we have (where N is the number of lottery tickets required to reach indifference):

$$EU(Gamble) = \$100P(E) = EU(Lottery) = \frac{\$1000000}{N} \qquad (10.2)$$

Hence, $P(E) = 10000/N$. Lotteries can be used analogously to elicit unclear utilities for an outcome state by manipulating the probability of reaching that state until the expert is indifferent between the proposed gamble and some lottery ticket with known value and probability of winning.

Our VE software tool provides a number of useful automated functions that facilitate probability elicitation. One function normalizes CPTs, allowing users to specify ratios in place of probabilities.[7] For example, if the expert thinks that someone has three times the chance of getting cancer as not getting cancer (say if they are a smoker), they can specify

$$P(Cancer = T|Smoker = T) : P(cancer = F|Smoker = F)$$

---

[7]Note that some other software packages have this feature.

as 3:1, which the tool translates into the probabilities 0.75 and 0.25, respectively.

VE can also perform a maximum entropy fill of CPTs, where the remaining probabilities are filled in uniformly with the probability remaining after subtracting supplied probabilities from 1. This means the expert need only provide probabilities for combinations about which s/he is confident. For example, if the variable A has states a_1, a_2 and a_3, and the probability for $A = a_1$ given some combination of values for its parent variables, $P(A = a_1|B = T, C = F)$ is set to 0.95, then $P(A = a_2|B = T, C = F)$ and $P(A = a_3|B = T, C = F)$ will both be set to 0.025 automatically.

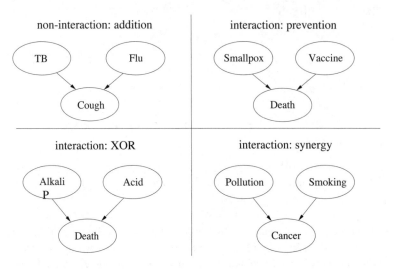

FIGURE 10.21: Different qualitative causal relationships.

10.4.5   Local structure

When parameterizing the relation between parents and a child node, the possibility of there being "local structure" was discussed in Chapter 6 in the context of automated parameter learning. It is also of interest in the elicitation process, of course. There are various *kinds* of causal interaction, such as those displayed in Figure 10.21. A classic example of interaction between causes is **XOR**, where each cause cancels the other out. The alkali/acid case (§6.4.1) is an example of this: *one might ingest alkali, and die; one might instead ingest acid, and die; but if one ingests both alkali and acid together, then one may well not die.*

Other causal interactions include **prevention**, where one causal factor intervenes to stop another, such as a *Vaccine* preventing *Smallpox* leading to *Death*. And again there is the possibility of **synergy**, where the effects are reinforced by the occurrence of both causes beyond the mere addition of the effects independently. All of these relationships can be looked for explicitly during an elicitation process. For example, **Q:** *"Given that* Acid *and* Alkali *are independently causes of* Death*, when taken jointly what happens to the risk?"*

A: *"It is decreased."*
Modeling: in this case, the causal interaction is clearly an XOR type.

These kinds of interaction imply that the probabilities associated with one or more of the possible instantiations of the parents are *independent* of the probabilities associated with the other parent instantiations. For example, knowing what happens when you ingest Acid but not Alkali tells you little or nothing about what happens when you ingest both.

Local structure is the opposite situation: there is some structure across the different parent instantiations that allows you to infer some probabilities from the others. We have already discussed some different models of non-interaction (local structure) in Chapter 6, namely noisy-or, logit models and classification tree models, all of which allow a more compact specification of the CPT under non-interaction. In our original noisy-or example of *Flu*, *TB* and *SevereCough* (in Figures 6.4 and 10.21), this relationship would be identified by negative answers to the questions:
Q: *"Does having* TB *change the way that* Flu *causes a* Severe Cough?"*
A: *"No."*
Q: *"Similarly, does* Flu *change the way that* TB *causes a* Severe Cough?"*
A: *"No."*

Assuming that *Flu* and *TB* have been identified as causes of *Severe Cough*, these answers imply that the probability of not having the symptom is just the product of the independent probabilities that every cause present will fail to induce the symptom. (This is illustrated in Table 6.2 and explained in the surrounding text.) Given such a noisy-or model, we only need to elicit three probabilities: namely, the probability that *Flu* will fail to show the symptom of *Severe Cough*, the probability that *TB* will fail to show the symptom and the background probability of not having the symptom.

Local structure, clearly, can be used to advantage in either the elicitation task or the automated learning of parameters (or both).

One method for eliciting local structure is "elicitation by partition" (Heckerman, 1991, Geiger and Heckerman, 1991, Mahoney and Laskey, 1999). This involves dividing the joint states of parents into subsets such that each subset shares the conditional probability distribution for the child states. In other words, we **partition** the CPT, with each subset being a **partition element**. The task is then to elicit one probability distribution per partition element. Suppose, for example, that the probability of a high fever is the same in children, but not adults, for both the flu and measles. Then the partition for the parent variables *Flu*, *Measles* and *Age* and effect variable *Fever* would produce two partition elements for adults and one for children.

Note that this elicitation method directly corresponds to the use of classification trees and graphs in automated parameter learning (see §6.4.3). So, one way of doing partitioning is by building that corresponding classification tree by hand. Figure 10.22 illustrates this possibility for a simple network.

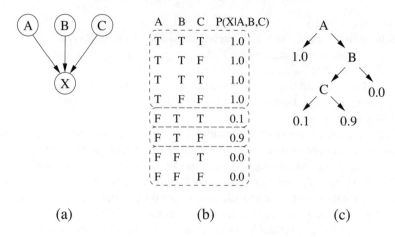

	A	B	C	P(X|A,B,C)
	T	T	T	1.0
	T	T	F	1.0
	T	T	T	1.0
	T	F	F	1.0
	F	T	T	0.1
	F	T	F	0.9
	F	F	T	0.0
	F	F	F	0.0

(a)   (b)   (c)

FIGURE 10.22: Local CPT structure: (a) the Bayesian network; (b) the partitioned CPT; (c) classification tree representation of the CPT.

10.4.6   Case-based evaluation

Case-based evaluation runs the BN model on a set of test cases, evaluating the results via expert judgment or some kind of prior knowledge of the domain. Test cases can be used either for **component testing**, where fragments of the BN are checked, or for **whole-model testing**, where the behavior of the full BN is tested. The results of interest for the test will depend on the domain, sometimes being the decision, or a diagnosis, or again a future prediction. Ideally, cases would be generated to test a wide variety of situations, possibly using the **boundary-value analysis** technique of software engineering: choosing inputs at the extreme upper and lower values of available ranges to ensure reasonable network performance across the largest possible set of conditions. The usability of this approach will depend upon the experts having enough experience to make useful judgments in all such cases.

Case-base evaluation tends to be informal, checking that expert opinion is being properly represented in the network. If a great many cases can be brought together, then more formal statistical methods can be applied. These are discussed immediately below, as validation methods.

10.4.6.1   Explanation methods

If a BN application is to be successfully deployed and used, it must be accepted first by the domain experts and then by the users.[8] As we have already discussed (§5.7 and §10.3.4), the independence-dependence relations encoded in the BN structure are not always obvious to someone who is not experienced with BNs. Also, the outcomes of probabilistic belief updating are often unintuitive. These factors have motivated the development of methods for the **explanation** of Bayesian networks.

---

[8]Of course the domain experts may also be the intended users.

Some of these methods focus on the explanation *of* the Bayesian network. Others provide explanations of the conclusions drawn about the domain *using* the Bayesian network — that is, how the inference has led from available evidence to the conclusions. This latter is how our NAG would like to be employed eventually (§5.7). The term "explanation" has also been used to describe the provision of the "most probable explanation"(MPE) (§3.7.1) of the evidence (e.g., Lacave et al., 2000). We view that more as an inferential matter, however.

NESTOR (Cooper, 1984), one of the first BN diagnostic tools, included two kinds of explanations: (1) a comparison of how two diagnoses account for the evidence and (2) a critique of a hypothesis. In both cases it generated a verbal explanation, with an accompanying visualization. INSITE (Suermondt, 1992) was an early explanation system that identified the findings that influence a certain variable and the paths through which the influence flows. It then generates visual cues (shading and highlighting paths) and verbal explanations of the probabilistic influences. Díez, in his expert system DIAVAL (1997), developed a method for explaining both the model and the reasoning. This was a menu-based system, where the explanation focused on one node or link at a time, but supported the user navigating through the network. GRAPHICAL-BELIEF (Madigan et al., 1997) is an explanatory system that provides a graphical visualization of how evidence propagates through a BN. Elvira (see §B.3) is a BN package that offers both verbal explanations (about nodes and arcs) as well as graphical explanations, such as the automatic coloring of arcs to indicate positive and negative influences.

Explanation methods have also been used for teaching purposes. For example, BANTER (Haddawy et al., 1997) is a BN tutoring shell, based on INSITE's methods, for tutoring users (such as medical students) in the evaluation of hypotheses and selection of optimal diagnostic procedures. In BANTER, the use of the BN by the system to perform its reasoning is completely invisible to the user. B2 (Mcroy et al., 1997) is an extension of BANTER that generates both graphical and verbal explanations of the reasoning in a more consistent way than BANTER.

In addition to supporting a deployed system or *being* part of the application itself, such explanation methods can be useful during the iterative evaluation of the KEBN process. They have been shown to support the detection of possible errors in the model (Haddawy et al., 1997) and can be used by the domain expert to verify the output of the model (Suermondt, 1992, Madigan et al., 1997, Druzdzel, 1996).

10.4.7 Validation methods

Validating a Bayesian network means to confirm that it is an accurate representation of the domain or process being modeled. Much of what we have discussed above, concerning comparing the behavior of a network with the judgments of experts, could be considered a kind of validation. However, we will reserve the term for more statistically oriented evaluation methods and specifically for methods suited to testing the correctness of a Bayesian network when a reasonably large amount of data describing the modeled process is available.

First there is a general consideration. The Bayesian networks are (usually) built or learned under an explicitly causal interpretation. But much validation work in the literature concerns testing the probability distribution represented by the network against some **reference distribution**. By a reference distribution (or, reference model, etc.) we mean a known distribution which is being used as the source for sample data; typically, this is a Bayesian network which is being used to artificially generate sample data by simulation. Now, since Bayesian networks within a single Markov equivalence class can be parameterized so as to yield identical probability distributions, testing their probability distributions against a reference distribution fails to show that you have the right causal model. For example, the Kullback-Leibler divergence (§3.6.5) from the true distribution to the learned distribution will not help distinguish between the different causal models within a single Markov equivalence class. For a discussion of alternative measures for this case, see §9.8.

One way of dealing with this problem is to collect experimental data, rather than simply take joint observations across the variables. In that case, we can represent the causal interventions explicitly and distinguish between causal models. There are many reasons why this option may be unavailable, the main one being that collecting such experimental data may be prohibitively expensive or time consuming.

A more readily available option is to analyze the problem and identify a subset of nodes which are characteristically going to be the query nodes in application and another subset which are going to be evidence nodes. Sample data can then be used to see how well the learned network predicts the values of the query nodes when the evidence nodes take the values observed in the sample. This can be done whether the query nodes are answers to diagnostic queries (i.e., causes of the evidence nodes) or are causally downstream from the evidence nodes. As described so far, this evaluation also does not take into account the causal structure being learned. It will often turn out that the restricted subnetwork being examined has few Markov equivalent subnetworks even when the full network does. Alternatively, you can examine causal structure directly, penalizing errors (as we discussed in §9.8.1). In one of our studies, we tested learned models restricted to leaf nodes only against the reference model's leaf nodes; if our method mislearned causal structure it would misidentify leaves, thus leading to a higher reported error (Neil and Korb, 1999).

In §7.5 we described a number of methods for evaluating Bayesian networks against data. Ideally this should be *real* data, sampled from the real process to be modeled, rather than against data simulated from reference models. Much of the research literature, however, employs artificial data, sampled from a reference model. It is important to be clear about the difference in intent behind these two procedures. If we are testing one machine algorithm against another (e.g., K2 against CaMML), then sampling from an artificially constructed Bayesian network and then seeing whether K2 or CaMML can learn from the sample data something like the original network makes good sense. In that case we are testing *algorithms* against each other. But if we are interested in testing or validating a *model* for a real process, then we presumably do not know in advance what the true model is. If we did, we would already be done. So, the practical validation problem for KEBN is to test some models, constructed by hand or learned by our machine learning algorithms, against real data reporting the history of the process to be modeled. All of the methods in §7.5

are applicable to evaluating against the real data, except for the Kullback-Leibler divergence (§3.6.5), which is used to motivate our information reward measure.

10.4.8 Sensitivity analysis

Another kind of evaluation is to analyze how sensitive the network is to changes in parameters or inputs; this is called **sensitivity analysis**. The network outputs may be either the posterior probabilities of the query node(s) or (if we are building a decision network) the choice of action. The changes to be tested may be variations in the evidence provided, or variations in the network parameters — specifically conditional probability tables or utilities. We will look at each of these types of changes in turn.

10.4.8.1 Sensitivity to evidence

Earlier in §10.3.4, we saw how the properties of d-separation can be used to determine whether or not evidence about one variable may influence belief in a query variable. It is also possible to measure this influence. Given a metric for changes in belief in the query node (which we address just below), we can, for example, rank evidence nodes for either the maximum such effect (depending on which value the evidence node takes) or the average such effect. It is also possible, of course, to rank sets of evidence nodes in the same way, although the number of such sets is exponential in the number of evidence nodes being considered. In either case, this kind of sensitivity analysis provides guidance for the collection of further evidence. An obvious application area for this is medical diagnosis, where there may be multiple tests available; the clinician may like to perform the test that most decreases the uncertainty of the diagnosis.

So, how to quantify this uncertainty? What metric of change should we employ? We would like to drive query node probabilities close to 0 and 1, representing greater certainty. **Entropy** is the common measure of how much uncertainty is represented in a probability mass. The entropy of a distribution over variable X is (cf. Definition 9.2):

$$H(X) = -\sum_{x \in X} P(x) \log P(x) \qquad (10.3)$$

For continuous variables, we can simply substitute integration for summation. The goal, clearly, is to minimize entropy: in the limit, entropy is zero if the probability mass is concentrated on a single value.

A second measure used sometimes is **variance**:

$$Var(X) = \sum_{x \in X} P(x - \mu)^2 P(x) \qquad (10.4)$$

where μ is the mean, i.e., $\sum_{x \in X} x P(x)$. Variance is the classic measure of the dispersion of X around its mean. The greater the dispersion, the less is known; hence, we again aim to reduce this measure to zero.

TABLE 10.5

Entropy and variance measures for three distributions
(H is computed with natural logs).

P(Z=1)	P(Z=2)	P(Z=3)	P(Z=4)	H(Z)	Var(Z)
1	0	0	0	0.0	0.0
0.25	0.25	0.25	0.25	1.39	1.25
0	0.5	0.5	0	0.69	0.25

Whether using entropy or variance, the metric will actually be computed for the query node *conditional upon* whatever evidence nodes are being tested for sensitivity — that is, for the distribution $P(X|\mathbf{E})$ rather than $P(X)$. To provide a feel for these measures, Table 10.5 compares the entropy and variance for a few distributions over a quaternary variable Z.

The main BN software packages provide a facility to quantify the effect of evidence using some such measure.

Lung cancer example

Let us re-visit our lung cancer example from Figure 2.1 to see what these measures can tell us. *Cancer* is our query node, and we would like to know observations of which single node will most reduce uncertainty about the cancer diagnosis. Table 10.6 shows the output from the Netica software's "sensitivity to findings" function, which provides this information. Each pair of rows tells us, if evidence were obtained for the node in column 1:

- The minimum (column 3) and maximum (column 4) posterior probabilities for $C = T$ or $C = F$ (column 2); and
- The reduction in entropy , both in absolute terms (column 5) and as a percentage (column 6).

The first set of results assumes no evidence has been acquired, with the prior $Bel(C = T) = 0.012$. As we would expect, all the "min" beliefs are less than the current belief and "max" beliefs above. Obtaining evidence for C itself is a degenerate case, with the belief changing to either 0 or 1. Of the other nodes, X (*XRay*) has the most impact on C, then S, D and P respectively. Netica's output does not tell us which particular observed value actually gives rise to the minimum and maximum new belief; however, this is easy to work out.

Suppose that the medical practitioner orders the test having the most potential impact, namely an X-ray, which then returns a positive result. The second set of results in Table 10.6 shows the new sensitivity to findings results (with values for $C = F$ omitted). In this example the relative ordering of the remaining nodes is the same — S,D and P — but in general this need not be the case.

These results only consider observations one at a time. If we are interested in the effect of observations of multiple nodes, the results can be very different. To compute the entropy reduction of a pair of observations each possible combination of values

TABLE 10.6

Output from Netica's sensitivity to findings function for the Cancer Example, with *Cancer* as the query node.

Node	value of C	min Bel(C)	max Bel(C)	Entropy Reduction	(%)
No evidence. Bel(C=T)=0.012					
C	T	0	1	0.0914	100%
	F	0	1		
X	T	0.001	0.050	0.0189	20.7%
	F	0.950	0.999		
S	T	0.003	0.032	0.0101	11.0%
	F	0.968	0.997		
D	T	0.006	0.025	0.0043	4.7%
	F	0.975	0.994		
P	T	0.001	0.029	0.0016	1.7%
	F	0.971	0.990		
Evidence $X = Pos$. Bel(C=T)=0.050					
C	T	0	1	0.2876	100%
S	T	0.013	0.130	0.0417	14.5%
D	T	0.026	0.103	0.0178	6.2%
P	T	0.042	0.119	0.0064	2.2%
Evidence $X = Pos, S = T$. Bel(C=T)=0.129					
C	T	0	1	0.5561	100%
D	T	0.069	0.244	0.0417	7.5%
P	T	0.122	0.3391	0.0026	0.47%
Evidence $X = Pos, S = T, D = T$. Bel(C=T)=0.244					
C	T	0	1	0.8012	100%
P	T	0.232	0.429	0.0042	0.53%

of both nodes must be entered and the entropy reductions computed. Netica does not provide for this explicitly in its GUI, but it is easily programmed using the Netica API.

10.4.8.2 Sensitivity to changes in parameters

Sensitivity analysis can also be applied to the parameters, checking them for correctness and whether more precision in estimating them would be useful. Most such sensitivity analyses are one-dimensional, that is, they only vary one parameter at a time. For many decision problems, decisions will be unaffected by the precision of either the model or the input numbers. This phenomenon is known as a **flat maximum**, as it marks the point beyond which small changes in probabilities or utilities are unlikely to affect the final decision (Covelo et al., 1986). Once a flat maximum is observed, no more effort should be spent refining the model or its inputs. However, even when a network may be insensitive to a change in one of its parameters, it may well still be sensitive to changes in combinations of parameters. As always, testing all possible combinations of parameters is exponentially complex.

It is plausible that the elicitation process can be supported by interactively performing sensitivity analyses of the network during construction, starting with initial rough assessments (Coupé et al., 2000). A sensitivity analysis indicates which probabilities require a high level of accuracy and which do not, providing a focus for subsequent elicitation efforts.

One can approach sensitivity analysis of BNs either empirically (e.g., Kipersztok and Wang, 2001, Coupe et al., 1999) or theoretically (e.g., Laskey, 1993, Kjærulff and van der Gaag, 2000). The empirical approach examines the effects of varying the parameters on a query node. It requires first identifying reasonable ranges for each parameter, then varying each parameter from its lowest to its highest value while holding all the other variables fixed. The changes in the posterior probability of the query node may be quantified with a measure such as entropy (see above). Or when we are concerned with decision making, we can identify at what point (if any) changes in a parameter can change the decision.

However, such a straightforward analysis can be extremely time consuming, especially on large networks. Coupé and Van der Gaag (2002) address this difficulty by first identifying a "sensitivity set" of variables given some evidence. These are those variables which can potentially change, meaning the remaining variables can be eliminated from further analysis. The sensitivity set can be found using an adapted d-separation algorithm (see Woodberry et al., 2004a).

The theoretical approach instead determines a function expressing the posterior probability of the query node in terms of the parameters. For example, Laskey (1993) computes a partial derivative of the query node with respect to each parameter. In addition to not requiring additional assessment by the expert, often this function can be computed efficiently by a modification of the standard inference algorithms. Coupé and Van der Gaag demonstrated that the posterior probability of a state given evidence under systematic changes to a parameter value can be given a functional representation, either linear or hyperbolic.

The two approaches can be used together, by first having a theoretical approach identify parameters with potentially high impact, then reasonable ranges can be assessed for these, allowing one to compute the sensitivity empirically.

For some time there was a belief common in the BN community that BNs are generally insensitive to inaccuracies in the numeric value of their probabilities. So the standard complaint about BNs as a method that required the specification of precise probabilities ("where will the numbers come from?") was met with the rejoinder that "the numbers don't really matter." This belief was based on a series of experiments described in Pradhan et al. (1996), which was further elaborated on in Henrion et al. (1996). Clearly, however, the inference that insensitivity is general is wrong, since it is easy to construct networks that are highly sensitive to parameter values (as one of the problems in §10.12 shows). Indeed, more recent research has located some real networks that are highly sensitive to imprecision in particular parameters (Kipersztok and Wang, 2001, van der Gaag and Renooij, 2001, Chan and Darwiche, 2003). The question of how to identify such sensitive parameters, clearly an important one from a knowledge engineering perspective, is a current research topic.

There is as yet only limited support in current BN software tools for sensitivity analysis. In § 11.4.4, we describe our own implementation, used in the Goulburn Catchment ERA.

10.5 Stage 3: Decision structure

Since the 1970s there have been many good software packages for decision analysis. Such tools support the elicitation of decisions/actions, utilities and probabilities, building decision trees and performing sensitivity analysis. These areas as well covered in the decision analysis literature (see for example, Raiffa's *Decision Analysis* Raiffa, 1968, an excellent book!).

The main differences between using these decision analysis tools and knowledge engineering with decision networks are: (1) the scale — decision analysis systems tend to require tens of parameters, compared to anything up to thousands in KEBN; and (2) the structure, as decision trees reflect straight-forward state-action combinations, without the causal structure, prediction and intervention aspects modeled in decision networks.

We have seen that the KE tasks for ordinary BN modeling are deciding on the variables and their values, determining the network structure, and adding the probabilities. There are several additional KE tasks when modeling with decision networks, encompassing decision/action nodes, utility (or value) nodes and how these are connected to the BN.

First, we must model what decisions can be made, through the addition of one or more decision nodes. If the decision task is to choose only a single decision at any one time from a set of possible actions, only one decision node is required. A good deal can be done with only a single decision node. Thus, a single *Treatment* decision node with options {*medication, surgery, placebo, no-treatment*} precludes consideration of a combination of surgery and medication. However, combinations of actions can be modeled within the one node, for example, by explicitly adding a *surgery-medication* action. This modeling solution avoids the complexity of multiple decision nodes, but has the disadvantage that the overlap between different actions (e.g., medication and surgery-medication) is not modeled explicitly.

An alternative is to have separate decision nodes for actions that are not mutually exclusive. This can lead to new modeling problems, such as ensuring that a "no-action" option is possible. In the treatment example, the multiple decision node solution would entail 4 decision nodes, each of which represented the positive and the negative action choices, e.g., {*surgery, no-surgery*}. The decision problem becomes much more complex, as the number of combinations of actions is $2^4 = 16$. Another practical difficulty is that many of the current BN software tools (including Netica) only support decision networks containing either a single one-off decision node or multiple nodes for sequential decision making. That is, they do not compute optimal *combinations* of decisions to be taken at the same time.

10.6 Stage 4: Utilities (preferences)

The next KE task for decision making is to model the utility of outcomes. The first stage is to decide what the unit of measure ("utile") will mean. This is clearly domain specific and in some cases fairly subjective. Modeling a monetary cost/benefit is usually fairly straightforward. Simply adopting the transformation $1 = 1 utile provides a linear numeric scale for the utility. Even here, however, there are pitfalls. One is that the utility of money is not, in fact, linear (as discussed in §4.2): the next dollar of income undoubtedly means more to a typical student than to a millionaire.

When the utile describes subjective preferences, things are more difficult. Requesting experts to provide preference orderings for different outcomes is one way to start. Numeric values can be used to fine tune the result. As noted previously, hypothetical lotteries can also be used to elicit utilities. It is also worth noting that the domain expert may well be the wrong person for utility elicitation, either in general or for particular utility nodes. It may be that the value or disvalue of an outcome arises largely from its impact on company management, customers or citizens at large. In such cases, utilities should be elicited from those people, rather than the domain experts.

It is worth observing that it may be possible to get more objective assessments of value from social, business or governmental practice. Consider the question: What is the value of a human life? Most people, asked this question, will reply that it is impossible to measure the value of a human life. On the other hand, most of *those* people, under the right circumstances, will go right ahead and measure the unmeasurable. Thus, there are implicit valuations of human life in governmental expenditures on health, automobile traffic and air safety, etc. More explicitly, the courts frequently hand down judgments valuing the loss of human life or the loss of quality of life. These kinds of measurement should not be used naively — clearly, there are many factors influencing such judgments — but they certainly can be used to bound regions of reasonable valuations. Two common measures used in these kinds of domains are the **micromort** (a one in a million chance of death) and the **QALY**, a quality-adjusted life year (equivalent to a year in good health with no infirmities).

The knowledge engineer must also determine whether the overall utility function consists of different value attributes which combine in an additive way.

Q: *"Are there different attributes that contribute to an overall utility?"*
Modeling: add one utility node for each attribute.

Finally, the decision and utility nodes must be linked into the graph structure. This involves considering the *causal* effects of decisions/actions, and the *temporal* aspects represented by information links and precedence links. The following questions probe these aspects.
Q: *"Which variables can decision/actions affect?"*
Modeling: add an arc from the action node to the chance node for that variable.

Q: *"Does the action/decision itself affect the utility?"*
Modeling: add an arc from the action node to the utility node.
Q: *"What are the outcome variables that there are preferences about?"*
Modeling: add arcs from those outcome nodes to the utility node.
Q: *"What information must be available before a particular decision can be made?"*
OR **Q:** *"Will the decision be contingent on particular information?"*
Modeling: add information arcs from those observation nodes to the decision node.
Q: *"(When there are multiple decisions) must decision D1 be taken before decision D2?"*
Modeling: add precedence arcs from decision node D1 to decision node D2.

10.6.1 Sensitivity of decisions

In general, rather than estimating raw belief changes in query nodes, we are more interested in what evidence may *change* a decision. In particular, if an observation *cannot* change the prospective decision, then it is an observation not worth making — at least, not in isolation.

In medical diagnosis, the decision at issue may be whether to administer treatment. If the BN output is the probability of a particular disease, then a simple, but ad hoc, approach is to deem treatment appropriate when that probability passes some threshold. Suppose for our cancer example that threshold is 50%. Then any one observation would not trigger treatment. Indeed, after observing a positive X-ray result and learning that the patient is a smoker, the sensitivity to findings tells us that evidence about shortness of breath still won't trigger treatment (since max(Bel(C=T)) is 0.244).

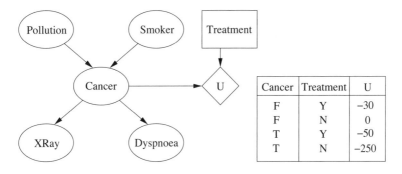

FIGURE 10.23: A decision network for the cancer example.

This whole approach, however, is ad hoc. We are relying on an intuitive (or, at any rate, unexplicated) judgment about what probability of cancer is worth acting upon. What we really want is to decide for or against treatment based upon a consideration of the expected value of treatment. The obvious solution is to move to decision networks, where we can work out the expected utility of both observations and treatments. This is exactly the evaluation of "test-act" combinations that we looked at in §4.4, involving the **value of information**.

Figure 10.23 shows a decision network for the extended treatment problem. The effect of different observations on the expected utilities for *Treatment* are shown in Table 10.7. We only consider the observations that can increase the belief in cancer and hence may change the treatment decision. We can see that evidence about any single node or pair of nodes will not change the decision, because in all cases the expected utility for not treating remains greater than the expected utility for treating the patient. However, once we consider the combinations of three evidence nodes, the decision can change when *X-Ray = pos*.

TABLE 10.7
Expected utilities for *Treatment* in the extended lung cancer problem (highest expected utility in **bold**).

X=pos	S=T	D=T	P=high	EU(Treat=Y)	EU(Treat=N)
No evidence				-30.23	**-2.908**
•				-31.01	**-12.57**
	•			-30.64	**-8.00**
		•		-30.50	**-6.22**
			•	-30.58	**-7.25**
•	•			-32.59	**-32.37**
	•	•		-31.34	**-16.71**
		•	•	-31.22	**-15.19**
•		•		-32.06	**-25.73**
	•		•	-31.00	**-12.50**
•			•	-32.37	**-29.62**
•	•	•		**-34.88**	-60.94
•	•		•	**-33.83**	-47.87
•		•	•	**-34.51**	-56.38
	•	•	•	-32.05	**-25.59**
•	•	•	•	**-36.78**	-84.78

The work we have done thus far is insufficient, however. While we have now taken into account the expected utilities of treatment and no treatment, we have not taken into account the costs of the various observations and tests. This cost may be purely monetary; more generally, it may combine monetary costs with non-financial burdens, such as inconvenience, trauma or intrusiveness, as well as some probability of additional adverse medical outcomes, such as infections caused by examinations or biopsies.

Let us work through such reasoning with the simple, generic decision network shown of Figure 10.24. Here there is a single disease node and a single test findings node. The utility is a function of the disease and the treatment decision. We would also like to take into account the option for running the test, represented by another decision node, which also has an associated cost, C. (A cost of $C = 5$ is used for the calculations in the remainder of this section.)

If we do not run the test, the probability of the disease is just the prior (0.30) and

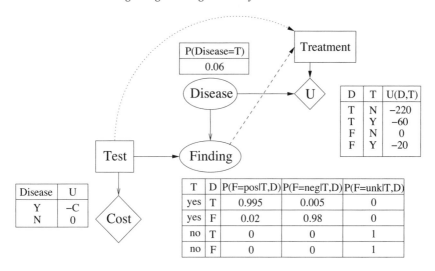

FIGURE 10.24: A decision network for disease treatment.

the expected utilities of the treatment options are:

$$
\begin{aligned}
EU(Treatment = Y) &= P(Disease = T) \times U(T = Y, D = T) \\
&\quad + P(Disease = F) \times U(T = Y, D = F) \\
&= 0.06 \times -60 + 0.94 \times -20 = -22.4 \\
EU(Treatment = N) &= P(Disease = T) \times U(T = N, D = T) \\
&\quad + P(Disease = F) \times U(T = N, D = F) \\
&= 0.06 \times -220 + 0.94 \times 0 = -13.2
\end{aligned}
$$

So, without a test, the decision would be *not* to have treatment.

If, on the other hand, we run the test and get a value for the findings (positive or negative), we obtain the expected utilities of Table 10.8. The decision tree evaluation for this problem is shown in Figure 10.25, using the method described in §4.4.3.

TABLE 10.8

Expected utilities given different test findings (highest EU given in **bold**, indicating decision table entry).

EU	Treatment	
Finding	Y	N
positive	**-55.42**	-172.31
negative	-25.01	**-5.07**

Hence the **expected benefit** of running the test, the *increase* in our expected utility, is the difference between the expected utility of running the test and the expected

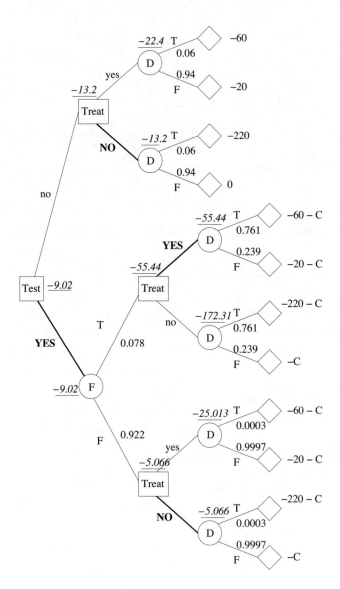

FIGURE 10.25: A decision tree evaluation of the disease test-treatment decision network.

utility of our decision without the test, namely: $EU(Test = Y) - EU(Treatment = N)$ = -9.02 − (-13.20) = 4.18.

What if there are several tests that can be undertaken? In order to determine the best course of action, ideally we would look at all possible combinations of tests, including doing none. This will clearly often be impractical. In such cases, a greedy search may be used, where at each step, the single test with the highest expected benefit is chosen (if this is greater than no test). The price of practicality is that greedy search can result in a non-optimal sequence, if the outcome of a rejected test would have made another more informative.

10.6.1.1 Disease treatment example

Before we leave this topic, let us briefly look at an example of how a decision may be sensitive to changes in parameters, using the generic disease "test-treatment" sequential decision network from Figure 10.24. Figure 10.26 shows how the expected utilities for the test decision (y-axis) vary as we change a parameter (x-axis), for three different parameters.

(a) **The test cost, C:** While EU(Test=N) does not change, we can see that the test decision will change around the value of C=9.

(b) **U(Disease=F,Treatment=Y):** In this case, the change in decision from "no" to "yes" occurs between -5 and -6.

(c) **The prior, P(Disease=T):** Here, the decision changes between 0.03 and 0.04.

10.7 Modeling example: missing car

We will now run through the modeling process for an example we have used in our BN courses, the missing car problem (Figure 10.27).[9] This is a small problem, but illustrates some interesting modeling issues.

The first step is to highlight parts of the problem statement which will assist in identifying the Bayesian network structure, which we have already done in Figure 10.27. For this initial analysis, we have used the annotations:

- possible situations ($\rightarrow$ nodes and values)
- **words connecting situations** ($\rightarrow$ graphical structure)
- indication of evidence ($\rightarrow$ observation nodes)
- focus of reasoning ($\rightarrow$ query nodes)

The underlined sections suggest five Boolean nodes, shown in the table in Figure 10.27, consisting of two query nodes and three observation nodes.

There are two possible **explanations**, or **causes**, for the car being missing, suggesting causal arcs from *CarStolen* and *SamBorrowed* to *MissingCar*. Signs of

---

[9]The networks developed by our students are available from the book Web site.

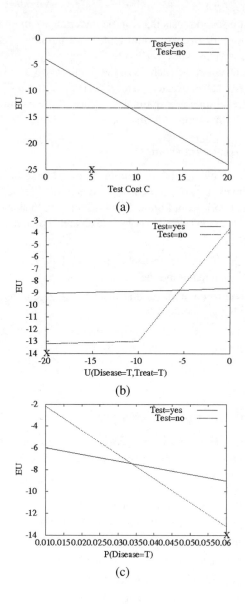

FIGURE 10.26: Sensitivity analysis showing effect of varying a parameter on the test decision, for three parameters: (a) Cost of test, C; (b) U(Disease=T,Treatment=yes); (c) P(Disease=T). Parameter value in model marked with X on the x-axis.

Marked-Up Problem Statement

John and Mary Nguyen arrive home after a night out to find *that their second car is not in the garage.* Two **explanations** *occur to them:* **either** the car has been stolen **or** *their daughter* Sam has borrowed the car *without permission. The Nguyens know that* **if** *the car was stolen,* **then** *the garage will probably* show *signs of forced entry. The Nguyens also know that Sam has a busy social life, so* **even if** *she didn't borrow the car, she* **may be** out socializing. Should they be worried about their second car being stolen and notify the police, or has Sam just borrowed the car again?

Preliminary Choice of Nodes and Values

Type	Description	Name	Values
Query	Has Nguyen's car been stolen?	*CarStolen*	*T,F*
	Has Sam borrowed the car?	*SamBorrowed*	*T,F*
Observation	See 2nd car is missing	*MissingCar*	*T,F*
	See signs of forced entry to garage?	*ForcedEntry*	*T,F*
	Check whether or not Sam is out?	*SamOut*	*T,F*

FIGURE 10.27: Missing car problem: preliminary analysis.

forced entry are a likely **effect** of the car being stolen, which suggests an arc from *CarStolen* to *ForcedEntry*. That leaves only *SamOut* to connect to the network. In the problem as stated, there is clearly an association between Sam being out and Sam borrowing the car; which way should the arc go? Some students add the arc *SamOut → SamBorrowed* because that is the direction of the inference or reasoning. A better way to think about it is to say "Sam borrowing the car *leads to* Sam being out," which is both a causal and a temporal relationship, to be modeled by *SamBorrowed → SamOut*.

The BN constructed with these modeling choices is shown in Figure 10.28(a). The next step of the KEBN process is to determine the parameters for this model; here we underline text about the "numbers."

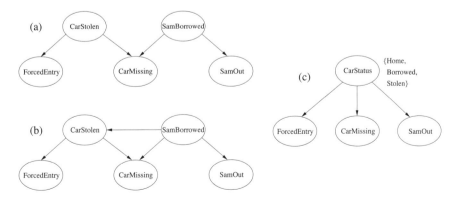

FIGURE 10.28: Alternative BNs for the missing car problem.

Additional problem information

The Nguyens know that <u>the rate</u> of car theft in their area is about <u>1 in 2000</u> each day, and that if the car was stolen, there is <u>a 95% chance</u> that the garage will show signs of forced entry. There is nothing else worth stealing in the garage, so it is reasonable to assume that if the car isn't stolen, the garage <u>won't show</u> signs of forced entry. The Nguyens also know that Sam borrows the car without asking <u>about once a week</u>, and that even if she didn't borrow the car, there is <u>a 50% chance</u> that she is out.

There are two root nodes in the model under consideration (Figure 10.28(a)). The prior probabilities for these nodes are given in the additional information:

- $P(CarStolen)$ is the rate of car theft in their area of "1 in 2000," or 0.0005
- $P(SamBorrowed)$ is "once a week," or ≈ 0.143

The conditional probabilities for the *ForcedEntry* node are also straightforward:

- $P(ForcedEntry = T | CarStolen = T) = 0.95$ (from the "95% chance")
- $P(ForcedEntry = T | CarStolen = F) = 0$

In practice, it might be better to leave open the possibility of there being signs of forced entry even though the car hasn't been stolen (something else stolen, or thieves interrupted), by having a very high probability (say 0.995) rather than the 1; this removes the possibility that the implemented system will grind to a halt when confronted with a combination of impossible evidence (say, $CarStolen = F$ and $ForcedEntry = T$). There is a stronger case for

- $P(SamOut = T | SamBorrowed = T) = 1$

The new information also gives us

- $P(SamOut = T | SamBorrowed = F) = 0.5$

So the only CPT parameters still to fill in are those for the *MissingCar* node. But the following probabilities are clear:

- $P(MissingCar = T | CarStolen = T, SamBorrowed = F) \approx 1$
- $P(MissingCar = T | CarStolen = F, SamBorrowed = T) \approx 1$

If we want to allow the possibility of another explanation for the car being missing (not modeled explicitly with a variable in the network), we can adopt

- $P(MissingCar = T | CarStolen = F, SamBorrowed = F) = \alpha > 0$

Finally, only one probability remains, namely

- $P(MissingCar = T | CarStolen = T, SamBorrowed = T)$

Alert! Can you see the problem? How is it possible for Sam to have borrowed the car *and* for the car to have been stolen? These are mutually exclusive events!

The first modeling solution is to add an arc *SamBorrowed* → *CarStolen* (or vice versa)(see Figure 10.28(b)), and make this mutual exclusion explicit with

- $P(CarStolen = T | SamBorrowed = T) = 0$

In which case, it doesn't matter what numbers are put in the CPT for $P(Missing Car = T | CarStolen = T, SamBorrowed = T)$. While this "add-an-arc" modeling solution "works," it isn't very elegant. Although Sam's borrowing the car will prevent it from being stolen, it is equally true that someone's stealing the car will prevent Sam's borrowing it!

A better solution is to go further back in the modeling process and re-visit the choice of nodes. The Nguyens are actually interested in the state of their car, whether it has been borrowed by their daughter, stolen, or is safe at home. So instead of two Boolean query nodes, we should have a single query node *CarStatus* with possible values {*home, borrowed, stolen*}. This simplifies the structure to that shown in Figure 10.28(c). A disadvantage of this simplified structure is that it requires additional parameters about other relationships, such as between signs of forced entry and Sam borrowing the car, and the car being stolen and Sam being out.

Decision making

Recall that the original description concluded with *"Should they [the Nguyens] be worried about their second car being stolen and notify the police?"* They have to *decide* whether to notify the police, so *NotifyPolice* becomes the decision node. In this problem, acting on this decision doesn't affect any of the state variables; it might lead to the Nyugens obtaining more information about their car, but it will not change whether it has been stolen or borrowed. So there is no arc from the decision node to any of the chance nodes. What about their preferences, which must be reflected in the connections to the utility node?

Additional information about Nguyen's preferences

If the Nguyen's car is stolen, they want to notify the police as soon as possible to increase the chance that it is found undamaged. However, being civic-minded citizens, they don't want to waste the time of the local police if Sam has borrowed it.

This preference information tells us that utility depends on both the decision node and the *CarStatus* node, shown in Figure 10.29 (using the version in Figure 10.28(c)). This means there are four combinations of situations for which preferences must be elicited. The problem description suggests the utility ordering, if not the exact numbers, shown in Table 10.9.[10] The decision computed by this network as evidence is added sequentially, is shown in Table 10.10. We can see that finding the car missing isn't enough to make the Nguyen's call the police. Finding signs of forced entry changes their decision; however more information — finding out that Sam is out — once again changes their decision. With all possible information now available, they decide not to inform the police.

---

[10]Note that the utilities incorporate unmodeled probabilistic reasoning about getting the car back.

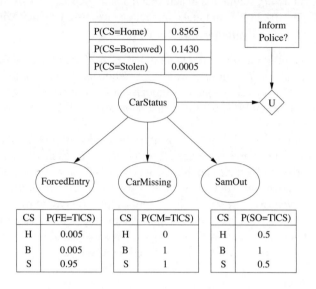

FIGURE 10.29: A decision network for the missing car problem.

TABLE 10.9
Utility table for the missing car problem.

CarStatus	Inform Police	Outcome utility	
		Qualitative	Quantitative
borrowed	*no*	neutral	0
home	*no*	neutral	0
borrowed	*yes*	poor	-20
home	*yes*	poor	-20
stolen	*yes*	bad	-40
stolen	*no*	terrible	-80

TABLE 10.10
Expected utilities of *InformPolice* decision node for the missing car problem as evidence is added sequentially.

Evidence	EU(*InformPolice*=Y)	EU(*InformPolice*=N)	Decision
None	-20.01	**-0.04**	No
CarMissing=T	-20.07	**-0.28**	No
ForcedEntry=T	**-27.98**	-31.93	Yes
SamOut=F	-24.99	**-19.95**	No

10.8   Incremental modeling

Common Modeling Mistake 17 *Trying to build the full model all at once*

It is nearly impossible to build a complex BN in one go, that is, including all the variables right from the start. A good rule of thumb is to limit the size of any initial model to 10-15 nodes. There are a number of well-known general problem-solving strategies that can be applied to BN knowledge engineering.

In this subsection, we'll use the Goulburn Catchment ERA BN (Figure 5.13) to illustrate ways to break up the modeling.

10.8.1   Divide-and-conquer

There are many ways to divide a problem up into more tractable sub-problems; here are some concrete suggestions.

Divide into sub-components One way is to identify clear sub-components of the system being modeled, which become sub-networks in the BN. The Goulburn Catchment BN is an obvious example of this; its sub-components are the water quality, hydraulic habitat, structural habitat, biological interaction and species diversity.

Single target variable. The problem can be broken-down by looking at one target variable at a time. For example, in Goulburn Catchment, when looking only at the *FutureAbundance* (and not considering *FutureDiversity* in the first instance, the Species Diversity section of the model would not be included.

Reduce scope. We can also reduce the scope of the problem by taking out a variable (which is usually a root node) and building a smaller BN for *one* of its values. In the Goulburn Catchment example, this could be done by focusing on a single site. This leads to many nodes having one fewer parent and hence smaller CPTs. In some cases, the structure in the rest of the network becomes simpler; for example, in Goulburn Catchment, 5 sites (*G_Murch*, *G_Shep*, *G_McCoy*, *G_Undera*, *G_Echuca*) have no barrier on them, meaning the *Barrier* node can be removed for this sub-problem.

Single time frame. The problem can be split up and each time frame considered in turn. In the Goulburn Catchment, this applies at two levels. First, there are two seasonal variations within the year that affect the hydraulic habitat, Winter-spring and Summer-autumn; these could be considered separately. Second, each time frame over which the future abundance and diversity is being considered. Note that this is a special case of reducing the scope in the Goulburn Catchment BN, as *TimeScale* is a root node with 2 values, *OneYear* and *FiveYears*.

10.8.2   Top-down vs. Bottom-up

Another way to do incremental modeling is to vary the level of abstraction. For example, it can be useful to consider a "big picture" overview of the system – the most abstract level. Alternatively, sometimes it is necessary to look at all the details of at least part of the system to fully understand it; this is the least abstract, or most concrete view. The BN is then constructed using hierarchical components, with each level in the hierarchy becoming progressing less abstract. OOBNs, as discussed earlier in § 4.7, are exactly intended to support such modeling.

A **top-down approach** means looking at the problem at the highest level of abstraction first, then looking in ever more increasing detail at each of the pieces. Each of the sub-systems would be fleshed out in turn, as a separate subnetwork, through all the building and evaluating stages. (Note that for the Goulburn Fish BN, this means only Stages 1 and 2.) The components are then joined together incrementally – one at a time. Each time a new subnetwork is connected into the main model, there will be another iteration through the KEBN process. The focus when joining sub-networks is on the interactions *between* the components – adding and evaluating arcs and possibly new intermediate nodes, before doing the additional parameterization required.

In a **bottom-up approach**, the modeling starts at a detailed, concrete level. The specific target variables are usually identified first, then the most relevant direct influences are identified in turn. The process is done incrementally, with an iteration of the process done with just a few new variables added each time.

In practice, top-down and bottom-up modeling are usually combined.

10.9   Adaptation

Adaptation in Bayesian networks means using some machine learning procedure to modify either the network's structure or its parameters over time, as new data arrives. It can be applied either to networks which have been elicited from human domain experts or to networks learned from some initial data. The motivation for applying adaptation is, of course, that there is some uncertainty about whether or not the Bayesian network is correct. The source of that uncertainty may be, for example, the expert's lack of confidence, or perhaps a lack of confidence in the modeled process itself being stable over time. In any case, part of the discipline of KEBN is the ongoing collection of statistics; so the opportunity of improving deployed networks by adaptation ought to be available.

We have already seen elements of adaptation. In Chapter 6 we saw how the Multinomial Parameterization Algorithm 6.1 requires the specification of priors for the parameters, including an equivalent sample size which implicitly records a degree of confidence in those parameters. Also, §9.6.3 shows how prior probabilities for arc structure can be fed into the learning process. And, indeed, this section relies upon an understanding of those sections. The difference is one of intent and degree: in those chapters we were concerned with specifying some constraints on the learning of a causal model; here we are concerned with modifying a model already learned (or elicited).

10.9.1 Adapting parameters

We describe how to modify multinomial parameters given new complete joint observations. This can be done using the Spiegelhalter-Lauritzen Algorithm 6.1 under the same global and local assumptions about parameter independence. Recall that in that process the prior probabilities for each parameter are set with a particular equivalent sample size. If the parameters have actually been learned with some initial data set, then subsequent data can simply be applied using the same algorithm, starting with the new equivalent sample size and Dirichlet parameters set at whatever the initial training has left. If the parameters have been elicited, then you must somehow estimate your, or the expert's, degree of confidence in them. This is then expressed through the equivalent sample size, the larger the sample size the greater the confidence in the parameter estimates and the slower the change through adapting them with new data.

Suppose for a particular Dirichlet parameter α_i the expert is willing to say she gives a 90% chance to the corresponding probability $p_i = 0.2$ lying within the interval $[.1, .3]$. The expected value (i.e., the mean) of the Dirichlet distribution over state i corresponds to the estimated parameter value:

$$\mu_i = \frac{\alpha_i}{s} = 0.2$$

where $s = \sum_j \alpha_j$ is the equivalent sample size. The variance over state i of the Dirichlet distribution is:

$$\sigma_i^2 = \frac{\alpha_i(s - \alpha_i)}{s^2(s+1)} = \frac{\mu_i(1 - \mu_i)}{s+1}$$

The 90% confidence interval of this distribution lies within 1.645 standard deviations of the mean and is

$$\mu_i \pm \sigma_i = \mu_i \pm 1.645\sqrt{\frac{\mu_i(1 - \mu_i)}{s+1}}$$

We can solve this algebraically for values of α_i and s that yield the mean and 90% confidence interval. A solution is an equivalent sample size $s = 15$ and a parameter value α_i of 3.

Of course, when this procedure is applied independently to the different Dirichlet parameters for a single parent instantiation the results may not be fully consistent. If the expert reports the above interval for the first state of the binary child variable and $[0.4, 0.6]$ for the second, then the latter will lead to an equivalent sample size of $s = 24$ and $\alpha_2 = 12$. Since the equivalent sample size applies to all of the values of the child variable for a single instantiation, it cannot be both 15 and 24! The sample size must express a common degree of confidence across all of the parameter estimates for a single parent instantiation. So, the plausible approach is to compromise, for example, taking an average of the equivalent sample sizes, and then finding numbers as close to the estimated means for each state as possible. Suppose in this case, for example, we decide to compromise with an equivalent sample size of 20. Then the original probabilities for the two states, 0.2 and 0.5, yield $\alpha = (4, 10)$, which does not work. Normalizing (with round off) would yield instead $\alpha = (6, 14)$.

When parameters with confidence intervals are estimated in this fashion, and are not initially consistent, it is of course best to review the results with the expert(s) concerned.

Fractional updating is what Spiegelhalter and Lauritzen (1990) call their technique for adapting parameters when the sample case is missing values, i.e., for incomplete data. The idea is simply to use the Bayesian network as it exists, applying the values observed in the sample case and performing Bayesian propagation to get posterior distributions over the unobserved cases. The observed values are used to update the Dirichlet distributions for those nodes; that is, a 1 is added to the relevant state parameter for the observed variable. The posteriors are used to proportionally update those variables which were unobserved; that is, $p' < 1$ is added to the state parameter corresponding to a value which takes the posterior p'. The procedure is complicated by the fact that a unique parent instantiation may not have been observed, when the proportional updating should be applied across all the possible parent instantiations, weighted by their posterior probabilities. This procedure unfortunately has the drawback of overweighting the equivalent sample size, resulting in an artificially high confidence in the probability estimates relative to new data.

Fading refers to using a time decay factor to underweight older data exponentially compared to more recent data. If we fade the contribution of the initial sample to determining parameters, then after sufficient time the parameters will reflect only what has been seen recently, allowing the adaptation process to track a changing process. A straightforward method for doing this involves a minor adjustment to the update process of Algorithm 6.1 (Jensen, 2001, pp. 89-90): when state i is observed, instead of simply adding 1 to the count for that state, moving from α_i to $\alpha_i + 1$, you first discount all of the counts by a multiplicative decay factor $d \in [0, 1]$. In other words the new Dirichlet distribution becomes $D[d\alpha_1, \ldots, d\alpha_i + 1, \ldots, d\alpha_\tau]$. In the limit, the Dirichlet parameters sum to $\frac{1}{1-d}$, which is called the **effective sample size**.

10.9.2 Structural adaptation

Conceivably, rather than just modifying parameters for an existing structure, as new information comes to light we might want to add, delete or reverse arcs as well. Jensen and Nielsen report that "no handy method for incremental adaptation of structure has been constructed" (Jensen and Nielsen, 2007, p.214). They suggest the crude, but workable, approach of accumulating cases and rerunning structure learning algorithms in batch mode periodically.

However, the Bayesian approach allows for structural adaptation, at least in principle. If we let what has previously been learned be reflected in our prior probabilities over structures, then new data will update our probability distributions over causal structure. That simply is Bayesian theory. To be sure, it is also Bayesian updating without model selection, since we need to maintain a probability distribution over all of the causal structures that we care to continue to entertain. A straightforward implementation of such inference will almost immediately be overwhelmed by computational intractability.

However, this approach is straightforward with CaMML. As §9.6.3 describes (and O'Donnell et al., 2006), there is a wide range of alternatives for specifying prior probabilities for CaMML's search, including specifying degrees of confidence in those probabilities which effectively act as equivalent sample sizes. These priors can put probabilities on directed or undirected arc structure, as well as the more common partitioning of variables into tiers. In other words, Jensen and Nielsen's comment is out of date (and, indeed, was so when published).

10.10   Summary

There is an interplay between elements of the KEBN process: variable choice, graph structure and parameters. Both prototyping and the spiral process model support this interplay. Various BN structures are available to compactly and accurately represent certain types of domain features. While no standard knowledge engineering process exists as yet, we have sketched a framework and described in more detail methods to support at least some of the KE tasks. The integration of expert elicitation and automated methods, in particular, is still in early stages. There are few existing tools for supporting the KEBN process, although some are being developed in the research community, including our own development of VE and Matilda.

10.11   Bibliographic notes

An excellent start on making sense of software engineering is Brooks' *Mythical Man-Month* (1995). Sommerville's text (2010) is also worth a read. You may also wish to consult a reference on software quality assurance in particular, which includes detailed advice on different kinds of testing (Ginac, 1997).

Laskey and Mahoney (2000) also adapt the ideas of prototyping and the spiral model to the application of Bayesian networks. Bruce Marcot (1999) has described a KEBN process for the application area of species-environment relations. Linda van der Gaag's group at Utrecht have been leaders in developing KEBN methods, including the use of ontologies in the early stages of the structural modeling, qualitative probability elicitation and combining qualitative and quantitative assessment. Boneh's KEBN process was formalized in her Ph.D. thesis on the use of BNs in the weather forecasting domain, which we supervised. However the ideas evolved during our collaboration on several BN research projects, including the decimals intelligence tutoring system (§11.3), Matilda, the Goulburn Catchment ERA (§11.4), and the Cardiovascular Risk Assessment project (§11.5).

One introduction to parametric distributions is that of Balakrishnan and Nevzorov (2003). Mosteller et al.'s *Statistics by Example* (1973) is an interesting and readily accessible review of common applications of different probability distributions. An excellent review of issues in elicitation can be found in Morgan and Henrion's *Uncertainty* (1990), including the psychology of probability judgments.

Lacave and Díez (2002) provide an excellent review of explanation methods for Bayesian networks.

10.12   Problems

Problem 1

The elicitation method of partitioning to deal with local structure is said in the text to directly correspond to the use of classification trees and graphs in Chapter 6. Illustrate this for the *Flu*, *Measles*, *Age*, *Fever* example by filling in the partitioned CPT with hypothetical numbers. Then build the corresponding classification tree. What do partition elements correspond to in the classification tree?

Problem 2

Consider the BN that you constructed for your own domain in Problem 5, Chapter 2. (Or if you haven't completed this problem, use an example network from elsewhere in this text, or one that comes with a BN software package.)

1. Identify the type of each node: query, observation, context, controllable.
2. Re-assess your choice of nodes and values, particularly if you discretized a continuous variable, in the light of the discussion in §10.3.1.
3. Label each arc with the relationship that is being modeled, using the relationships described in §10.3.3 or others you might think of.
4. Use Matilda to investigate the dependence and independence relationships in your BN.

5. Can you map the probabilities in the CPTs into a qualitative scale, as in §10.4.2?

Problem 3

Using some of methods from this chapter, reengineer one of the BN applications you developed for one of the problems from Chapter 5. That is, redo your application, but employ techniques from this chapter. Write up what you have done, with reference to specific techniques used. Also, describe the difference use of these techniques has made in the final BN.

Problem 4

Start with the decision network for *Julia's manufacturing* problem from §4.10. In case it isn't already, parameterize it so that the decision (between production, marketing, further development) is relatively insensitive to the exact probability of a high vs. low product quality. Then reparameterize it so that the decision is sensitive to product quality. Comment on these results.

Problem 5

In Table 10.9 we gave a utility function for the missing car problem, for the decision network shown in Figure 10.29, with expected utility results in Table 10.10. Investigate the effects on the decision in the following situations:

1. The same relative ordering of outcome preferences is kept, but the quantitative numbers are all made positive.
2. The preference ordering is changed so the highest priority is not wasting police time.
3. The rate of car theft in the area is much higher, say 1 in 500.

Problem 6

In §10.4.8.1 we described how it is possible to compute the entropy reduction due to evidence about a pair of observations. Write code to do this using the Netica API.

11

KEBN Case Studies

11.1 Introduction

In this chapter we describe four applications of Bayesian networks. These are, in particular, applications we were involved in developing, and we choose them because we are familiar with the history of their development, including design choices illustrating KEBN (or not), which otherwise often go unreported.

11.2 Bayesian poker revisited

In §5.5, we described a Bayesian decision network for the game of 5-card stud poker. The details of the network structure and betting strategies used by the Bayesian Poker Player (BPP) are a more recent version of a system that has been evolving since 1993. Here we present details of that evolution as an interesting case study in the knowledge engineering of Bayesian networks.

11.2.1 The initial prototype

The initial prototype developed by Jitnah (1993) used a simple polytree structure. This was largely for pragmatic reasons, as the message passing algorithms for polytrees were simple to implement, and at that time, BN software tools were not widely available.[1] The original network structure is shown in Figure 11.1(a). The polytree structure meant that we were making the (clearly incorrect) assumption that the final hand types are independent.

In the initial version, the nodes representing hand types were initially given values which divided hands into the nine categories. The size of the model was also limited by compressing the possible values for the *OPP_Action* node to {*pass/call, bet/raise, fold*}. This produced a level of play comparable only to a weak amateur. Since any busted hands, for example, were treated as equal to any other, BPP would bet inappropriately strongly on middling busted hands and inappropriately weakly

---

[1] Early non-commercial tools included IDEAL (Srinivas and Breese, 1989) and Caben (Cousins et al., 1991a); see §B.2 for a brief history of BN software development.

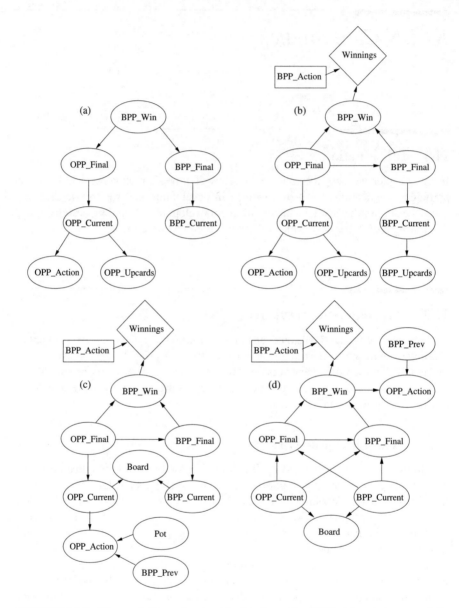

FIGURE 11.1: The evolution of network structures for BPP: (a) polytree used for both 1993 and 1999 version; (b) 2000 decision network version; (c) Texas Hold'em structure 2003; (d) Improved opponent modeling 2007.

on Ace-high hands. The lack of refinement of paired hands also hurt its performance. Finally, the original version of BPP was a hybrid system, using a BN to estimate the probability of winning, and then making the betting decision using randomized betting curves, which were hand constructed. Bluffing was introduced as a random low probability (5%) of bluffing on any individual bet in the final round. The probabilities for the *OPP_Action* node were adapted over time based on observations of opponents' actions. Information about the opponent's betting actions in different situations was used to learn the *OPP_Action* CPT.

11.2.2   Developments 1995-2000

During 1995, alternative granularities for busted and pair hand types were investigated (Thomson, 1995). At this stage, we introduced an automatic opponent that would estimate its winning probability by dealing out a large sample of final hands from the current situation, and then use simple rules for betting.

In the 1999 version (Korb et al., 1999) we kept the same polytree structure, but opted for a modest refinement of the hand types, moving from the original 9 hand types to 17, by subdividing busted hands and pairs into low (9 high or lower), medium (10 or J high), queen, king and ace. In this version, the parameters for the betting curves were found using a stochastic greedy search. This version of BPP was evaluated experimentally against: a simple rule-based system; a program which depends exclusively on hand probabilities (i.e., without opponent modeling); and human players with experience playing poker.

A number of individual changes to the system were investigated individually during 2000 (Carlton, 2000). These changes included:

- Expansion of the hand type granularity from 17 to 24, by explicitly including all 9 different pair hands.
- Allowing *OPP_Action* to take all four different action values, {*fold, pass, call, raise*}.
- Adding an arc from *BPP_Final* to *OPP_Final*, to reflect the dependence between them.
- Extending the BN to a decision network, in particular changing the decision method from one based on probability betting curves to one based on expected winnings.
- Improving the bluffing by taking into account the opponent's belief in BPP winning and making bluffing consistent throughout the last round when undertaken.

The network structure for this version, previously shown in Figure 5.14 is shown again in Figure 11.1(b) for easy comparison.

11.2.3   Adaptation to Texas Hold'em, 2003

In 2003, we are modified BPP to play Texas Hold'em Poker, to allow us to compete against other automated opponents online (Boulton, 2003). In Texas Hold'em each

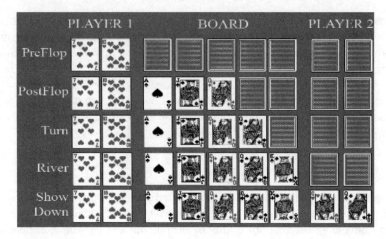

FIGURE 11.2: Example game of Texas Hold'em poker: player 2 wins with 4 queens. (Taylor, 2007, Fig. 2.1.) (Reproduced with permission.)

player is dealt 2 cards face down and a betting round commences (the PreFlop). The dealer then places 3 shared community or board cards in the middle of the table, then another round of betting occurs (PostFlop). Two more additional cards are placed in the center, with a betting round after each card (Turn, then River). Each player combines their hidden (hole) cards with the 5 board cards to form their best poker hand. An example game of Texas Hold'em is shown in Figure 11.2; here, player 2 wins with four queens. Table 11.1 shows how the probability of different hands varies between 5 card stud and Texas Hold'em.

TABLE 11.1

Poker hand types: weakest to strongest with prior probabilities in both Five Card Stud and Texas Hold'em.

Hand Type	Example	Probability 5 card stud	Texas Hold'em
Busted	A♣ K♠ J◇ 10◇ 4♡	0.5015629	0.1728400
Pair	2♡ 2◇ J♠ 8♣ 4♡	0.4225703	0.4380000
Two Pair	5♡ 5♣ Q♠ Q♣ K♣	0.0475431	0.2351900
Three of a Kind	7♣ 7♡ 7♠ 3♡ 4◇	0.0211037	0.048 3400
Straight (sequence)	3♠ 4♣ 5♡ 6◇ 7♠	0.0035492	0.0479900
Flush (same suit)	A♣ K♣ 7♣ 4♣ 2♣	0.0019693	0.029 9000
Full House	7♠ 7◇ 7♣ 10◇ 10♣	0.0014405	0.0255000
Four of a Kind	3♡ 3♠ 3◇ 3♣ J♠	0.0002476	0.0018800
Straight Flush	3♠ 4♠ 5♠ 6♠ 7♠	0.0000134	0.0003600

The change from 5 card stud to Texas Hold'em required the addition of a node to represent the shared board cards, as shown in Figure 11.1(c). Other structural changes were made to improve learning and play included making *OPP Action* a child node of a new *BPP Prev* node, and the *Pot* so that BPP's previous action and the pot jointly condition the opponent's behavior.

11.2.4 Hybrid model, 2003

One of the obvious disadvantages to BPP's play is the so-called "hand abstraction" used to contain the complexity of the decision network. In our various versions of 5-card stud BPP we tinkered with different abstractions, from 9 to the current 25 hand types (which still represented over 133 million card combinations). The hand abstraction causes some information to be lost and therefore introduces inaccuracy into the model; in particular, subtle differences between similar hands, that are crucial when determining which hand will win, may be lost. A balance between model performance and computational requirements is required. In 2003, we analyzed the magnitude of the inaccuracy in this trade-off.

Two measures of the hand abstraction inaccuracy are considered. Firstly, a theoretical measure, the Kullback-Leibler divergence (see § 7.5.4), is considered as a measurement of inaccuracy between the *BPP_Win* node probability distributions output by two different models. Secondly, a more practical measure is observed by playing different models against each other, and also against the automated Texas Hold'em player poki developed by Billings et al.[2]

In addition, a so-called *Combinatorial* model was designed as a base case against which other models could be tested. Each position in a game is evaluated by considering 100,000 random decks of cards. This number of trials was chosen as the largest number that would allow acceptable playing speed using the resources available. The hand is dealt out to completion using these decks and frequency counts are compiled for winning, tied, and losing outcomes. This information is used to calculate an accurate probability that the agent will hold the strongest hand if the game should proceed to a showdown. No information is lost in the hand representation for this model, meaning that it does not suffer from the hand abstraction problems effecting the BPP network. Also, this method does not consider information concerning the opponent's actions. When comparing BPP to the Combinatorial model, no evidence is added about the opponent's actions, which allows a direct comparison of the effects of hand abstraction alone.

Five models with differing granularities of hand abstraction (33, 25, 18, 15, 13) were compared to determine the effect that the number of states has on the accuracy of the network (see Boulton, 2003 for details of this experiment). The busted and pair categories were focused on due to the high frequency with which they occur, thus making this experiment more sensitive. Each model was compared with the combinatorial model in each of the four betting rounds to gauge how they performed through out the entire course of a hand. In each category, the KL divergence was calculated by averaging the results from 100,000 random hands. These results are shown graphically in Figure 11.3.

From Figure 11.3, it is clear that hand abstraction has a much larger influence

---

[2]Note that all results described here were obtained against the 2003 version of Poki.

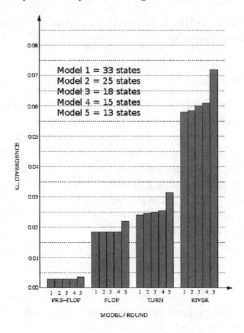

FIGURE 11.3: The KL divergence between the BN model and the Combinatorial model for 5 different abstractions, at each betting round.

in the later betting rounds, while being almost irrelevant at the beginning of a hand. Secondly, it appears that initial decreases in granularity cause a significant increase in accuracy, while further decreases have less effect. This is especially true when it is considered that model 1 (33 states) requires a relatively greater increase in state numbers for a much smaller gain in accuracy. Our current 25 hand types seems a reasonable compromise. Table 11.1 indicates that hands with two pairs becomes much more common when playing the game of Texas Hold'em. This suggests that this hand type might make a good candidate for further segmentation when future increases in state numbers are considered.

While these theoretical results confirm that hand abstraction does indeed play an important role, they do not quantify the actual disadvantage that it will cause under practical playing conditions. The performance of an automated agent in actual game play is clearly the gold standard for testing successful strategies. To this end, the primitive Bayesian Model with no opponent modeling was played against the Combinatorial Model over 20,000 random hands to quantify the practical financial loss that was occurring. As these two models differ only in the presence of hand abstraction, the losses observed by the Bayesian Model can be directly credited to this factor. Model 2 (25 states) was used for this experiment. This model was observed to lose 0.49 ± 0.12 small betting units per hand against the Combinatorial model.

To further evaluate this effect, both models were evaluated against the Billing's agent over the online poker server. Due to the speed of play over the Internet, smaller numbers of hands could be played. Over 6000 hands, the Bayesian Model lost 0.43 ± 0.17 small betting units per hand. Over the same number of hands, the Combinatorial model lost 0.33 ± 0.15 small betting units per hand. It is interesting to note that this difference (0.1 small betting units per hand) is smaller than the losses observed when the two models are played against each other. One possible cause for this discrepancy might be the learning techniques incorporated into the Billing's agent. These techniques may be more efficient against the Combinatorial model, thus reducing the gap between them. These results demonstrate that hand abstraction provides a significant financial disadvantage in situations of actual game play and is a worthwhile area in which to invest further experimental resources. We now present one possible solution to this problem.

Both the Combinatorial model and the decision network model have advantages and disadvantages when implemented as a poker playing agent. The Combinatorial model, while being able to provide accurate predictions of final hand probabilities, is poorly equipped to implement an effective opponent modeling strategy. The decision network model, while providing the potential for intuitive and transparent opponent modeling, suffers from the effects of hand abstraction, meaning that it is unable to predict the final hand distributions accurately. The Hybrid model aims to meld these two techniques in an attempt to provide accurate mathematical predictions of final hand types while still allowing effective opponent modeling.

First, in the hybrid model, the priors for each final hand are adjusted to the result from the Combinatorial model. This required the separation of the link between the two final hand nodes, but it was felt that this introduced only a small inaccuracy. In a similar fashion, this technique was intended to bias the results of the BN towards that of the Combinatorial model. By itself, however, this doesn't work, as, intuitively, information from the various sources of available evidence are effectively introduced twice, causing an overly extreme belief to be computed. In self-play trials, this model fared significantly worse than the unaltered BN.

In order to rectify this problem, additional uncertain, or so-called "*virtual*" evidence was entered into the two final hand nodes. This uncertain evidence represented the difference between the original BN probabilities and those produced by the Combinatorial roll outs. This was done by using the ratio by which the distribution estimated by the combinatorial method differed from the original BN estimate. These values were then normalized and re-introduced to the network as likelihood evidence (see § 3.4).

The Hybrid model was evaluated against both the Combinatorial model and the original decision network model (Figure 11.1). The Hybrid model was able to win 0.50 ± 0.12 small betting units per hand from the BN model over 20,000 hands. This was a similar result to that achieved by the Combinatorial model. As expected, the Hybrid model lost a small amount against the Combinatorial model (0.02 small betting units per hand), although this was not statistically significant. The Hybrid model shows that BNs are capable of being combined with more accurate techniques to almost eliminate the effects of hand abstraction.

11.2.5   Improved opponent modeling 2005-2007

In 2007, Mascaro produced a version of BPP available for online play (BPP, 2010) with further structural changes, Figure 11.1 (d). The key different was that *OPP Action* didn't depend on the opponent's current hand, but on whether or not they would win. Taylor (2007) extended that BPP's opponent model, by adding a node *Opp Conservative* as another parent of *Opp Action*, to distinguish between the betting actions of conservative versus aggressive opponents.

11.2.6   Ongoing Bayesian poker

Finally, as this text is being written, our poker project continues (McGain, 2010). Possible directions of ongoing research include the following.

Ideally, the program should also take advantage of the difference between BPP's upcards and its full hand. The point is that when one's strong cards are showing on the table, there is no reason to bet coyly; on the contrary, it is advantageous to make opponents pay for the opportunity of seeing any future cards by betting aggressively. On the other hand, when one's strongest card is hidden, aggressive betting can drive opponents out of the game prematurely. This could be done by using the BN model from both BPP and the opponent's side to obtain different user model views of the strength of BPP. When the opponent's view is "strong" but BPP's view is weak, BPP should bet heavily. When both views are strong, BPP should bet more softly to keep the opponent playing.

Just as for BPP, the conservativeness or aggressiveness of an opponent's play, which is learned and captured by recalibrating the matrices relating *OPP Current* and *OPP Action*, does not fully describe the bluffing behavior of the opponent. A plausible extension would be to add an opponent bluffing node which is a parent of *OPP Action* and a child of *OPP Current* and *BPP Current* (since the latter gives rise to BPP's upcards and behavior, even though they are not explicitly represented).

These online gaming environments will also require extensions to handle multi-opponent games. In multiple opponent games it will be more important to incorporate the interrelations between what is known of different player's hands and the node representing their final hands.

We also anticipate using a dynamic Bayesian network (DBN) to model more effectively the interrelation between rounds of play; more details of this modeling problem are left as a homework problem (see Problem 5.10).

Finally, we are aware that the lack of non-showdown information is likely to introduce some selection bias into the estimates of the conditional probabilities, but we have not yet attempted to determine the nature of this bias.

11.2.7   KEBN aspects

There are several points to make about this case study, in terms of the overall KEBN process.

All the network structures have been hand-crafted, while the parameters have been either generated (by dealing out large numbers of poker hands) or learnt during playing sessions. This combination of elicitation and learning is fairly typical of BN projects to date, where the amount of expert domain knowledge (and lack of data) means that hand-crafting the structure is the only option.

Our domain expert on this project was one of the authors (who has had considerable poker playing experience), which meant we didn't have to deal with the KE difficulties associated with using an expert who is unfamiliar with the technology.

As we are interested in poker as a challenging application for our BN research agenda, BPP has never been deployed in a "live" environment where BPP and its opponents are playing for money. This has undoubtedly limited the validation and field testing phases of our proposed lifecycle model (see Figure 10.1), and there has certainly been no industrial use.

On the other hand, there *has* been an iterative development cycle, with continual refinement of the versions, following the incremental prototype model. We started with sufficient assumptions to allow the development of a simple but working initial model and gradually increased the complexity, adding variables, increasing the number of values, adding arcs (moving from a polytree to a graph) and extending a BN into a decision network.

The management of versions has been adequate, allowing subsequent versions of BPP to be tested against earlier versions. We did not document changes and the rationale for them as part of a formal KEBN process, but fortunately a series of research reports and papers served that purpose.

11.3 An intelligent tutoring system for decimal understanding

In this section we present a case study in the construction of a BN in an intelligent tutoring system (ITS) application,[3] specifically decimal understanding, for the target age range of Grades 5 to 10 (Nicholson et al., 2001, Stacey et al., 2003).

To understand the meaning and size of numbers written using a decimal point involves knowing about place value columns, the ordering of numbers by size and the value of digits. Although this is a topic plagued by misconceptions, very often students don't know they harbor them. Our education domain experts had been working in this area for some time, gathering and analyzing data on students' thinking about decimals. An ITS consisting of computer games involving decimals was being designed and developed as a useful learning tool to supplement classroom teaching. Our domain experts needed an underlying reasoning engine that would enable the ITS to diagnose and target an individual's wrong way of thinking about decimals; here we describe the development of a BN to do this.

We begin with the background domain information that was available at the start of the project (§11.3.1), and then we present the overall architecture of the ITS ar-

---

[3]There have been a number of other successful ITS systems that use Bayesian networks, e.g., (Conati et al., 1997, 2002, Mayo and Mitrovic, 2001, VanLehn and Niu, 2001).

chitecture (§11.3.2). Next we give a detailed description of the both expert elicitation phase, including evaluations (§11.3.3), and the investigation that was undertaken using automated methods (§11.3.4). Finally, we describe results from field trials and draw some conclusions from the case study as a whole.

11.3.1   The ITS domain

Students' understanding of decimal numeration has been mapped using a short test, the Decimal Comparison Test (DCT), where the student is asked to choose the larger number from each of 24 pairs of decimals (Stacey and Steinle, 1999). The pairs of decimals are carefully chosen so that from the patterns of responses, students' (mis)understanding can be diagnosed as belonging to one of a number of classifications. These classifications have been identified manually, based on extensive research (Stacey and Steinle, 1999, Resnick et al., 1989, Sackur-Grisvard and Leonard, 1985, Stacey and Steinle, 1998). The crucial aspects are that misconceptions are prevalent, that students' behavior is very often highly consistent and that misconceptions can be identified from patterns amongst simples clues.

About a dozen misconceptions have been identified (Stacey and Steinle, 1999), labeled vertically in Table 11.2. This table also shows the rules the domain experts originally used to classify students based on their response to 6 types of DCT test items (labeled horizontally across the top of the table): H = High number correct (e.g., 4 or 5 out of 5), L = Low number correct (e.g., 0 or 1 out of 5), with '.' indicating that any performance level is observable for that item type by that student class other than the combinations seen above.

TABLE 11.2
Response patterns expected from students with different misconceptions.

Coarse Class	Fine Class	Description	Item type (with sample item)					
			1 0.4 0.35	2 5.736 5.62	3 4.7 4.08	4 0.452 0.45	5 0.4 0.3	6 0.42 0.35
A	ATE	apparent expert	H	H	H	H	H	H
	AMO	money thinker	H	H	H	L	H	H
	AU	unclassified A	H	H	.	.	.	.
L	LWH	whole number thinking	L	H	L	H	H	H
	LZE	zero makes small	L	H	H	H	H	H
	LRV	reverse thinking	L	H	L	H	H	L
	LU	unclassified L	L	H	.	.	.	.
S	SDF	denominator focused	H	L	H	L	H	H
	SRN	reciprocal negative	H	L	H	L	L	L
	SU	unclassified S	H	L	.	.	.	.
U	MIS	misrule	L	L	L	L	L	L
	UN	unclassified	.	.	.	.	.	.

Most misconceptions are based on false analogies, which are sometimes embellished by isolated learned facts. For example, many younger students think 0.4 is smaller than 0.35 because there are 4 parts (of unspecified size, for these students) in the first number and 35 parts in the second. However, these "whole number thinkers" (LWH, Table 11.2) get many questions right (e.g., 5.736 compared with 5.62) with the same erroneous thinking. So-called 'reciprocal thinking' students (SRN, Table 11.2) choose 0.4 as greater than 0.35 but for the wrong reason, as they draw an analogy between fractions and decimals and use knowledge that 1/4 is greater than 1/35.

The key to designing the DCT was the identification of "item types." An item type is a set of items which a student with any misconception should answer consistently (either all right or all wrong). The definition of item types depends on both the mathematical properties of the item and the psychology of the learners. In practice, the definition is also pragmatic — the number of theoretically different item types can be very large, but the extent to which diagnostic information should be squeezed from them is a matter of judgement. The fine misconception classifications had been "grouped" by the experts into a coarse classification — L (think longer decimals are larger numbers), S (shorter is larger), A (correct on straightforward items (Types 1 & 2)) and U (other). The LU, SU and AU "catch-all" classifications for students who on their answers on Type 1 and 2 items behave like others in their coarse classification, but differ on other item types. These and the UNs may be students behaving consistently according to an unknown misconception, or students who are not following any consistent interpretation.

The computer game genre was chosen to provide children with an experience different from, but complementary to, normal classroom instruction and to appeal across the target age range (Grades 5 to 10). The system offers several games, each focused on one aspect of decimal numeration, thinly disguised by a story line.

In the "Hidden Numbers" game students are confronted with two decimal numbers with digits hidden behind closed doors; the task is to find which number is the larger by opening as few doors as possible. Requiring similar knowledge to that required for success on the DCT, the game also highlights the place value property that the most significant digits are those to the left. The game "Flying Photographer" requires students to "photograph" an animal by clicking when an "aeroplane" passes a specified number on a numberline. The "Number Between" game is also played on a number line, but particularly focuses on the density of the decimal numbers; students have to type in a number between a given pair. Finally, "Decimaliens" is a classic shooting game, designed to link various representations of the value of digits in a decimal number. These games address several of the different tasks required of an integrated knowledge of decimal numeration based on the principles of place value. It is possible for a student to be good at one game or the diagnostic test, but not good at another; emerging knowledge is often compartmentalized.

11.3.2 ITS system architecture

The high-level architecture of our system is shown in Figure 11.4. The BN is used to model the interactions between a student's misconceptions, their game playing abilities and their performance on a range of test items. The BN is initialized with a generic model of student understanding of decimal numeration constructed using the

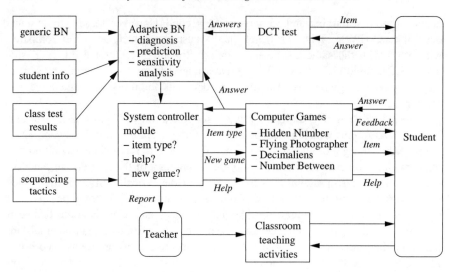

FIGURE 11.4: Intelligent tutoring system architecture.

DCT results from a large sample of students (Steinle and Stacey, 1998). The network can also be tailored to an individual student using their age or results from a short online DCT. During a student's use of the system, the BN is given information about the correctness of the student's answers to different item types encountered in the computer games.

Given the model, and given evidence about answers to one or more item types, the Bayesian belief updating algorithm then performs diagnosis; calculating the probabilities that a student with these behaviors has a particular misconception. Changes in the beliefs in the various misconceptions are in turn propagated within the network to perform prediction; the updating algorithm calculates the new probabilities of a student getting other item types right or wrong. After each set of evidence is added and belief updating performed, the student model stored within the network is updated, by changing the misconception node priors. Updating the network in this way as the student plays the game (or games) allows changes in students' thinking and skills to be tracked. The identification of the misconception with the highest probability provides the best estimate of the students' current understanding.

The information provided by the BN is, in turn, used by a controller module, together with the specified sequencing tactics (see below), to do the following: select items to present to the student; or decide whether additional help presentation is required; or decide when the user has reached expertise and should move to another game. The controller module also makes a current assessment of the student available to the teacher and reports on the overall effectiveness of the adaptive system. The algorithm for item type selection incorporates several aspects, based on the teaching model. Students are presented with examples of all item types at some stage during each session. Students meeting a new game are presented with items that the

BN predicts they are likely to get correct to ensure they have understood the rules and purposes of the game. If the probabilities of competing hypotheses about the student's misconception classification are close, the system gives priority to diagnosis and proposes a further item to be given to the user. By employing the Netica BN software's "sensitivity to findings" function (see §10.4.8.1), the user is presented with an item that is most likely to distinguish between competing hypotheses. Finally, the network selects items of varying difficulty in an appropriate sequence for the learner. The current implementation of the system allows comparison of different sequencing tactics: presenting the hard items first, presenting the easy items first and presenting easy and hard items alternately. Note that the terms easy and hard are relative to the individual student and based on the BN's prediction as to whether they will get the item correct.

The ITS also incorporates two forms of teaching into the games: clear feedback and visual scaffolding. For example, in the Flying Photographer, where students have to place numbers on number lines, feedback is given after each placement (see Figure 11.5): if the position is incorrect a sad red face appears at the point, whereas the correct position is (finally) marked by a happy green face. In this game, the visual scaffolding marks intermediate numbers on the interval; for example, when a student has twice failed to place a number correctly on the interval [0,1], the intermediate values 0.1, 0.2, 0.3, ..., 0.9 appear for this and the next item.

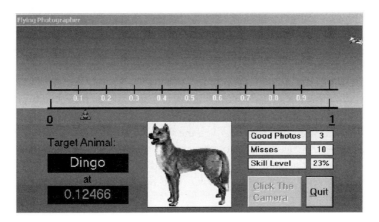

FIGURE 11.5: A screen from the ITS "Flying Photographer" game showing visual feedback (triangle) and scaffolding (marked intervals).

11.3.3 Expert elicitation

In this section we describe the elicitation of a fragment of the decimal misconception BN from the education domain experts. This fragment was the first prototype structure built, involving only the diagnosis of misconceptions through the DCT; the game fragments were constructed in later stages of the project. This exposition is given sequentially: the identification of the important variables and their values; the identification and representation of the relationships between variables; the parameterization of the network; and finally the evaluation. In practice, of course, there was considerable iteration over these stages.

11.3.3.1   Nodes

We began the modeling with the main "output" focus for the network, the representation of student misconceptions. The experts already had two levels of classifications for the misconceptions, which we mapped directly onto two variables. The *coarseClass* node can take the values {*L, S, A, UN*}, whereas the *fineClass* node, incorporating all the misconception types identified by the experts, can take the 12 values shown in column 1 of Table 11.2. Note that the experts considered the classifications to be mutually exclusive; at any one time, the student could only hold one misconception. While the experts knew that the same student could hold different misconceptions at different times, they did not feel it was necessary, at least initially, to model this explicitly with a DBN model.

Each different DCT item type was made an observation variable in the BN, representing student performance on test items of those types; student test answers are entered as evidence for these nodes. The following alternatives were considered for the possible values of the item type nodes.

- Suppose the test contains N items of a given type. One possible set of values for the BN item type node is {0,1, ..., N}, representing the number of the items the student answered correctly. The number of items may vary for the different types and for the particular test set given to the students, but it is not difficult to adjust the BN. Note that the more values for each node, the more complex the overall model; if N were large (e.g., > 20), this model may lead to complexity problems.

- The item type node may be given the values {*High, Medium, Low*}, reflecting an aggregated assessment of the student's answers. For example, if 5 such items were presented, 0 or 1 correct would be considered *low*, 2 or 3 would be *medium*, while 4 or 5 would be *high*. This reflects the expert rules classification described above.

Both alternatives were evaluated empirically (see §11.3.3.5 below); however the H/M/L option was used in the final implementation.

11.3.3.2   Structure

The experts considered the coarse classification to be a strictly deterministic combination of the fine classification; hence, the *coarseClass* node was made a child of the *fineClass* node. For example, a student was considered an *L* if and only if it was one of an *LWH, LZE, LRV* or *LU*. For the DCT test item types, the *coarseClass* node was not necessary; however, having it explicitly in the network helped both the expert's understanding of the network's reasoning and the quantitative evaluation. In addition, including this node allowed modeling situations where all students with the same coarse misconception exhibit the same behavior (e.g., in the Flying Photographer game).

The *type* nodes are observation nodes, where entering evidence for a type node should update the posterior probability of a student having a particular misconception. As discussed in §2.3.1, such diagnostic reasoning is typically reflected in a BN structure where the class, or "cause," is the parent of the "effect" (i.e., evidence) node. Therefore an arc was added from the *fineClass* node to each of the *type* nodes. No connections were added between any of the *type* nodes, reflecting the experts' intuition that a student's answers for different item types are independent, given the subclassification.

A part of the expert elicited BN structure implemented in the ITS is shown in Figure 11.6. This network fragment shows the *coarseClass* node (values $\{L,S,A,UN\}$), the detailed misconception *fineClass* node (12 values), the item *type* nodes used for the DCT, plus additional nodes for the Hidden Number and Flying Photographer games. The bold item type nodes are those corresponding to the DCT test data set and hence used subsequently for evaluation and experimentation.

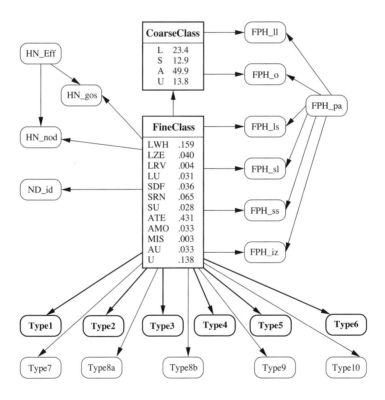

FIGURE 11.6: Fragment of the expert elicited BN currently implemented. Bold nodes are those discussed here.

The "HN" nodes relate to the Hidden Numbers game, with evidence entered for the number of doors opened before an answer was given (*HN_nod*) and a measure of the "goodness of order" in opening doors (*HN_gos*). The root node for the Hidden Number game subnet reflects a player's game ability, specifically their door opening "efficiency" (*HN_eff*). The "FPH" node relate to the Flying Photographer game. The node *FPH_ls* records evidence when students have to place a long number, which is small in size, such as 0.23456. As noted above, LWH students are likely to make an error on this task. The other nodes perform similar functions, with the root node *FPH_pa* reflecting a player's overall game ability.

The BNs for the other two games are separate BNs as student performance on these provides information about abilities other than the 12 misconceptions; however their construction was undertaken using similar principles.

11.3.3.3 Parameters

The education experts had collected data that consisted of the test results and the expert rule classification on a 24 item DCT for over 2,500 test papers from students in Grades 5 and 6. These were then pre-processed to give each student's results in terms of the 6 test item types; 5,5,4,4,3,3 were the number of items of these type 1 to 6 respectively. The particular form of the pre-processing depends on the item type values used: with the 0-N *type* node values, a student's results might be 541233, whereas with the H/M/L values, the same student's results would be represented as $H\,H\,L\,M\,H\,H$.

The expert rule classifications were used to generate the priors for the sub-classifications. The priors are found to vary slightly between different classes and teaching methods, and quite markedly between age groups; however these variations are ignored for current purposes. All the CPTs of the item types take the form of

- $P(Type = Value | Classification = X)$

As we have seen from the domain description, the experts expect particular classes of students to get certain item types correct and others wrong. However we do need to model the natural deviations from such "rules," where students make a careless error. We model this uncertainty by allowing a small probability of a careless mistake on any one item. For example, students who are thinking according to the LWH misconception are predicted to get all 5 items of Type 2 correct. If, however, there is a probability of 0.1 of a careless mistake on any one item, the probability of a score of 5 is $(0.9)^5$, and the probability of other scores follows the binomial distribution; the full vector for P(Type2|Subclass=LWH) is (0.59,0.33,0.07,0.01,0.00,0.00) (to two decimal places). When the item type values H/M/L are used, the numbers are accumulated to give the vector (0.92,0.08,0.00) for H, M and L.

The experts considered that this mistake probability was considerably less than 0.1, of the order of 1-2%. We ran experiments with different probabilities for a single careless mistake (*pcm*=0.03, 0.11 and 0.22), with the CPTs calculated in this manner, to investigate the effect of this parameter on the behavior of the system. These numbers were chosen to give a combined probability for *HIGH* (for 5 items) of 0.99, 0.9 and 0.7 respectively, numbers that our experts thought were reasonable.

Much more difficult than handling the careless errors in the well understood behavior of the specific known misconceptions is to model situations where the experts do not know how a student will behave. This was the case where the experts specified '.' for the classifications *LU, SU, AU* and *UN* in Table 11.2. We modeled the expert not knowing what such a student would do on the particular item type in the BN by using 0.5 (i.e., 50/50 that a student will get each item correct) with the binomial distribution to produce the CPTs.

11.3.3.4 The evaluation process

During the expert elicitation process we performed the following three basic types of evaluation. First was *case-based evaluation* (see §10.4.6) , where the experts "play" with the net, imitating the response of a student with certain misconceptions and review the posterior distributions on the net. Depending on the BN parameters, it was often the case that while the incorporation of the evidence for the 6 item types from the DCT test data greatly increased the BN's belief for a particular misconception, the expert classification was not the BN classification with the highest posterior, because it started with a low prior. We found that it was useful to the experts if we also provided the ratio by which each classification belief had changed (although the highest posterior is used in all empirical evaluations). The case-based evaluation also included sequencing, where the experts imitate repeated responses of a student, update the priors after every test and enter another expected test result. The detection of the more uncommon classifications through repetitive testing built up the confidence of the experts in the adaptive use of the BN.

Next, we undertook **comparative evaluation** between the classifications of the BN compared to the expert rules on the DCT data. It is important to note here that the by-hand classification is only a best-guess of what a student is thinking — it is not possible to be certain of the "truth" in a short time frame. As well as a comparison grid (see next subsection), we provided the experts with details of the records where the BN classification differed from that of the expert rules. This output proved to be very useful for the expert in order to understand the way the net is working and to build confidence in the net.

Finally, we performed **predictive evaluation** (see §7.5.1) which considers the prediction of student performance on individual item type nodes rather than direct misconception diagnosis. We enter a student's answers for 5 of the 6 item type nodes, then predict their answer for the remaining one; this is repeated for each item type. The number of correct predictions gives a measure of the predictive accuracy of each model, using a score of 1 for a correct prediction (using the highest posterior) and 0 for an incorrect prediction. We also look at the predicted probability for the actual student answer. Both measures are averaged over all students.

We performed these types of evaluation repeatedly throughout the development of the network, which showed the effect changes in structure or parameters may have had on the overall behavior. The iterative process halted when the experts felt the behavior of the BN was satisfactory.

11.3.3.5 Empirical evaluation

Table 11.3 is an example of the comparison grids for the classifications that were produced during the comparison evaluation phase. The vertical list down the first column corresponds to the expert rules classification, while the horizontal list across the top corresponds to the BN classification, using the highest posterior; each entry in the grid shows how many students had a particular combination of classifications from the two methods. The grid diagonals show those students for whom the two classifications are in agreement, while the "desirable" changes are shown in *italics* and undesirable changes are shown in **bold**. Note that we use the term "match," rather than saying that the BN classification was "correct," because the expert rule classification is not necessarily ideal.

TABLE 11.3

Comparison grid: expert rule (vertical) vs expert elicited BN (horizontal) classification; type node states 0-N, *pcm*=0.11. Desirable re-classifications are *italicized*, while undesirable ones are in **bold**.

	lwh	lze	lrv	lu	sdf	srn	su	ate	amo	mis	au	un
lwh	386	0	0	0	0	0	0	0	0	0	0	0
lze	0	98	0	0	0	0	0	0	0	0	0	0
lrv	**10**	0	0	0	0	0	0	0	0	0	0	0
lu	*6*	*9*	0	54	0	0	0	0	0	0	0	**6**
sdf	0	0	0	0	83	0	**4**	0	0	0	0	0
srn	0	0	0	0	0	159	0	0	0	0	0	0
su	0	0	0	0	2	22	40	**3**	0	0	0	**2**
ate	0	0	0	0	0	0	0	1050	0	0	0	0
amo	0	0	0	0	0	0	0	0	79	0	0	0
mis	0	0	0	0	0	0	0	0	0	6	0	0
au	*9*	0	0	0	0	0	0	*63*	*8*	0	0	**1**
un	*43*	*6*	*0*	*15*	*35*	*14*	*11*	*119*	*26*	*2*	*0*	66

Further assessment of these results by the experts revealed that when the BN classification does not match the expert rules classification, the misconception with the second highest posterior often did match. The experts then assessed whether differences in the BN's classification from the expert rules classification were in some way desirable or undesirable, depending on how the BN classification would be used. They came up with the following general principles which provided some general comparison measures.

1. It is **desirable** for expert rule classified LUs to be re-classified as another of the specific Ls, similarly for AUs and SUs, and it was desirable for Us to be re-classified as anything else (because this is dealing with borderline cases that the expert rule really can't say much about).

2. It is **undesirable** for (a) specific classifications (i.e., not those involving any kind of "U") to change, because the experts are confident about these classifications, and (b) for any classification to change to UN, because this is in some sense throwing away information (e.g., LU to UN loses information about the "L-like" behavior of the students).

TABLE 11.4
Results showing fine classification summary comparison of various models compared to the expert rules (match, desirable and undesirable changes), together with accuracy of various models predicting student item type answers (using two different measures).

Method	Type values		Match	Des. change	Undes. change	Avg Pred. Accuracy	Avg Pred. Prob.
Expert	0-N	0.22	77.88	20.39	1.72	0.34	0.34
BN		0.11	82.93	15.63	1.44	0.83	0.53
		0.03	84.37	11.86	3.78	0.82	0.70
	H/M/L	0.22	80.47	18.71	0.82	0.89	0.69
		0.11	83.91	13.66	2.42	0.89	0.80
		0.03	90.40	6.48	3.12	0.88	0.83
SNOB	24 DCT		79.81	17.60	2.49		
	0-N		72.06	16.00	11.94		
	H/M/L		72.51	17.03	10.46		
learned	0-N	Avg	95.97	2.36	1.66	0.83	0.74
parameters	H/M/L	Avg	97.63	1.61	0.75	0.89	0.83
CaMML	0-N	Avg	86.51	5.08	8.41	0.83	0.72
constr.	H/M/L	Avg	83.48	8.12	8.34	0.88	0.79
CaMML	0-N	Avg	86.15	5.87	7.92	0.83	0.74
uncons.	H/M/L	Avg	92.63	4.61	2.76	0.89	0.83

Many classification comparison grids were obtained through varying the probability of a careless mistake (*pcm*=0.22, 0.11 and 0.03) and the item type values (0-N vs H/M/L). The percentages for match, desirable and undesirable change were calculated for each grid.

Considerable time was spent determining the following factors that might be causing the differences between the BN and the expert rule classifications. First, the expert rules give priority to the type 1 and type 2 results, whereas the BN model gives equal weighting to all 6 item types. Second, the expert elicited BN structure and parameters reflects both the experts' good understanding for the known fine classifications, and their poor understanding of the behavior of "U" students (LU, SU, AU and UN). Finally, as discussed earlier, some classes are broken down into fine classifications more than others, resulting in lower priors, so the more common classifications (such as ATE and UN) tend to draw in others.

Table 11.4 (Set 1) shows a summary of the expert BN fine classification, varying the type values and probability of careless mistake, in terms of percentage of matches (i.e., on the grid diagonal), desirable changes and undesirable changes and the two prediction measures (see §7.5.1), averaged over all predicted item types, for all students. The experts considered the undesirable change percentages to be quite low, especially considering that they felt some of these can be considered quite justified.

Overall the experts were satisfied that the elicited network performs a good classification of students' misconceptions and captures well the different uncertainties in the experts' domain knowledge. In addition, they were reassured by the fact that its performance appeared quite robust to changes in parameters such as the probability of careless mistakes or the granularity of the evidence nodes.

11.3.4   Automated methods

The next stage of the project involved the application of certain automated methods for knowledge discovery to the domain data, for each main task in the construction process.

1. We applied a classification method to student test data.
2. We performed simple parameter learning based on frequency counts to the expert BN structures; and
3. We applied an existing BN learning program, CaMML (see §9.5).

In each case we compared the performance of the resultant network with the expert elicited networks, providing an insight into how elicitation and knowledge discovery might be combined in the BN knowledge engineering process.

11.3.4.1   Classification

The first aspect investigated was the classification of decimal misconceptions. We applied the SNOB probabilistic classification program (Wallace and Dowe, 2000), based on the information theoretic Minimum Message Length (MML) (Wallace and Boulton, 1968) to the data from 2437 testpapers in different forms:

1. The raw data from the 24 DCT items
2. The data pre-processed from 24 items into the 6 item types

   (a) Using the values 0-N

   (b) Using the values H/M/L

Using the SNOB's most probable class for each student, we constructed comparison grids comparing the SNOB classification with the expert rule classification.

Given the raw data of 24 DCT items, SNOB produced 12 classes, 8 of which corresponded closely to the expert classifications (i.e., had most members on the grid diagonal). Two classes were not found (LRV and SU). Of the other 4 classes, 2 were mainly combinations of the AU and UN classifications, while the other 2 were mainly UNs. SNOB was unable to classify 15 students (0.6%). The percentages of match, desirable and undesirable change are shown in Table 11.4 (set 2, row 1). They are comparable with the expert BN 0-N and only slightly worse than the expert BN H/M/L results.

The comparison results using the data pre-processed into 6 item types (values 0-N and H/M/L) were not particularly good.

- For O-N type values, SNOB found only 5 classes (32 students = 1.3% not classified), corresponding roughly to some of the most populous expert classes (LWH, SDF, SRN, ATE and UN), subsuming the other expert classes.
- For H/M/L type values, SNOB found 6 classes (33 students = 1.4% not classified), corresponding roughly to 5 of the most populous expert classes (LWH, SDF, SRN, ATE, UN), plus a class that combined MIS with UN. The match results are shown in Table 11.4 (set 2, rows 2 and 3).

Clearly, summarizing the results of 24 DCT into 6 item types gives a worse classification; one explanation of this was that many pairs of the classes are distinguished by student behavior on just one item type, and SNOB might consider these differences to be noise within one class.

The overall good performance of the classification method shows that automated knowledge discovery methods may be useful in assisting experts to identify suitable values for classification type variables, particularly when extensive analysis of the domain is not available.

11.3.4.2 Parameters

Our next investigation was to learn the parameters from the data, while keeping the expert elicited network structure. The data was randomly divided into five 80%-20% splits for training and testing; the training data was used to parameterize the expert BN structures (using the the Spiegelhalter-Lauritzen Algorithm 6.1 in Netica), while the test data was given to the resultant BN for classification. The match results (averaged over the 5 splits) for the fine classification comparison of the expert BN structures (with the different type values, 0-N and H/M/L) with learned parameters are shown in Table 11.4 (set 3), together with corresponding prediction results (also averaged over the 5 splits).

Clearly, learning the parameters from data, if it is available, gives results that are much closer to the expert rule classification than using the parameters elicited from the experts. The trade-off is that the network no longer makes changes to the various "U" classifications, i.e., it doesn't shift LUs, SUs, AUs and UNs into other classifications that may be more useful in a teaching context. However it does mean that expert input into the knowledge engineering process can be reduced, by doing the parameter learning on an elicited structure.

11.3.4.3 Structure

Our third investigation involved the application of CaMML (see §9.5) to learn network structure. In order to compare the learned structure with that of the expert elicited BN, we decided to use the pre-processed 6 type data; each program was given the student data for 7 variables (the fine classification variable and the 6 item types), with both the 0-N values and the H/M/L values. The same 5 random 80%-20% splits of the data were used for training and testing. The training data was given as input to the structural learning algorithm and then used to parameterize the result networks using Netica's parameter learning method.

We ran CaMML once for each split (a) without any constraints and (b) with the ordering constraint that the classification should be an ancestor of each of the type nodes. This constraint reflects the general known causal structure. Each run produced a slightly different network structure, with some having the *fineClass node* as a root, some not. Two measures of network complexity were used: (i) ratio of arcs/nodes, which varied from 1.4 to 2.2 and (ii) the total number of probabilities in the CPTs, which varied from about 700 to 144,000. The junction-tree cost was not used as a measure, although it probably should have been!

The percentage match results comparing the CaMML BN classifications (constrained and unconstrained, O-N and H/M/L) are also shown in Table 11.4 (sets 4 and 5), together with the prediction results. The undesirable changes include quite a few shifts from one specific classification to another, which is particularly bad as far as our experts are concerned. The variation between the results for each data set 1-5 was much higher than for the variation when learning parameters for the expert BN structure, no doubt reflecting the difference between the network structure learned for the different splits. However we did not find a clear correlation between the complexity of the learned network structures and their classification performance.

Our experts also looked at the learnt structures during this phase of the project, but they did not have an intuitive feel for how these structures were modeling the domain. In particular, the change in the direction of the arc between the classification node and (one or more of) the item type nodes did not reflect the causal model we had introduced them to during the elicitation phase. Also, there were many arcs *between* item type nodes, and they could not explain these dependencies with their domain knowledge. Some time later, one of the experts investigated one of these learnt structures using Matilda (see §10.3.4), as part of Matilda's evaluation process. By using this KEBN support tool, the expert gained some understanding of the dependencies captured in the seemingly non-intuitive structure. Overall, however, the lack of an adequate explanation for the learnt structures, together with the unacceptable undesirable re-classifications, meant that none of the learnt structures were considered for inclusion in the implemented system.

11.3.5   Field trial evaluation

All components of the complete system were field-tested. The games were trialed with individual students holding known misconceptions and their responses and learning have been tracked (McIntosh et al., 2000). This has refined the design of the games and of the visual scaffolding and led to the decision to provide the feedback and visual scaffolding automatically.

The complete system has also been field tested with 25 students in Grades 5 and 6, who had persistent misconceptions after normal school instruction (Flynn, 2002, Helm, 2002). Students worked with a partner (almost always with the same misconception) for up to 30 minutes, without adult intervention. The observer recorded their conversations, which were linked to computer results and analyzed to see where students learned or missed learning opportunities and how cognitive conflict was involved. Long term conceptual change was measured by re-administering the DCT about three weeks later. Students played in pairs so that the observer could, without intervention, monitor their thinking as revealed by their conversations as they played the games.

Ten students tested as experts on the delayed post-test, indicating significant progress. Seven students demonstrated improvement while eight retained their original misconception. There were some instances where students learned from the visual scaffolding of the help screens, but active teacher intervention seems required for most students to benefit fully from these. Very frequently, students learned by observing and discussing with their partners, but they did not always learn the same things at the same time. This means that the computer diagnosis was not necessarily meaningful for both students so that the item type selection may not perform as designed for either student. This disadvantage needs to be weighed against the benefits of working with a partner.

Feedback provided by the games provoked learning in two ways. In some instances students added new information to their conceptual field, without addressing misconceptions (e.g., learned that 0 in the tenths column makes a number small, without really changing basic whole number thinking). In other instances the feedback provoked cognitive conflict and sometimes this was resolved within the session, resulting in a significant change from a misconception to expertise, maintained at the delayed post-test. The item type selection was set to alternate between "easy" and "hard" items for these field trials but this experiment indicated that it gave too many easy items. The real-time updated diagnosis by the system of the student's thinking patterns was (generally) consistent with the observer's opinion. Discrepancies between classifications and the delayed post-test were tracked to known limitations of the DCT, which could not diagnose an unusual misconception prevalent in that class.

On balance, our experts consider the ITS to be a useful supplement to class instruction.

11.3.6   KEBN aspects

The combination of elicitation and automated methods used in this case study was only possible because we had both the involvement of experts with a detailed understanding of the basis of the misconceptions and how they relate to domain specific activities and an extensive data set of student behavior on test items in the domain.

Building the student model from misconceptions (cf. Sleeman, 1984, Hagen and Sonenberg, 1993), rather than in terms of gaps in correct pieces of domain knowledge (cf. Conati et al., 1997), is a little unusual and is only viable because of the nature of the domain and the extensive research on student understanding in the domain discussed in §11.3.1.

This application is relatively complex in that the BN is used in several ways, namely for diagnosis, prediction and value of information.

The automated techniques were able to yield networks which gave quantitative results comparable to the results from the BN elicited from the experts. However, the experts preferred to use the elicited BN in the implemented system, based on their qualitative evaluation of the undesirable re-classifications.

The quantitative results do provide a form of 'validation' of the learning techniques and suggests that automated methods can reduce the input required from domain experts. In any case, these results build confidence in the elicited BNs. It also supports the reciprocal conclusion regarding the validity of manual construction when there is enough expert knowledge but no available data set.

In addition, we have seen that the use of automated techniques can provide opportunities to explore the implications of modeling choices and to get a feel for design tradeoffs — some examples of this were reported above in both the initial elicitation stage and the discovery stage (e.g., 0-N *vs.* H/M/L).

The actual process of undertaking a quantitative evaluation was tedious, requiring programming in the BN software API to run sets of experiments and collecting and collating results. Much data massaging was required, with all the automated methods requiring different input data formats and producing different output.

While the evaluation consisted of the standard set of case-based, comparative and predictive evaluations, domain specific criteria (such as the division into desirable and undesirable classification changes) were developed during the investigation process and the experts refined their requirements.

The implemented system contains a mechanism for fully recording sessions (with both user inputs and the BNs outputs), which could be invaluable for ongoing evaluation and maintenance of the system. However the system will be deployed via the distribution of CD-ROM to interested math teachers, and there is no incentive or mechanism for them to send this data back to us, the system developers.

11.4　Goulburn Catchment Ecological Risk Assessment

In §5.4 we described the BN developed to model the effects of management actions and associated environmental conditions on the native fish community in the Goulburn Catchment, in Victoria. Here we describe the knowledge engineering process used in this case study, particularly the parameterization (Stage 2 in Boneh's KEBN process as described in §10.2.3. This case study was originally published in a series of papers (Woodberry et al., 2004a,b, Pollino et al., 2007).

11.4.1　Conceptual modeling

A initial stakeholder workshop (Workshop 1), including catchment managers and scientists and conducted in the region, identified native fish communities as being at risk. A second workshop was then held with experts including fish ecologists and local catchment managers, to identify the main biological processes, the human-related activities impacting on the native fishcommunities, and the linkages between them. The experts collaboratively developed a conceptual model (shown in Figure 11.7) of the relationships between physical, chemical and biological factors, and the study query variables (*FutureAbundance* and *FutureDiversity*). Note that the color of the links (not visible in this black and white version) indicate different types of relationships. This workshop identified the need for the model to be flexible, representing multiple spatial scales and at least two temporal scales (1 year and 5 years). The resultant conceptual model is shown in Figure 11.7.

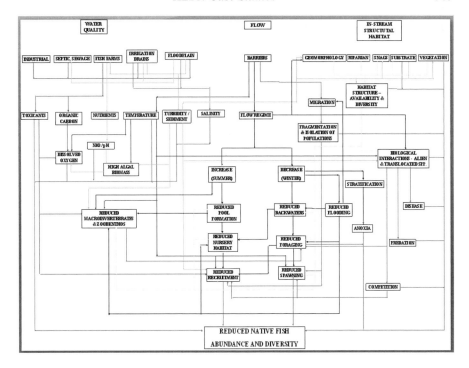

FIGURE 11.7: Goulburn Catchment ERA: initial conceptual model.

11.4.2 The KEBN process used for parameterization

Figure 11.8 shows the detailed KEBN Stage process undertaken for this case study, that is the estimation and evaluation of the probabilities. The diagrams illustrates possible flows (indicated by arrows) through the different KE processes (rectangular boxes), which are executed either by humans (the domain expert and the knowledge engineer, represented by clear boxes) or computer programs (shaded boxes). Major choice points are indicated by hexagons.

Stage 1 of the KEBN process is summarized here in **Structural Development and Evaluation** box. This was done iteratively, using the conceptual model as a starting point, resulting in the BN structure (shown in Figure 5.13); this process is described in (Pollino et al., 2004).

Once a BN structure has been established, the next step is **parameter estimation**, involving specifying the CPTs for each node. Figure 11.8 shows that the parameter estimates can be elicited from domain experts (path 1),[4] or learned from data (2) or, as proposed here, generated from a combination of both sources (an example is shown in path 3). In early prototypes the parameter estimates need not be exact, and uniform distributions can be used if neither domain knowledge nor data are read-

---

[4]This can also include the domain literature as a source of parameter estimates.

ily available. A detailed description of the parameter estimation process is provided below.

The second major aspect of quantitative knowledge engineering is **quantitative evaluation**. Evaluative feedback can be generated using either domain experts or data or both, as we have done here. When data is available, several measures can be used to evaluate BNs, including predictive accuracy, expected value computations and information reward. The domain expert evaluation techniques used include elicitation reviews and model walkthroughs. We also used sensitivity analysis (see §10.4.8 to evaluate the BN. This involves analyzing how sensitive the network is, in terms of changes in updated probabilities of some query nodes to changes in parameters and inputs. Measures for these were computed automatically using BN tools (shown as **Sensitivity to Parameters** and **Sensitivity to Findings** processes, in Figure 11.8), but were then evaluated by the domain expert in conjunction with the knowledge engineer.

11.4.3   Parameter estimation

Variables were split into two groups: those with data were initially given uniform probability distributions; the remainder (see Figure 5.13) were first fully elicited from domain experts. The domain experts were asked to report their confidence in these estimates, which was categorized as either low or high. We note that the experts tended to be more confident estimating variables pertaining to the physical and chemical relationships in the system and less so with the biological relationships. In this study, the elicited confidence applied to the node, i.e., to the whole CPT, rather than to individual parameters; this need not be the case in general.

We dealt with the problem of incomplete data by incorporating expert knowledge before automated learning. The combination of elicitation and data-based parameterization requires the elicited information to be weighted relative to the data available, e.g., via the equivalent sample sizes of §6.2.2. In Figure 11.8 this is done in the **Assign Expert Experience** process, where an experience weighting is assigned to the expert parameter estimates, based on the confidence in the estimates obtained during expert elicitation. These are then treated as equivalent to the size of a hypothetical initial data sample (the equivalent sample size, or ESS).

After incorporating the data in parameter estimation, the next step is to compare the new with the original parameterization. In Figure 11.8 we consider this to be an automated process, **Assess Degree of Changes**. As mentioned above, during parameter elicitation an acceptable range of values can also be elicited. Any parameters estimated from the data to be outside this range should be flagged for attention. In this study, however, we used an alternative method for comparing the parameterizations, namely the Bhattacharyya distance (Bhattacharyya, 1943) between the two probability distributions. This distance is computed for each possible combination of parent values; higher distances between conditional distributions trigger further attention. The domain expert must then assess whether these flagged parameter refinements obtained after automated learning are acceptable (in the **Accept Changes**

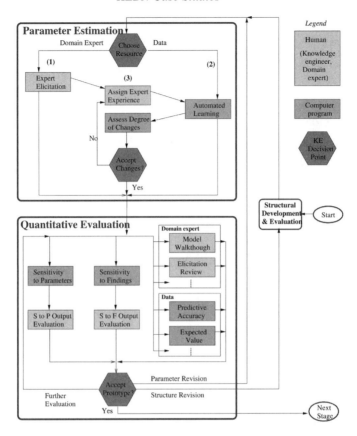

FIGURE 11.8: Knowledge engineering process applied to ERA case study. [With kind permission of Springer Science+Business Media (Woodberry et al., 2004b) and reprinted from Parameterisation of a Bayesian network for use in an ecological risk management case study. Pollino et al. *Environmental Modeling and Software 22*(8), 1140–1152, Copyright (2007), with permission from Elsevier.]

decision point in Figure 11.8). If not, we undertook an iterative investigation of different mappings of the expert experience into equivalent sample sizes.

Of the L&S, expectation maximization (EM) and gradient descent (GD) automated learning methods available in the Netica BN software (Norsys, 2010), the EM method was selected, since the L&S method was not very useful with many parent instantiations missing in the data and the GD method was susceptible to local maxima. Automated learning trials were then carried out using EM in order to investigate the effects of different weightings of expert elicited CPTs. A pre-trial with the L&S method was used for comparative purposes.

A series of trials were conducted (see Table 11.5) using the EM method to refine the parameters of all nodes. Each trial used a series of experience weightings. Each of these trial EM parameterizations was compared, using the Bhattacharyya distance, with the L&S BN, and an assessment was made as to whether the degree

TABLE 11.5

Nodes whose CPTs were first expert elicited, with the different experience weightings used for trials of the EM automated learning method.

Node	H=10,M=5 Trial No.			H=20,M=10 Trial No.			Combined Trial No.	
	1	2	3	1	2	3	4	5
WaterQuality (P)	10	15	18	20	25	25	25	24
HydraulicHabitat (P)	10	15	18	20	25	25	25	24
StructuralHabitat (P)	10	7	4	20	15	10	1	1
BiologicalPotential (B)	5	2	4	10	5	5	5	5
TemperatureModification (P)	10	5	1	20	15	10	1	1
CommunityChange (B)	5	1	1	10	5	1	1	1
FloodplainInundation (P)	10	5	1	20	15	10	1	1
PotentialRecruitment (B)	5	1	3	10	5	2	3	3
Connectivity (P)	10	10	10	20	17	14	12	12
MigratorySPP (B)	5	10	15	10	15	15	15	16
CurrentDiversity (B)	5	1	1	10	7	4	1	2
FutureAbundance (B)	5	2	4	10	7	4	5	6
FutureDiversity (B)	5	5	5	10	5	5	5	5
Remaining Nodes	0	0	0	0	0	0	0	0

Note: P indicates variable part of physical process, B part of biological.

of change was acceptable. If the change was deemed unacceptably large, the ESS was increased, while if there was no or minor changes, the ESS was decreased. This assessment process was iterated, comparing the new EM parameterization with the L&S parameterization and setting a new ESS value, W_{i+1}, using Algorithm 11.1. If an ESS value adjustment was too great and changes went from too large to too small (or vice versa), a smaller, reverse direction, ESS change of 2 was used. As the trials progressed, smaller "tweaks" were used.

Algorithm 11.1 1¡*Adjusting ESS (with ESS changes from ERA case study)*

   Loop *until* **ESS** *values converge*
   Parameterize *network with current* **ESS** *values*
      Switch
         Case *changes unrealistic:* $W_{i+1} \leftarrow W_i$+upLarge (5)
         Case *would allow greater changes :* $W_{i+1} \leftarrow W_i$+upSmall (3)
         Case *little OR no change:* $W_{i+1} \leftarrow W_i$+downLarge (5)
         Case *insignificant change:* $W_{i+1} \leftarrow W_i$-downSmall (3)
         Case *changes become unrealistic:* $W_{i+1} \leftarrow W_{i-1}$+bounceup (2)
         Case *changes disappear:* $W_{i+1} \leftarrow W_{i-1}$-bouncedown (2)
         Case *final trials, small adjustments only:* $W_{i+1} \leftarrow W_{i-1}\pm$tweak (1)
         Case *changes acceptable:* $W_{i+1} \leftarrow W_i$
   End Loop

11.4.4 Quantitative evaluation

After parameterization of the BN, the second major aspect of quantitative knowledge engineering is evaluation, which guides further iterations of BN development.

As the ERA data was limited, automated evaluation was of little use. The available data were split randomly so that 80% of data were used for model training and 20% used for testing. Predictive accuracy was used to test the situations that were available in the data; data was entered for the non-query variables, and the model was used to predict the outcomes of the two query nodes, given the evidence data available for other nodes in the cases. The error rate of the query nodes, *FutureAbundance* and *FutureDiversity*, were only 5.8% and 0%, respectively. This was not very informative as there were only 10 high *FutureAbundance* cases and no high *Future-Diversity* cases. Thus, they were equal to the percentage of high cases for each query node.

Other, less formal, data evaluation was conducted as follows. Current abundance and diversity at each of the sites under existing environmental conditions was plotted against model output probabilities of *FutureAbundance* (see Figure 11.9) and *FutureDiversity* (see Figure 11.10). The trends between real data and BN predictions are well maintained throughout the catchment. Output from both the L&S and EM learned models are included for comparative purposes, showing a general improvement in the predictive output of the EM model and hence our automated learning methodology.

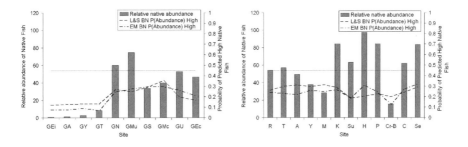

FIGURE 11.9: Relative abundances of native fish at (a) Goulburn river sites (b) Tributary sites, observed (> 1970 fisheries data) vs. predicted *FutureAbundance=high* (> 54) at the same sites.

11.4.4.1 Evaluation by domain expert

A semi-formal model walkthrough was conducted with management and ecology experts. Overall, the model received positive feedback. The model was regarded as being a reasonable representation of a complex environment, but the need for routine updating was emphasized. The authors also recognize the need for more data to improve the robustness of model predictions, and to reduce parameter uncertainties. A model deficiency was the inability to differentiate between the responses of dif-

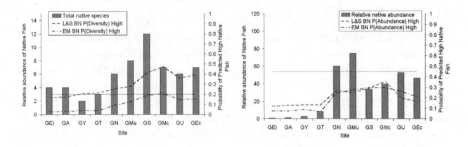

FIGURE 11.10: Total number of native fish species at (a) Goulburn river sites (b) Tributary sites, (> 1970 fisheries data) vs. predicted *FutureDiversity=high* (> 3) at the same sites.

ferent native fish groups, particularly as different groups can respond differently to environmental conditions, such as flow.

11.4.4.2   Sensitivity to findings analysis

We implemented an algorithm to computer the sensitivity to findings, as described in § 10.4.8, shown in Algorithm 11.2. Our algorithm computes and displays both the entropy of a specified query node and the ranked mutual information values for a specified set of interest nodes, given a set of evidence for some other observed nodes. The user can subsequently investigate how changes to the evidence will affect the entropy and MI measures. This process allows the domain expert to identify whether a variable is either too sensitive or insensitive to other variables in particular contexts, which in turn may help identify errors in either the network structure or the CPTs.

Algorithm 11.2 Sensitivity to Findings
> *loop*
>> **Compute** *entropy of query node and mutual information values of interest nodes*
>> **Display** *entropy of query node*
>> **Rank** *and* **Display** *mutual information values*
>> **Prompt** *user for action*
>> **if** *Action = Set Node Finding* **then**
>>> **Prompt** *user for node name and value*
>>> **Enter** *node finding*
>> **else if** *Action = Back up* **then**
>>> **Remove** *last node finding*
>> **else if** *Action = Save Report* **then**
>>> **Save** *sensitivity analysis results to file*
>>> **break Loop**
>> **end if**
> *end loop*

TABLE 11.6

Sensitivity to findings analysis (using Netica) performed on *FutureAbundance*.

	No Evidence		*WaterQuality = low*		*WaterQuality = high*	
Entropy of *FutureAbundance*	0.762080		0.379315		0.878575	
Node	MI	Rank	MI	Rank	MI	Rank
FutureDiversity	0.086744	1	0.039833	1	0.097424	1
WaterQuality	0.055666	2	-	-	-	-
HydraulicHabitat	0.031075	3	0.005350	2	0.029836	3
BiologicalPotential	0.030830	4	0.000189	21	0.056507	2
Site	0.027754	5	0.002104	3	0.006124	4
Barrier	0.022200	6	0.001866	6	0.002228	11
Type	0.022190	7	0.001890	5	0.003686	6
Temperature	0.021858	8	0.001827	7	0.000999	18
AvrSummer	0.019413	9	0.001919	4	0.003406	7
Min Summer	0.011461	10	0.001400	9	0.003406	8
AvrWinter	0.009974	11	0.001524	8	0.002454	9
MaxWinter	0.008584	14	0.001235	10	0.002357	10
CurrentAbundance	0.004246	16	0.000148	22	0.004836	5

Application of the sensitivity to findings algorithm showed that the query variable, *FutureAbundance*, in the absence of other evidence, is most sensitive to *FutureDiversity*, followed by *WaterQuality (WQ)* (see Table 11.6). When findings for *WQ* were entered into the network, the sensitivity measures changed, and the ranking of variables changed. When $WQ = low$, evidence about other variables have less impact on the *FutureAbundance* node, demonstrating the dominating effect of a low water quality in the system. Alternatively, when $WQ = high$, some of the remaining variables became more influential. We also looked at site specific cases; for example, (Pollino et al., 2007, Table 5) shows the sensitivity analysis for Eildon, as site where the native fish communities are highly stressed. These observations agreed with the domain expert's understanding of the system.

11.4.4.3 Sensitivity to parameters analysis

We implemented a form of sensitivity to parameters, given in Algorithm 11.3, following Coupé and Van der Gaag (2002), as described in § 10.4.8. When a particular evidence instantiation is set, our algorithm identifies the type of sensitivity function for the parameters by checking whether the query node has any observed descendant nodes. Once the sensitivity function is determined for a parameter, its coefficients can be computed (see Algorithm 11.4). If the plotted sensitivity function does not behave as the domain expert expects (its slope, direction or range is unexpected), then this could indicate errors in the network structure or CPTs.

Algorithm 11.3 Sensitivity to Parameters

loop
   Prompt *user for action*
   Switch
        Case *Set Interest Node*
           Prompt *user for node name*
        Case *Set Test Node Parameter*
           Prompt *user for node name*
           Prompt *user for parent set (CPT row)*
           Prompt *user for test node state (CPT column)*
        Case *Set Evidence Set*
           Prompt *user for evidence set*
           Set *node findings*
        Case *Display Plot*
           if test node in sensitivity set **then**
               Compute *sensitivity function coefficients*
               Display *sensitivity function to screen*
           end if
        Case *Find Sensitive Parameters*
           if test node in sensitivity set **then**
               for each parameter in test node **do**
                   Compute *sensitivity function coefficients*
                   Compute *sensitivity value at parameter value*
                   Display *sensitivity* **And** *parameter values*
               end for
           end if
end loop

Algorithm 11.4 Algorithm to find Sensitivity Functions

Loop *2 times for linear, 3 times for hyperbolic*
   Set *new normalized probability distribution in test node*
   Compile *network*
   Get *node belief in interest node*
   Store *parameter and belief pair*
   Restore *old probability distribution*
End Loop
Solve *for coefficients of sensitivity function*

The revised normalized probability distribution of the test node is set by first selecting a new value, P_{new} for the parameter under investigation, P_j. The remaining parameters, P_i, are normalized to retain relative values by the updating function,

$$P_i \Leftarrow P_i \times \frac{1 - P_{new}}{1 - P_j}, i \neq j \qquad (11.1)$$

before the parameter under study is updated, $P_j \Leftarrow P_{new}$

The most sensitive parameter identified in the ERA model was $P(FutureAbundance = low|Time\_Scale = one\_year, WQ = low, StrucHab = low, BiolPoten = low, OverallFlow = xxtIncrease)$ for the posterior probability $P(FutureAbundance|Site = G\_Eild, Time\_Scale = one\_year)$ (see Figure 11.11). This observation agreed with the domain expert's evaluation of the system, in the magnitude, direction and range of the visual representation of the function.

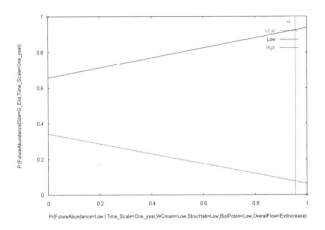

FIGURE 11.11: Linear functional representation of parameter sensitivity (Woodberry et al., 2004a). Y-axis is the posterior probability, X-axis is the particular parameter (CPT entry).

11.4.5 Conclusions

The development of quantitative decision-support systems in catchment management is of high priority as they enable more robust, defensible and tractable decisions. For managing ecological assets, it is preferable that models be integrative, representing the range of hazards that can potentially harm that asset, allowing risks and management actions to be identified and prioritized. However, given that both the understanding of many complex ecological systems is limited, and the existing modeling technologies for describing complex and variable systems are poor, progress has been limited. In this study, a BN was developed to assess the suitability of using the modeling approach in an ecological risk management case study. The model needed to characterise a complex ecological system and, common to many catchments, there were high uncertainties associated with the lack of data and poor knowledge of the quantitative relationships between variables in the model.

The BN parameterization process described in the study enabled expert knowledge and data to be combined using a robust, iterative approach. The current model prototype can be used to inform future management decisions at multiple spatial scales, while taking into account associated uncertainties. To test the robustness of this model further, it is essential that the model be further field-tested to determine its accuracy pre- and post-management interventions or system changes.

11.4.6   KEBN aspects

In the course of this case study, we developed and formalized a methodology for combining expert elicitation and data for parametering the BN. In many ecological applications, including our case study, information sources are often poorly documented, poorly understood, and generally incomplete. Although other causal network structures (e.g. Borsuk et al., 2004, Ticehurst et al., 2007) have been developed using such information sources, unlike this study, parameter estimates for a variable were obtained from only one source (i.e. experts or data). To parameterize the native fish BN model, we directed our efforts towards combining multiple information sources, each with associated uncertainties, and undertaking an iterative process to derive acceptable parameter estimates.

A suite of evaluative methods was used to investigate the uncertainties and inaccuracies in model structure, relationships and outputs. This process enabled a more targeted approach to the identification of parameters that needed to be accurately quantified and also to recommendations for targeted monitoring and studies to collect further information and data.

11.5   Cardiovascular risk assessment

In § 5.3 we described how we knowledge engineered BNs for predicting Coronary Heart Disease (CHD) from the medical literature. In this section we first describe how we applied the causal discovery program, CaMML (see § 9.5) to learn from the Busselton data, and compare the knowledge engineered BNs with the BNs learned by CaMML and other BN learners from the literature. We then describe TakeHeart II, our clinical decision support tool, which provides an interface to the CHD BNs.

11.5.1   Learning CHD BNs

In our learning experiments, we used two versions of CaMML (§ 9.5), one which learns linear path models and another which learns discrete BNs. Although linear CaMML uses numerical variables, for testing using the Weka machine learning environment (Witten and Frank, 2005), we had to discretize the variables; we used the Weka-provided MDL discretizer.

Rather than looking only at CaMML's best model, which may have the highest posterior probability and yet be highly unlikely, we looked at a "summary model" showing each arc that appeared more than, say, 10% of the time, with more frequent arcs drawn more boldly. We can then be very confident of arcs that appear nearly all (or none) of the time, though not always their direction.

First we ran discrete CaMML on the whole of the Busselton-PROCAM dataset. Figure 11.12 shows three summary graphs showing arc frequencies for three methods of handling missing data: (a) modeling missing explicitly; (b) imputing mean (or mode) values; and (c) removing missing values. Thicker arcs indicate higher frequencies. Although there are variations with different treatment of missing values, we note that CHD10 is *never* directly connected to SBP, Diabetes, or Smoking, but is always linked to Age, and also to HDL and LDL except when modeling missing.

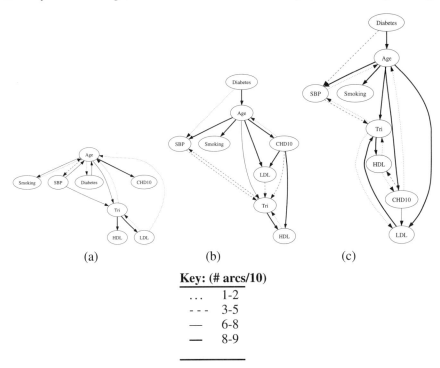

(a)   (b)   (c)

Key: (# arcs/10)

...	1-2
---	3-5
—	6-8
—	8-9

FIGURE 11.12: Three summary graphs showing arc frequencies for BNs produced by CaMML from the Busselton-PROCAM dataset. (a) Modeling missing; (b) Imputing mean (or mode) values; (c) Removing missing. (Twardy et al., 2006, Fig.5. With permission.)

The linear model learnt (with missing cases removed) is shown in Figure 11.13(a), and a summary graph in Figure 11.13(b). Age has a strong positive effect on CHD10, and HDL has a moderate negative effect. Although the causal directions must be viewed skeptically, we can see that SBP has a mixed effect: high SBP "raises" Age, which raises CHD10, but high SBP also "raises" HDL, thereby lowering CHD10. Here, as in the models learn by discrete, Smoking and Diabetes are related to CHD10 *only* through other variables.

We also ran CaMML over the complete Busselton variable set (for the 1978 cohort), which adds Height, Weight, DBP, SmokeAmt, Drinker, and AlcAmt to the vari-

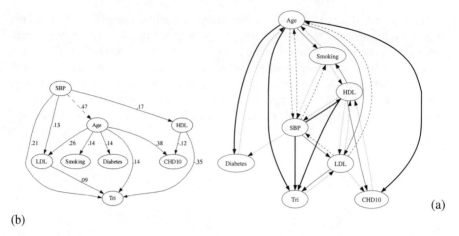

(a)

(b)

FIGURE 11.13: The PROCAM Busselton path model, as found by Linear CaMML. (a) the best model (weights are path coefficients, dashed indicates dubious arc); (b) the summary graph. (Twardy et al., 2006, Fig.6. With permission.)

ables of Busselton-PROCAM. To allow for some direct comparison with the smaller models, we use data for males only, omitting Sex ($N \approx 1820$).

Figure 11.14 shows the results for discrete CaMML, removing all missing cases. Looking at the summary graph and the best model (Figure 11.14(a) & (b)), we see that given Age and Total Cholesterol, the new variables do not help much for predicting CHD10. However, in the best model, drinking raises HDL (good) cholesterol. When we intervene on Drinking (to prevent additional correlations from the "back paths" through Age) we find the probability of high HDL ($> 1.3 \ mmol/L$) is 0.4 for nondrinkers, but 0.6 for drinkers. However, in this model, that will have no effect on anything else, because there are no variables "downstream" of HDL.

11.5.1.1 Experimental methodology

In our experimental evaluation, we use 10-fold stratified cross-validation We present the results using two metrics: ROC curves (§ 7.5.3) and Bayesian Information Reward (BIR) (§ 7.5.8).

In the first set of experiments, we compare the knowledge-engineered BNs – Busselton, PROCAM-German (German priors) and PROCAM-adapted (Busselton priors) (see § 5.3) – and the BN learned by CaMML when removing missing data – CaMML-remove (see § 9.5). Note that this BN is learned from the entire data set, just as the Busselton and PROCAM models were learned on their respective full datasets.

In the second set of experiments, we compared CaMML against the standard machine-learning algorithms provided by Weka: Naive Bayes, J48 (C4.5), AODE, logistic regression, and an artificial neural network (ANN) (all run with default settings). For comparison, we also included Perfect (always guesses the right answer

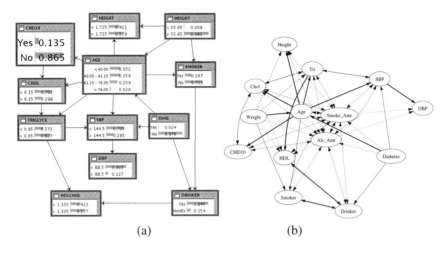

(a)   (b)

FIGURE 11.14: The results from discrete CaMML on the whole Busselton dataset, Males only. (a) the "best" model (b) the summary graph. (Twardy et al., 2006, Fig.7. With permission.)

with full confidence) and Prior (always guesses the training prior), which should set the upper and lower bounds on reasonable performance, respectively.

11.5.1.2   Results

Figure 11.15 (a) and (b) show the ROC curves and BIR results for the BNs on the Busselton-PROCAM data. The two PROCAM BNs score the same AUC (.845); CaMML does just as well, though we know that it uses only two or three variables to predict CHD10! And although the Busselton BN no longer matches the transformed variables in the Busselton-PROCAM dataset, it still manages a respectable AUC of .828.

Figure 11.16(a) and (b) shows how CaMML and standard machine learners fare on the Busselton-PROCAM data. Logistic regression wins with AUC of .844 and a BIR clearly above 0. None of the other learners did as well on ROC. The next best AUC was Naive Bayes at .828. However, Naive Bayes does very poorly on BIR, with scores indistinguishable from the prior. CaMML and AODE also do well on BIR (other methods of handling missing values averaged about the same, with slightly less variance). They perform about the same on AUC, scoring about .81. J48 did quite poorly.

We repeated the previous experiment, but used the original Busselton dataset, males only. The results were much as before, only the Busselton BN now does better, and the other models do worse, so the Busselton and Procam BNs get AUC ≈ .83.

In conclusion, we found that the PROCAM BN (German priors) does as well as a logistic regression model of the Busselton data, which is otherwise the best model. They had the same AUC, with about the same curve. This means they ranked cases in roughly the same order. However, we also found that the PROCAM BN did just

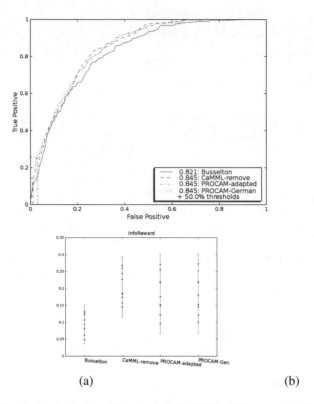

(a)   (b)

FIGURE 11.15: Evaluating the elicited and CaMML BNs on the Busselton-PROCAM dataset: (a) ROC and AUC; (b) BIR cross-validation results. (Twardy et al., 2006, Fig.8. With permission.)

as well as the logistic regression on BIR and related metrics, which suggests that it *is* well-calibrated, regardless of the fact that the individual regression coefficients might be different.

11.5.2   The clinical support tool: TakeHeart II

Monash University's Epidemiological Modeling Unit developed the original Take Heart program (McNeil et al., 2001) in conjunction with BP Australia's Occupation Health Services Department. Take Heart estimated CHD10 risk for approximately 900 BP Australia employees, starting from 1997 and extending for over two and a half years. Take Heart's epidemiological model used equations from the Multiple Risk Factor Intervention Trial (MRFIT) study (Kannel et al., 1986), adjusted so that the background risk factor probabilities fit the Australian population. It used a Microsoft Access database to store cases, and Access macros for the calculations.

Figure 11.17 shows the architecture for TakeHeart II, divided into the BN construction phase (on the left) and its use in a clinical setting (right). The construction

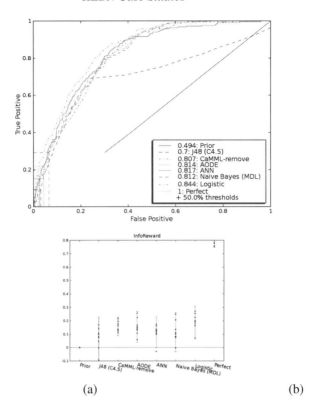

(a)   (b)

FIGURE 11.16: Evaluating CaMML BNs and the standard machine learners on the Busselton-PROCAM dataset: (a) ROC and AUC; (b) BIR cross-validation results. (Twardy et al., 2006, Fig.9. With permission.)

phase depicts the general process described earlier in this chapter, with the BN built using a combination of models from the epidemiological literature, expert knowledge and data. It also includes an adaptation phase, where a BN (built using any method) can be adapted for a different population (such as re-parameterizing the PROCAM model from the Busselton dataset, described in § 5.3.4).

TakeHeart II was implement in the special-purpose scripting language *Modular Bayesian Network* (ModBayes). This links a BN to graphical controls such as drop-down menus, buttons and check-boxes; as well as visualisations such as the chart and risk displays of Figure 11.18. It also manages a case file associated with the BN. ModBayes allows speedy organisation of the controls and displays on screen: someone knowledgeable in the scripting language could redesign the layout in minutes. In addition, it automatically integrates with the *Causal Reckoner* (Korb et al., 2004, Hope, 2008). When evidence is entered in the Causal Reckoner, it appears on Mod-Bayes's controls, and vice-versa, which enables the BN display to change according to the needs of the user. The scripting language itself is a dialect of Lisp, and hence

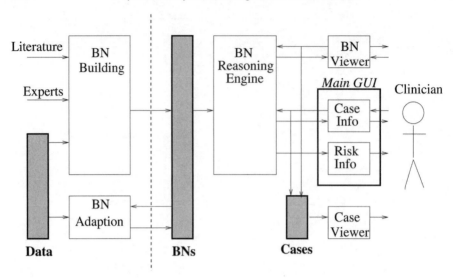

FIGURE 11.17: TakeHeart II architecture: construction and adaptation (left) provide BNs used for risk assessment in a clinical setting (right). Reproduced with permission from Wiley (Nicholson et al., 2008)[Fig.3.6].

is a full-featured programming language in its own right. Therefore the scripting language can be extended (by an advanced user) to create arbitrary BN interfaces.

TakeHeart II's main interface, shown in Figure 11.18 is divided into two sections. The top section is for case information about the patient being assessed, which is then entered as evidence in BN by the reasoning engine. The case title is an editable text field that becomes the case label on the risk assessment displays and on the case viewer.

TakeHeart II also makes a risk assessment of a CHD in the next 10 years. If age information has been entered, then TakeHeart displays the extrapolated 10 year risk for the current "case". The case is also saved (with its label) in the case database, allowing the clinician to modify it, for example to show the effect of a life-style change such as quitting smoking. The risk assessment then shows the risk for *both* cases (see Figure 11.19). In this example, the higher curve is her current projected risk, and the lower curve is her projected risk should she quit smoking; clearly, smoking has increased her risk of heart attack substantially. There is no formal limit to the number of cases that can be displayed together, however in practice the size of the screen will dictate a limit of 6-8 cases.

TakeHeart also provides risk assessment by *age*. Figure 11.20 shows a category plot of heart attack risk by age, again comparing smokers with non-smokers (the form was filled out the same as for Cassy, but with age left blank). Again, this provides a clear visualisation of the impact of smoking, almost doubling the risk for each age group.

How are the risk assessment values calculated? The category display is simple:

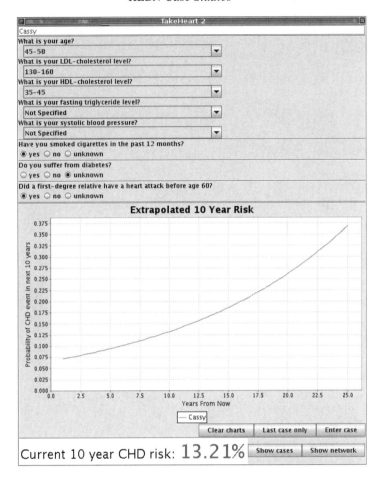

FIGURE 11.18: The main GUI for TakeHeart II. Reproduced with permission from Wiley (Nicholson et al., 2008)[Fig.3.7].

the risk for each age group, given the other risk factors, is simply calculated in turn. The extrapolated risk chart is more complex. An exponential curve ($y = e^{mx+c}$), is fitted to two points: the current risk and the risk for someone five years older (or five years younger if the person is in the oldest age range). The exponential distribution is chosen because it matches observed data in how risk increases with age (e.g., McNeil et al., 2001).

There are two additional displays provided with TakeHeart II. The first is Causal Reckoner's BN viewer, which is fully integrated into TakeHeart II: evidence entered in either the survey form or in the BN display is updated in the other. The BN viewer may also display non-input (i.e. "intermediate") nodes, whose values are not included in the main survey form. Finally, the Case Viewer allows the clinician to look at

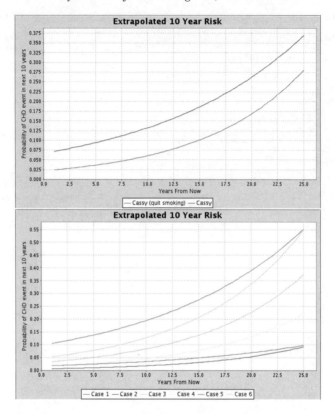

FIGURE 11.19: TakeHeart II's extrapolated 10-year for (above) 2 cases and (below) 6. (Cases reproduced with permission from Wiley (Nicholson et al., 2008) [Fig. 3.8].)

the history of cases entered into TakeHeart. Full details of TakeHeart II's features, including the extrapolation algorithm/method, are provided in Hope et al. (2006).

11.5.3 KEBN aspects

In § 5.3, we described the construction of two Bayesian networks for cardiovascular risk assessment. These networks were built using information available in the epidemiology literature, with only minor additional input from medical experts, avoiding much of the knowledge engineering "bottleneck" associated with expert elicitation. We have also justified our knowledge engineering choices, providing a case study that that may assist others develop similar BN applications in this way for risk prediction of other medical conditions.

We had to discretize our continuous variables. Often a tedious an uncertain step, we were fortunate in being able to use established diagnostic ranges, either from the models themselves or from other literature. Still, discretization introduces edge effects, and it would be better to perform continuous inference (perhaps via particle

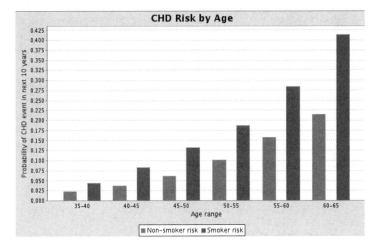

FIGURE 11.20: Heart attack risk by age, for a smoker and and a non-smoker. Reproduced with permission from Wiley (Nicholson et al., 2008) [Fig. 3.9].

filter sampling), and discretize only at the end, if necessary to match diagnostic categories. Netica cannot yet perform "native" continuous inference, but we should point out that this is a limit of our choice of tools, not the models or theory itself.

We then described our tool to support clinical cardiovascular risk assessment, TakeHeart II, which has a BN reasoning engine. The modular nature of the TakeHeart II architecture, and its implementation in the ModBayes scripting language, means that it can incorporate different BNs (for example changing to a BN that is a better predictor for a given population) without requiring major changes to the interface.

We are currently directing our efforts in two directions. First, at evaluating the TakeHeart II tool in a clinical setting, which may well lead to modifications of the interface. Second, at using expert knowledge to provide more direction to causal discovery process, hopefully yielding more informative and predictive models.

11.6 Summary

The four case studies demonstrate most of the main knowledge engineering features presented in the previous two chapters. The poker case study illustrates the spiral model of prototype development. All the poker structures were hand-crafted, and their evaluation was based in experimental use against both automated and human opponents. In the ITS case study both the structure and parameters were initially elicited from experts, with automated methods used to validate aspects of the model, using data that was only available for the DCT part of the system. In the ERA case study, the structure was elicited from experts in workshops, with a conceptual model developed before building the BN, while an iterative combination of elicitation and data was used to parameterize the model. Much more data was available for the CHD case study, which allowed comparison of networks built using models from the epidemiological literature, with BNs learned from from data, using CaMML and other BN learners. Consequently, the domain experts had a smaller role in this case study, compared to the other three, and evaluation consisted only of validation methods based on the data. Our experiences in these, and other, projects were useful in the development of our ideas for the KEBN process presented in this text. KEBN, in turn, should feed into our future Bayesian network projects, resulting in better focused application efforts.

A

Notation

This appendix provides a list of notational symbols and acronyms, together with page numbers where they are defined or first used.

$A \subset B$	A is a proper subset of B	p. 7
$A \subseteq B$	A is a proper subset of B or equal to B	p. 6
$A \backslash B$	set A with all elements of B removed	p. 60
$\emptyset$	the empty set	p. 6
$A \cup B$	the union of A and B	p. 6
$A \cap B$	the intersection of A and B	p. 6
$x \in A$	x is a member of set A	p. 4
$|A|$	the number of objects in set A	p. 9
$x \in [y, z]$	$y \le x \le z$	p. 4
$\{A_i\}$	a set of events indexed by i	p. 8
$\bigcup_i A_i$	the union of all $A_i \in \{A_i\}$	p. 8
$A \wedge B$	A and B	p. 37
$A \equiv B$	A is equivalent to B	p. 7
$\neg A$	not A	p. 5
$A \perp\!\!\!\perp B$	A is independent of B	p. 7
$A \perp\!\!\!\perp B \vert C$	A is independent of B given C	p. 7
$A \not\perp\!\!\!\perp C \vert B$	A is dependent upon B given C	p. 41
$A \perp C \vert B$	A is d-separated from B given C	p. 42
$A \not\perp C \vert B$	A is d-connected to B given C	p. 42
X_i	A variable	p. 8
$X_i = x$	The variable X_i takes value x	p. 35
x_i	ith state of variable X	p. 9
Ω_X	state space of variable X	p. 9
X_i^t	node for variable X_i at the tth time-slice	p. 113
$\overline{X}$	$\sum_{i=1}^{n} \frac{x_i}{n}$	p. 251
$\forall x$	for all x	p. 8
$N!$	N factorial	p. 247
α	normalizing constant	p. 56
$P(X)$	(prior, marginal) probability of X	p. 6
P'	estimated distribution P'	p. 75
$P(X = x)$	probability that X takes value x	p. 6
$P(X \vert E)$	probability of X given evidence E	p. 7
$Bel(X)$	posterior distribution over X	p. 12
$\widehat{Bel}(X)$	predicted belief for X	p. 116
$MB(h, e)$	degree of increased belief	p. 5
$MD(h, e)$	degree of increased disbelief	p. 5
$CF(h, e)$	certainty factor	p. 5

$Parents(X)$	set of parent nodes of X	p. 32
$\pi(X)$ or π_X	parent set of X	p. 257
$\phi(i)$	an instantiation of $\pi(X_i)$	p. 262
Φ_k	instantiations of parent set of X	p. 258
$A - B$	an undirected link	p. 242
$A \rightarrow B$	a directed link	p. 29
$E[X]$	the expected value of X	p. 213
$E[f(X)]$	the expected value of $f(X)$	p. 192
$U(O|A)$	the utility of outcome O, given action A	p. 15
$EU(A|E)$	the expected utility of action A, given evidence E	p. 15
$i \leftarrow i+1$	assignment (used in algorithms)	p. 76
$\int_a^b f(x)dx$	the integral of $f(x)$ from a to b	p. 9
$\sum_i f(x_i)$	the sum of $f(x_i)$ indexed by i	p. 8
$\prod_i f(x_i)$	the product of $f(x_i)$ indexed by i	p. 19
$C_j^n = \begin{pmatrix} n \\ j \end{pmatrix}$	The number of ways of taking a subset of j objects from a set of size n	p. 240
$\lambda(x)$	likelihood	p. 56
$\pi(X)$	parameter in message-passing algorithm	p. 59
$\lambda(X)$	parameter in message-passing algorithm	p. 59
$\pi_X(Y)$	message sent from parent Y to child X	p. 60
$\lambda_X(Y)$	message sent from child Y to parent X	p. 60
$E_{U_i \backslash X}$	evidence connected to U_i	p. 60
$E_{Y_j \backslash X}$	evidence connected to Y_j	p. 60
$u_1 \ldots u_n$	an instantiation of $U_1 \ldots U_n$	p. 60
$C(\mathbf{E})$	measure of conflict between evidence $\mathbf{E}$	p. 81
$max_{X_i \in P\mathbf{X}}(f(X_i))$	returns the X_i that gives maximum $f(.)$	p. 111
$\mathbf{E}_{\{1,t\}}$	evidence nodes from first to tth time-slice	p. 115
a_{ij}	linear coefficient for $X_j \rightarrow X_i$	p. 235
μ_i	mean of X_i	p. 235
σ_i	standard deviation of X_i	p. 235
p_{ij}	path coefficient for $X_j \rightarrow X_i$	p. 235
r_{ij}	sample correlation between X_j and X_i	p. 235
Φ_k	active path	p. 237
$v(\Phi_k)$	valuation of an active path	p. 237
θ_i	parameterization of model i	p. 244
$\rho_{XY \cdot S}$	partial correlation between X and Y, S fixed	p. 247
S_{XY}	sample covariance between X and Y	p. 250
S_X	sample standard deviation of X	p. 250
$r_{XY \cdot Z}$	sample partial correlation	p. 251
$D[\alpha_1, \ldots, \alpha_i, \ldots, \alpha_\tau]$	Dirichlet distribution with τ parameters	p. 188
$\vec{\theta}$	the vector $< \theta_1, \ldots, \theta_\tau >$	p. 189
α_{kjl}	number of matching sample cases	p. 258
$H(X)$	entropy of X	p. 262
$H(X,Y)$	mutual information between X and Y	p. 262
$N(0, \sigma_j)$	Normal (Gaussian) distribution	p. 269
χ^2	the χ^2 distribution	p. 290
IR_G	Good's information reward	p. 224
IR_B	Bayesian information reward	p. 225

Acronyms

A0DE	averaged zero-dependence estimators	p. 214
AUC	area under curve	p. 216
BIR	Bayesian information reward	p. 225
BN	Bayesian network	p. 29
BPP	Bayesian poker player	p. 157
CHD	cardiovascular heart disease	p. 82
CPT	conditional probability table	p. 32
D-map	Dependence-map	p. 33
DBN	dynamic Bayesian network	p. 112
DDN	dynamic decision network	p. 118
EB	expected benefit	p. 112
EM	expectation maximization	p. 194
ERA	environmental risk assessment	p. 138
EU	expected utility	p. 15
EW	expected winnings	p. 160
GA	genetic algorithm	p. 269
IC	Inductive Causation	p. 241
I-map	Independence-map	p. 33
ITS	intelligent tutoring system	p. 369
KEBN	knowledge engineering with Bayesian networks	p. 297
KL	Kullback-Leibler divergence	p. 78
LS	logic sampling	p. 75
LW	likelihood weighting	p. 76
MCMC	Markov chain Monte Carlo	p. 78
MDL	minimum description length	p. 259
MML	minimum message length	p. 259
MPE	most probable explanation	p. 80
NAG	Nice Argument Generator	p. 160
NB	Naive Bayes	p. 205
NRM	natural resources management	p. 138
OOBN	object-oriented Bayesian network	p. 120
PC	PC algorithm	p. 241
QPN	qualitative probabilistic network	p. 321
ROC	receiver-operator curve	p. 215
TAN	tree augmented naive Bayes	p. 208
TOM	totally ordered model	p. 267
VE	the Verbal Elicitor software package	p. 328

B

Software Packages

B.1 Introduction

The rapid development of Bayesian network research over the past 20-plus years has been accompanied by a proliferation of BN software tools. These tools have been built to support both these research efforts and the applications of BNs to an ever-widening range of domains. This appendix is intended as a resource guide to those software tools.

We begin the Appendix with a brief history of BN software development. We then provide a listing of software packages for Bayesian networks and graphical models (§B.3), based on the one built up and maintained by Kevin Murphy over a number of years[1]. We also note other Web sites for BN software and other resources. Most of these packages have a long list of features that we cannot even list here, so our survey is by no means exhaustive. Note that we have personal experience, through teaching, research projects or application development, with the following software: AgenaRisk, BayesiaLab, BNT, BUGS, CABeN, CaMML, Hugin, IDEAL, Netica, GeNIe/SMILE and TETRAD.

In general, all the packages with GUIs include the advanced GUI features (e.g., menu options, short-cut icons, drag-and-drop, online help) that have become the norm in recent years. In this resource review, we will generally ignore GUI aspects, unless there is some feature that stands out. Instead, we will concentrate on aspects of the functionality.

We make no attempt to give any sort of ranking of these packages; so our survey presents the packages alphabetically. The survey listing notes whether the products are free or commercial but available in a restricted form. Otherwise we do not make any comments on the cost of commercial products.

---

[1] Our thanks to Kevin Murphy for giving us permission to use this listing.

Src Is the source code included? **N=no**. If yes, what language? **J** = Java, **M** = Matlab, **L** = Lisp, **C**, **C++**, **C#**, **R**, **A** = APL, **P** = Pascal, **Ru** = Ruby, **F** = Fortran.

API Is an application program interface included?

N means the program cannot be integrated into your code, i.e., it must be run as a standalone executable. **Y** means it can be integrated.

Exec The **executable** runs on: **W** = Windows (95/98/2000/NT), **U** = Unix, **M** = MacIntosh, **-** = Any machine with a compiler, or Java Virtual machine.

GUI Is a Graphical User Interface included? **Y=Yes,N=No**.

D/C Are continuous-valued nodes supported (as well as discrete)? **G** = (conditionally) Gaussians nodes supported analytically, **Cs** = continuous nodes supported by sampling, **Cd** = continuous nodes supported by discretization, **Cx** = continuous nodes supported by some unspecified method, **D** = only discrete nodes supported.

DN Are decision networks/influence diagrams supported? **Y=Yes,N=No**.

DBN Are dynamic Bayesian networks/influence diagrams supported? **Y=Yes,N=No, T** means some modeling over time but not with DBNs.

OOBN Are Object-oriented Bayesian networks/influence diagrams supported? **Y=Yes,N=No, S=Not** full OO, but support for subnetworks.

Params Does the software functionality include parameter learning? **Y=Yes,N=No**.

Struct Does the software functionality include structure learning? **Y=Yes,N=No**.

IC means **Y**, using conditional independency tests (see §8.3)

K2 means **Y**, using Cooper & Herskovits' K2 algorithm (see §9.2)

D/U What kind of graphs are supported? **U** = only undirected graphs, **D** = only directed graphs, **UD** = both undirected and directed, **CG** = chain graphs (mixed directed/undirected).

Inf Which inference algorithm is used? (See Chapter 3)

JT = Junction Tree, VE = variable (bucket) elimination, CO = conditioning, PT = Pearl's polytree, E = Exact inference (unspecified), **MH = Metropolis Hastings, MC = Markov chain Monte Carlo (MCMC), GS = Gibbs sampling, IS = Importance sampling, S = Sampling, O = Other** (usually special purpose), **++ = Many** methods provided, **?** = Not specified, **N** = None, the program is only designed for structure learning from completely observed data.

NB: Some packages support a form of sampling (e.g., **likelihood weighting, MDMC**), in addition to their exact algorithm; this is indicated by **(+S)**.

Free Is a free version available? **O=Free** (though possibly only for academic use), **$** = Commercial (although most have free versions which are restricted in various ways, e.g., the model size is limited, or models cannot be saved, or there is no API.)

FIGURE B.1: Description of features used in our BN software survey (based on Murphy's), in Tables B.1 and B.2.

B.2 History

The development of the first BN software, beyond algorithm implementation, occurred concurrently with the surge of BN research in the late 1980s.

The original **HUGIN** shell Andersen et al. (1989) was initially developed by a group at the Aalborg University, as part of an ESPRIT project which also produced MUNIN system Andreassen et al. (1989). HUGIN's development continued through another Lauritzen-Jensen project called ODIN. Hugin Expert was established to start commercializing the Hugin Tool in 1989. The close connection between Hugin Expert and the Aalborg research group has continued, including co-location and personnel moving between the two.

The Lisp-based **IDEAL** (Influence Diagram Evaluation and Analysis in Lisp) test-bed environment was developed at Rockwell (Srinivas and Breese, 1989).[2]

Another early BN inference engine was **CABeN** (a Collection of Algorithms for Belief Networks; Cousins et al., 1991b), which contains a library of routines for different stochastic simulation inference algorithms. Lumina Decision Systems, Inc., was founded in 1991 by Max Henrion and Brian Arnold, which produces **Analytica**.

The development of **Netica**, now produced by Norsys Ltd, was started in 1992, by Brent Boerlage who had just finished a Masters degree at the University of British Columbia, where his thesis looked at quantifying and displaying "link strengths" in Bayesian networks (Boerlage, 1995). Netica became commercially available in 1995 and is now widely used.

Developed by Druzdzel's decision systems group, **GeNIe**'s GUI including DNs first appeared in 1998 and is the oldest free BN software tool produced by a research group that is still being actively supported and extended.

The Matlab **Bayes Net Toolbox (BNT)** was developed during Kevin Murphy's time at U.C. Berkeley as a Ph.D. student, where his thesis (Murphy, 2002) addressed DBN representation, inference and learning. He also worked on BNT while an intern at Intel, which lead to a C++ version, **PNL**. Now on the faculty at the University of British Columbia, Murphy is developing a new toolkit, **PMTk**, which will be available in late 2010.

JavaBayes was the first BN software produced in Java (1998) and is distributed under the GNU License. **MSBNx** was produced by the very strong Bayesian network research group at Microsoft, appearing in 2001. There was some attempt to have its XML-based BN format become a standard, but that hasn't eventuated. **AgenaRisk** is a commercial package launched in 2003 by Agena, a company founded by BN researchers Norman Fenton and Martin Neil. The **Samiam** software was developed by Adnan Darwiche to support his academic research and teaching.

There have been a number of Open source initiatives, most of which have had limited active lifespans. A promising exception appears to be **UnBBayes**, an initiative from the University of Brasilia started in 2001, that now has wide functionality and is still being supported and extended.

---

[2]One of the authors used it for her Ph.D. research.

Causal discovery software was first widely available via TETRAD II, which came with the book *TETRAD II: Tools for Discovery* (Scheines et al., 1994). This has since evolved into TETRAD IV. TETRAD contains a number of distinct programs, but the simplest, PC (named after Peter and Clark?), has been widely reimplemented, e.g., in Weka. Numerous other causal discovery programs were developed since the early 1990s, but are not readily available as stand-alone programs with easily understood interface. That description applies also to our CaMML, but we plan to change that by generating an open-source version and making CaMML available through Weka. Weka, by the way, is a program from Waikato University (Waikato Environment for Knowledge Analysis) and is the most widely used freeware environment for using and testing machine learning algorithms (Witten and Frank, 2005).

B.3   BN Software Package Survey

This survey, shown in Tables B.1 and B.2, is based on the one maintained by Kevin Murphy for many years, now hosted at `http://www.cs.ubc.ca/ murphyk/ Bayes/bnsoft.html`. We note the authors, and if the software is commercial but the company has links with particular institutions or BN researchers that we are aware of, those are also noted.

The survey covers basic feature information such as technical information about source availability, platforms, GUI and API, very high level functionality such as types of nodes supported (i.e., discrete and/or continuous), whether decision networks, dynamic Bayesian networks and OOBNs are supported, whether undirected graphs are supported. It includes whether the software allows learning (parameters and/or structure), describes the main inference algorithms (if this information is available) and indicates whether the software is free or commercial. The details of the meaning of each column are given in Figure B.1.

While we have attempted to validate the survey, by contacting developers, by looking at the documentation and in some cases by running the software, it is inevitable there will be errors. Thus we encourage developers or users to send corrections or updates to be incorporated in the an online version of this survey we are maintaining at

   `http://www.csse.monash.edu.au/bai/book/bnsoft.html,`

which gives links for each package. There we also describe some of the major software packages — those with the most functionality, or with a particular feature of interest — in more detail.[3] Most of these packages have a long list of features that we cannot even list here, so our survey is by no means exhaustive. We endeavor to point out particular features that relate to issues we have raised earlier in this text. Google's list of tools is available at:

---

[3]This description of features was included in this Appendix in B.4 – B.6 in the first edition of this text.

```
http://directory.google.com/Top/Computers/Artificial_
Intelligence/Belief_Networks/Software/
```

Wikipedia's has a listing of BN software at:

```
http://en.wikipedia.org/wiki/Bayesian_network#Software_
resources.
```

The Bayesian Network Repository contains examples of BNs, plus datasets for learning them:

```
http://compbio.cs.huji.ac.il/ğalel/Repository
```

TABLE B.1: Feature comparison of software packages (Part I)

Name	Authors	Src	API	Exec	GUI	D/C	DN	DBN	OOBN	Params	Struct	D/U	Infer	Free
Ace	Darwiche (UCLA)	N	Y	WUM	N	D	N	N	N	N	N	All?	AC	$
AgenaRisk	Agena	N	Y	WU	Y	Cd	N	?	Y	Y	Y	D	JT	0
AISpace (CISpace)	Poole et al. (UBC)	J	N	WU	Y	D	Y			N	N	D	VE	$
Analytica	Lumina	N	Y	W	Y	G, Cs, Cd	Y	Y	N	N	N	D	IS,G	0
B-course	U. Helsinki	N	N	WUM	Y	Cd	N	N	N	Y	Y	D	?	$
Banjo	Hartemink	J	Y	WUM	N	Cd	N	Y	N	N	Y	D	N	0
Bassist	U. Helsinki	C++	Y	U	N	G	N	?	?	Y	N	D	MH	0
BayesBuilder	Nijman (U. Nijmegen)	N	J	WUM	Y	D	N	N	N	N	N	D	JT	0
BayesiaLab	Bayesia Ltd	N	Y	WUM	Y	Cd	N	Y	?	Y	Y	CG	JT,G	$
Bayesware Discoverer	Bayesware	N	N	WUM	Y	Cd	N			Y	Y	D	?	$
bn4r	Bel, Dahl	Ru	Y	WUM	N	D	N	N	N	N	N	D	?	0
BNT	Murphy (U.C.Berkeley)	M/C	Y	WUM	N	G	Y	Y	N	Y	Y	UD	++	0
BucketElim	Rish (U.C.Irvine)	C++	Y	WU	N	D	N	N	N	Y	N	D	VE	0
BUGS	MRC/Imperial College	N	N	WU	Y	Cs	N	N	N	Y	N	D	GS	0
CaMML	Wallace,Korb (Monash)	N	N	WU	Y	D	N	N	N	Y	Y	D	N	0
Causeway/Siam	SAIC	N	N	W	Y	?	Y	?	Y	?	?	D		$
DBL Interacive	Smith (U.Qld)	N	N	I	Y	D	Y	N	N	?	?	D	?	$
DBNbox	Roberts et al	M	-	-	N	Y	N	Y	N	N	N	D	++	0
Dezide Advisor	Dezide	N	Y	W	Y	N	Y	N	S	Y	N	?	PT	
dlib	King, Davis	C++	Y	WUM	Y	D	N	N	N	N	N	D	JT,G	
Elvira	Collaboration	J	Y	WUM	Y	Cx	Y	N	N	Y	Y	U	JT,VE,IS	
GDAGsim	Wilkinson (U. Newcastle)	C	Y	WUM	N	G	N			N	N	D	E	0
GeNIe/SMILE	U. Pittsburgh	N	WU	WU	Y	D	Y	Y	S	N	N	D	JT	0

TABLE B.2: Feature comparison of software packages (Part II)

Name	Authors	Src	API	Exec	GUI	D/C	DN	DBN	OOBN	Params	Struct	D/U	Infer	Free
GMRFlib	Rue, H	C/F	Y	WUM	N	G	N	N	N	Y	Y	U	JT	0
GMTk	Bilmes (UW), Zweig (IBM)	N	Y	U	N	D	N			Y	Y	D	JT	0
Grappa	Green (Bristol)	R	-	-	N	D	N			N	N	D	JT	0
Hugin Expert	Hugin	N	Y	WUM	Y	G,Cd	Y	Y	Y	Y	CI	D,CG	JT	$
Hydra	Warnes (U.Wash.)	J	-	-	Y	Cs	N			Y	N	UD	MC	0
Java Bayes	Cozman (CMU)	J	Y	WUM	Y	D	N	N	N	N	N	D	VE	0
LibB	Friedman (Hebrew U)	N	Y	W	N	D	N			Y	Y	D	N	0
MASTINO	Mascherini	R	Y	WUM	N	G	N			Y	Y	D	SL	0
MIM	HyperGraph Software	P	Y	W	Y	G	N	N	N	Y	Y	UD,CG	JT	$
MSBNx	Microsoft	N	Y	W	Y	D	Y			N	N	D	JT	0
Netica	Norsys	N	Y	WUM	Y	Cs,Cd	Y	Y	Y	Y	Y	D	JT,S	$
Optimal Reinsertion	Moore, Wong (CMU)	N	N	WU	N	D	N			Y	Y	D	N	0
PMT	Pavlovic (BU,Rutgers)	M/C	-	-	N	D	N			Y	N	D	O	0
PNL	Eruhimov (Intel)	C++	-	-	N	D	N			Y	Y	UD	JT	0
ProBT	Probayes	N	Y	WUM	N	Cs,Cd	Y		D	Y	Y	D	++	
Pulcinella	IRIDIA	L	Y	WUM	Y	D	N			N	N	D	?	0
Quiddity	IET	N	Y	Y	T	?	Y	?	?	?	?	D	?	$
RISO	Dodier (U.Colorado)	J	Y	WUM	Y	G	N	N	N	N	N	D	PT	0
Sam Iam	Darwiche (UCLA)	N	N?	WU	Y	G?	Y	N	N	Y?	N	D	CO	0
Tetrad IV	CMU	N	N	WU	Y	G,Cx	N	N	N	Y	CI	UD	N	0
UnBBayes	Ladeira, Carvalho	J	Y	Y	Y	D	Y	Y	Y	Y	Y	D	++	0
Vibes /Infer.NET	Winn & Bishop (Cambridge)	C#	Y	WU	Y	Cx	N	N	N	Y	N	FG	++	0
WinMine	Microsoft	N	N	W	Y	Cx	N			Y	Y	UD	N	0

Bibliography

Abdelbar, A. M. and S. M. Hedetniemi (1998). The complexity of approximating MAP explanation. *Artificial Intelligence 102*, 21–38.

Abramson, B., J. Brown, W. Edwards, A. Murphy, and R.B. Winkler (1996). Hailfinder: A Bayesian system for forecasting severe weather. *International Journal of Forecasting 20*(1), 57–71. http://www.lis.pitt.edu/˜dsl/hailfinder/.

Abramson, B. and A. Finizza (1991, November). Using belief networks to forecast oil prices. *International Journal of Forecasting 7*(3), 299–315.

Acid, S., L. de Campos, J. Fernández-Luna, S. Rodríguez, J. María Rodríguez, and J. Luis Salcedo (2004). A comparison of learning algorithms for Bayesian networks: a case study based on data from an emergency medical service. *Artificial Intelligence in Medicine 30*(3), 215–232.

Adusei-Poku, K. (2005). *Operational risk management: Implementing a Bayesian network for foreign exchange and money market settlement.* Ph. D. thesis, University of Gottingen, German.

Aitkenhead, M. and I. Aalders (2009). Predicting land cover using GIS, Bayesian and evolutionary algorithm methods. *Journal of Environmental Management 90*(1), 236–250.

Albrecht, D. W., I. Zukerman, and A. E. Nicholson (1998). Bayesian models for keyhole plan recognition in an adventure game. *User Modeling and User-Adapted Interaction 8*(1-2), 5–47.

Aliferis, C., A. Statnikov, I. Tsamardinos, S. Mani, and X. Koutsoukos (2010a). Local causal and Markov blanket induction for causal discovery and feature selection for classification Part I: Algorithms and empirical evaluation. *Journal of Machine Learning Research 11*, 171–234.

Aliferis, C., A. Statnikov, I. Tsamardinos, S. Mani, and X. Koutsoukos (2010b). Local causal and Markov blanket induction for causal discovery and feature selection for classification Part II: Analysis and extensions. *Journal of Machine Learning Research 11*, 235–284.

Allen, D. and A. Darwiche (2008). Rc_link: Genetic linkage analysis using Bayesian networks. *International Journal of Approximate Reasoning 48*(2), 499–525.

Andersen, S., K. Olesen, F. Jensen, and F. Jensen (1989). HUGIN — a shell for building Bayesian belief universes for expert systems. In *Proceedings of the Eleventh International Joint Conference on Artificial Intelligence (IJCAI)*, Detroit, MI.

Anderson, J. (1983). *The Architecture of Cognition*. Cambridge, MA: Harvard University Press.

Anderson, K. M., P. M. Odell, P. W. Wilson, and W. B. Kannel (1991). Cardiovascular disease risk profiles. *American Heart Journal 121*, 293–298.

Anderson, S., D. Madigan, and M. Perlman (1997). A characterization of Markov equivalence classes for acyclic digraphs. *Annals of Statistics 25*, 505–541.

Andreassen, S., J. Benn, R. Hovorks, K. Olesen, and R. Carson (1994). A probabilistic approach to glucose prediction and insulin dose adjustment. *Computer Methods and Programs in Biomedicine 41*, 153–165.

Andreassen, S., F. V. Jensen, S. K. Andersen, B. Falck, U. Kjærulff, M. Woldbye, A. R. Sørensen, A. Rosenfalck, and F. Jensen (1989). MUNIN — an expert EMG assistant. In J. Desmedt (Ed.), *Computer-Aided Electromyography and Expert Systems*. Elsevier, Amsterdam.

Aristotle. *The Nicomachean Ethics*.

Asher, H. B. (1983). *Causal Modeling* (Second ed.). Beverly Hills, CA: Sage.

Assmann, G., P. Cullen, and H. Schulte (2002). Simple scoring scheme for calculating the risk of acute coronary events based on the 10-year follow-up of the Prospective Cardiovascular Münster (PROCAM) Study. *Circulation 105*(3), 310–315. http://circ.ahajournals.org/cgi/content/abstract/105/3/310.

Asuncion, A. and D. Newman (2007). UCI machine learning repository. http://www.ics.uci.edu/~mlearn/MLRepository.html.

Baars, B. (1988). *A Cognitive Theory of Consciousness*. Cambridge University Press, London.

Bacon, P., J. Cain, and D. Howard (2002). Belief network models of land manager decisions and land use change. *Journal of Environmental Management 65*(1), 1–23.

Baesens, B., T. Van Gestel, S. Viaene, M. Stepanova, J. Suykens, and J. Vanthienen (2003). Benchmarking state-of-the-art classification algorithms for credit scoring. *Journal of the Operational Research Society 54*, 627–635.

Balakrishnan, N. and V. Nevzorov (2003). *A Primer on Statistical Distributions*. New York: Wiley.

Banerjee, J., H.-T. Chou, J. Garza, W. Kim, D. woelk, N. Ballou, and H.-J. Kim (1987). Data model issues for object-oriented applications. *ACM Transactions on Office Informations Systems 5*(1), 3–27.

Bangso, O. and P. Wuillemin (2000). Top-down construction and repetitive structures representation in Bayesian networks. In *Proceedings of the Thirteenth International Florida Artificial Intelligence Research Society Conference*, pp. 282–286.

Barton, D., T. Saloranta, S. Moe, H. Eggestad, and S. Kuikka (2008). Bayesian belief networks as a meta-modeling tool in integrated river basin management–Pros and cons in evaluating nutrient abatement decisions under uncertainty in a Norwegian river basin. *Ecological Economics 66*(1), 91–104.

Bayes, T. (1764/1958). An essay towards solving a problem in the doctrine of chances. *Biometrika 45*, 296–315.

Beinlich, I., H. Suermondt, R. Chavez, and G. Cooper (1992). The ALARM monitoring system. In *Proceedings of the Second European Conference on Artificial Intelligence in Medicine*, pp. 689–693.

Bell, D. (1988). Disappointment in decision making under uncertainty. In H. R. D.E. Bell and A. Tversky (Eds.), *Decision making: Descriptive, Normative, and Prescriptive Interactions*, pp. 358–384. Cambridge Press, London.

Bhattacharyya, A. (1943). On a measure of divergence between two statistical populations defined by their probability distributions. *Bulletin of the Calcutta Mathematics Society 35*, 99–110.

Biedermann, A., F. Taroni, O. Delemont, C. Semadeni, and A. Davison (2005a). The evaluation of evidence in the forensic investigation of fire incidents. Part I: an approach using Bayesian networks. *Forensic Science International 147*(1), 49–57.

Biedermann, A., F. Taroni, O. Delemont, C. Semadeni, and A. Davison (2005b). The evaluation of evidence in the forensic investigation of fire incidents. Part II: Practical examples of the use of Bayesian networks. *Forensic Science International 147*(1), 59–69.

Billings, D., A. Davidson, J. Schaeffer, and D. Szafron (2002). The challenge of poker. *Artificial Intelligence 134*, 201–240.

Bishop, C. (2006). *Pattern Recognition and Machine Learning*. Springer Verlag.

Blalock, H. (1964). *Causal Inferences in Nonexperimental Research*. Chapel Hill: University of North Carolina Press.

Boehm, B. W. (1988). A spiral model of software development and enhancement. *IEEE Computer*, 61–72.

Boerlage, B. (1995). Link strengths in Bayesian Networks. Master's thesis, Department of Computer Science, University of British Columbia.

Bonafede, C. and P. Giudici (2007). Bayesian networks for enterprise risk assessment. *Physica A: Statistical Mechanics and its Applications 382*(1), 22–28.

Boneh, T. (2010). *Ontology and Bayesian decision networks for supporting the meteorological forecasting process*. Ph. D. thesis, Clayton School of Information Technology, Monash University.

Boneh, T., A. Nicholson, and L. Sonenberg (2006). Matilda: A visual tool for modeling with Bayesian networks. *International Journal of Intelligent Systems 21*(11), 1127–1150.

Borgelt, C. and R. Kruse (2002). *Graphical models: Methods for Data Analysis and Mining*. New York: Wiley.

Borsuk, M. (2008). Bayesian networks. In S. Jorgensen and B. Fath (Eds.), *Encyclopedia of Ecology*, pp. 307–317. Oxford, UK: Elsevier.

Borsuk, M., P. Reichert, and P. Burkhardt-Holm (2002). A Bayesian network for investigating the decline in fish catch in switzerland. In *Integrated Assessment and Decision Support. Proceedings of the 1st biennial meeting of the International Environmental Modeling and Software Society, Lugano, Switzerland*, pp. 108–113.

Borsuk, M., C. Stow, and K. Reckhow (2004). A Bayesian network of eutrophication models for synthesis, prediction, and uncertainty analysis. *Ecological Modeling 173*(2-3), 219–239.

Bouckaert, R. (1994). Probabilistic network construction using the minimum description length principle. Technical Report RUU-CS-94-27, Dept of Computer Science, Utrecht University.

Boudon, R. (1968). A new look at correlation analysis. In H. Blalock and A. Blalock (Eds.), *Methodology in Social Research*, pp. 199–235. McGraw-Hill, New York.

Boulton, D. (2003). Bayesian opponent modeling in poker. Honours thesis, School of Computer Science and Software Engineering, Monash University.

Boulton, D. and C. Wallace (1969). The information content of a multistate distribution. *Journal of Theoretical Biology 23*, 269–278.

Boutilier, C., N. Friedman, M. Goldszmidt, and D. Koller (1996). Context-specific independence in Bayesian networks. In *UAI96 – Proceedings of the Twelfth Conference on Uncertainty in Artificial Intelligence*, pp. 115–123.

Boyen, X. and D. Koller (1998). Tractable inference for complex stochastic processes. In *UAI98 – Proceedings of the Fourteenth Conference on Uncertainty in Uncertainty in Artificial Intelligence*, pp. 33–42.

BPP (2006-2010). Online version. http://www.csse.monash.edu.au/bai/poker/.

Bradford, J., C. Needham, A. Bulpitt, and D. Westhead (2006). Insights into protein-protein interfaces using a Bayesian network prediction method. *Journal of Molecular Biology 362*(2), 365–386.

Brightwell, G. and P. Winkler (1990). Counting linear extensions is #p-complete. Technical Report DIMACS 90-49, Dept of Computer Science, Rutgers.

Bromley, J., N. Jackson, O. Clymer, A. Giacomello, and F. Jensen (2005). The use of Hugin® to develop Bayesian networks as an aid to integrated water resource planning. *Environmental Modeling & Software 20*(2), 231–242.

Brooks, F. (1995). *The Mythical Man-Month: Essays on Software Engineering* (Second ed.). Reading, MA: Addison-Wesley.

Buchanan, B. and E. Shortliffe (Eds.) (1984). *Rule-Based Expert Systems: The MYCIN Experiments of the Stanford Heuristic Programming Project.* Reading, MA: Addison-Wesley.

Buntine, W. (1991). Theory refinement on Bayesian networks. In *UAI91 – Proceedings of the Seventh Conference on Uncertainty in Artificial Intelligence*, pp. 52–69.

Buntine, W. (1996). A guide to the literature on learning probabilistic networks from data. *IEEE Transactions on Knowledge and Data Engineering 8*, 195–210.

Burnside, E., D. Rubin, and R. Shachter (2004). Improving a Bayesian network's ability to predict the probability of malignancy of microcalcifications on mammography. In *International Congress Series*, Volume 1268, pp. 1021–026. Elsevier.

Cadwallader, P. L. (1978). Some causes of the decline in range and abundance of native fish in the Murray-Darling River system. In *The Murray-Darling River System Symposium, Proceedings of the Royal Society of Victoria*.

Cain, J. (2001). Planning improvements in natural resources management: Guidelines for using Bayesian networks to support the planning and management of development programmes in the water sector and beyond. Technical report, Centre for Ecology and Hydrology, Crowmarsh Gifford, Wallingford, Oxon, UK.

Carlton, J. (2000). Bayesian poker. Honours thesis, School of Computer Science and Software Engineering, Monash University. www.csse.monash.edu.au/ hons/projects/2000/Jason.Carlton/.

Carnap, R. (1962). *The Logical Foundations of Probability* (Second ed.). University of Chicago.

Cartwright, N. (1989). *Nature's Capacities and Their Measurement*. Oxford: Clarendon Press.

Cartwright, N. (2001). What is wrong with Bayes nets? *The Monist 84*, 242–64.

Caruana, R. and A. Niculescu-Mizil (2006). An empirical comparison of supervised learning algorithms. In *Proceedings of the 23rd International Conference on Machine Learning*, pp. 161–168.

Chaitin, G. (1966). On the length of programs for computing finite binary sequences. *Journal of the ACM 13*, 547–569.

Chan, H. and A. Darwiche (2003). When do numbers really matter? *Journal of Artificial Intelligence Research 17*, 265–287.

Charitos, T., L. van der Gaag, S. Visscher, K. Schurink, and P. Lucas (2009). A dynamic Bayesian network for diagnosing ventilator-associated pneumonia in ICU patients. *Expert Systems with Applications 36*(2), 1249–258.

Chee, Y., M. Burgman, and J. Carey (2005). Use of a Bayesian network decision tool to manage environmental flows in the Wimmera river, Victoria. Report No. 4, LWA/MDBC Project UMO43: Delivering Sustainability Through Risk Management, University of Melbourne, Australia.

Cheeseman, P. (1988). An inquiry into computer understanding. *Computational Intelligence 4*, 58–66.

Cheng, H. and G. Hadjisophocleous (2009). The modeling of fire spread in buildings by Bayesian network. *Fire Safety Journal 44*(6), 901–908.

Cheng, J. and M. Druzdzel (2000). AIS-BN: An adaptive importance sampling algorithm for evidential reasoning in large Bayesian networks. *Journal of Artificial Intelligence Research 13*, 155–188.

Cherniak, C. (1986). *Minimal Rationality*. Cambridge, MA: MIT Press.

Chickering, D. M. (1995). A tranformational characterization of equivalent Bayesian network structures. In *UAI95 – Proceedings of the 11th Conference on Uncertainty in Artificial Intelligence*, San Francisco, CA, pp. 87–98.

Chickering, D. M. (1996). Learning equivalence classes of Bayesian network structures. In *UAI96 – Proceedings of the Twelfth Conference on Uncertainty in Artificial Intelligence*, pp. 150–157.

Chihara, C. and R. Kennedy (1979). The Dutch book argument: Its logical flaws, its subjective sources. *Philosophical Studies 36*, 19–33.

Chow, C. and C. Liu (1968). Approximating discrete probability distributions with dependence trees. *IEEE Transactions on Information Theory 14*, 462–467.

Conati, C., A. Gertner, and K. VanLehn (2002). Using Bayesian Networks to manage uncertainty in student modeling. *User Modeling and User-Adapted Interaction 12*(4), 371–417.

Conati, C., A. Gertner, K. VanLehn, and M. Druzdzel (1997). On-line student modeling for coached problem solving using Bayesian Networks. In *UM97 – Proceedings of the Sixth International Conference on User Modeling*, pp. 231–242.

Cooper, G. (1984). *NESTOR: A Computer Based Medical Diagnostic Aid that Integrates Causal and Probabilistic Knowledge*. Ph. D. thesis, Medical Computer Science Group, Stanford University, Stanford, CA.

Cooper, G. (1997). A simple constraint-based algorithm for efficiently mining observational databases for causal relationships. *Data Mining and Knowledge Discovery 1*, 203–224.

Cooper, G., D. Dash, J. Levander, W. Wong, W. Hogan, and M. Wagner (2006). Bayesian methods for diagnosing outbreaks. In M. Wagner, A. Moore, and R. Aryel (Eds.), *Handbook of Biosurveillance*, pp. 273–288. Academic Press.

Cooper, G. and C. Yoo (1999). Causal discovery from a mixture of experimental and observational data. In *UAI99 – Proceedings of the Fifteenth Conference on Uncertainty in Artificial Intelligence*, pp. 116–125.

Cooper, G. F. (1990). The computational complexity of probabilistic inference using Bayesian belief networks. *Artificial Intelligence 42*, 393–405.

Cooper, G. F. and E. Herskovits (1991). A Bayesian method for constructing Bayesian belief networks from databases. In *UAI91 – Proceedings of the Seventh Conference on Uncertainty in Artificial Intelligence*, pp. 86–94.

Cooper, G. F. and E. Herskovits (1992). A Bayesian method for the induction of probabilistic networks from data. *Machine Learning 9*, 309–347.

Coupe, V., N. Peek, J. Ottenkamp, and J. Habbema (1999). Using sensitivity analysis for efficient quantification of a belief network. *Artificial Intelligence in Medicine 17*, 223–247.

Coupé, V. and L. van der Gaag (2002). Properties of sensitivity analysis of Bayesian belief networks. *Annals of Mathematics and Artificial Intelligence 36*, 323–356.

Coupé, V., L. van der Gaag, and J. Habbema (2000). Sensitivity analysis: an aid for belief-network quantification. *Knowledge Engineering Review 15*, 1–18.

Cousins, S., W. Chen, and M. Frisse (1991a). CABeN: A collection of algorithms for belief networks. Technical Report WUCS-91-25, Department of Computer Science, Washington University.

Cousins, S., W. Chen, and M. Frisse (1991b). CABeN: A collection of algorithms for belief networks. Technical Report WUCS-91-25, Department of Computer Science, Washington University.

Covelo, V., von D. Winterfeldt, and P. Slovic (1986). Risk communication: A review of the literature. *Risk Abstracts 3*(4), 171–182.

Cover, T. and J. Thomas (2006). *Elements of information theory* (2nd ed.). New York: Wiley.

Cowell, R., A. Dawid, T. Hutchinson, and D. Spiegelhalter (1991). A Bayesian expert system for the analysis of an adverse drug reaction. *Artificial Intelligence in Medicine 3*(5), 257–270.

Cowell, R. G., A. P. Dawid, S. L. Lauritzen, and D. Spiegelhalter (1999). *Probabilistic Networks and Expert Systems*. Statistics for Engineering and Information Science. New York: Springer Verlag.

D'Agostino, R., M. Russell, and D. Huse (2000). Primary and subsequent coronary risk appraisal: new results from the Framingham study. *American Heart Journal 139*, 272–81.

Dagum, P. and A. Galper (1993). Forecasting sleep apnea with dynamic network models. In *UAI93 – Proc of the Ninth Conference on Uncertainty in Artificial Intelligence*, pp. 65–71.

Dagum, P., A. Galper, and E. Horvitz (1992). Dynamic network models for forecasting. In D. Dubois, M. Wellman, B. D'Ambrosio, and P. Smets (Eds.), *UAI92 – Proceedings of the Eighth Conference on Uncertainty in Artificial Intelligence*, pp. 41–48.

Dagum, P. and M. Luby (1993). Approximating probabilistic inference in belief networks is NP-hard. *Artificial Intelligence 60*, 141–153.

Dai, H., K. B. Korb, C. S. Wallace, and X. Wu (1997). A study of casual discovery with weak links and small samples. In *IJCAI97 – Proceedings of the Fifteenth International Joint Conference on Artificial Intelligence*, pp. 1304–1309.

Daly, R., Q. Shen, and S. Aitken (2010). Learning Bayesian networks: Approaches and issues. *The Knowledge Engineering Review*. Forthcoming.

Davies, J. (1995). Mobile monitoring of lower-limb amputees and fall diagnosis in elderly patients. Thesis report. Department of Electrical and Computer Systems Engineering, Monash University.

Dawid, A., J. Mortera, and P. Vicard (2006). Representing and solving complex DNA identification cases using Bayesian networks. In *International Congress Series*, Volume 1288, pp. 484–491. Elsevier.

de Campos, L., J. Fernández-Luna, J. Gámez, and J. Puerta (2000). Ant colony optimization for learning Bayesian networks. *International Journal of Approximate Reasoning 31*, 291–311.

de Finetti, B. (1964). Foresight: Its logical laws, its subjective sources. In Kyburg and Smokler (Eds.), *Studies in Subjective Probability*. New York: Wiley. (Translation of "La prévision" Annales de l'Institut Henri Poincaré, 1937.)

de Laplace, P. S. (1820/1951). *A Philosophical Essay on Probabilities*. Dover.

de Melo, A. and A. Sanchez (2008). Software maintenance project delays prediction using Bayesian networks. *Expert Systems with Applications 34*(2), 908–919.

de Santa Olalla, F. M., A. Dominguez, F. Ortega, A. Artigao, and C. Fabeiro (2007). Bayesian networks in planning a large aquifer in Eastern Mancha, Spain. *Environmental Modeling & Software 22*(8), 1089–1100.

Dean, T. and K. Kanazawa (1989). A model for reasoning about persistence and causation. *Computational Intelligence 5*, 142–150.

Dean, T. and M. P. Wellman (1991). *Planning and Control*. San Mateo, CA.: Morgan Kaufman Publishers.

DeGroot, M. (1970). *Optimal Statistical Decisions*. New York: McGraw-Hill.

DeGroot, M. and M. Schervish (2002). *Probability and Statistics* (Third ed.). Reading, MA: Addison-Wesley.

Demographia. Demographic statistics. http://www.demographia.com.

Dempster, A., N. Laird, and D. Rubin (1977). Maximum likelihood from incomplete data via the EM algorithm. *Journal of the Royals Statistical Society, B 39*, 1–38.

Dietterich, T. (1998). Approximate statistical tests for comparing supervised classification learning algorithms. *Neural computation 10*(7), 1895–1923.

Díez, F., J. Mira, E. Iturralde, and S. Zubillaga (1997). Diaval, a Bayesian expert system for echocardiography. *Artificial Intelligence in Medicine 10*(1), 59–73.

Domingos, P. and M. Pazzani (1997). On the optimality of the simple Bayesian classifier under zero-one loss. *Machine learning 29*(2), 103–130.

Draper, D. (1995). *Localized partial evaluation of belief networks*. Ph. D. thesis, Department of Computer Science, University of Washington.

Druzdzel, M. (1996). Qualitative verbal explanations in Bayesian belief networks. *Artificial Intelligence and Simulation of Behaviour Quarterly 94*, 43–54.

Druzdzel, M. J. (1999, July 18-22). SMILE: Structural modeling, inference, and learning engine and GeNIe: A development environment for graphical decision-theoretic models. In *Proceedings of the Sixteenth National Conference on Artificial Intelligence (AAAI–99)*, Orlando, FL, pp. 902–903.

Duda, R. and P. Hart (1973). *Pattern classification and scene analysis*. New York: Wiley.

Duda, R. O., P. E. Hart, and N. J. Nilsson (1976). Subjective Bayesian methods for rule-based inference systems. In *AFIPS*, Volume 45, pp. 1075–1082.

Edwards, A. L. (1984). *An Introduction to Linear Regression and Correlation* (Second ed.). New York: W.H. Freeman.

Ellison, A. (1996). An introduction to Bayesian inference for ecological research and environmental decision-making. *Ecological Applications 6*, 1036–1046.

Erskine, W. D. (1996). Downstream hydrogeomorphic impacts of Eildon Reservoir on the mid-Goulburn River, Victoria. *Proceedings of the Royal Society of Victoria. 108*, 1–15.

Ettema, J., S. Østergaard, and A. Kristensen (2009). Estimation of probability for the presence of claw and digital skin diseases by combining cow-and herd-level information using a Bayesian network. *Preventive Veterinary Medicine 92*(1-2), 89–98.

Falzon, L. (2006). Using Bayesian network analysis to support centre of gravity analysis in military planning. *European Journal of Operational Research 170*(2), 629–643.

Feigenbaum, E. (1977). The art of artificial intelligence: Themes and case studies of knowledge engineering. In *Fifth International Conference on Artificial Intelligence – IJCAI-77*, San Mateo, CA, pp. 1014–1029. Morgan Kaufmann.

Fernández, A. and A. Salmerón (2008). BayesChess: A computer chess program based on Bayesian networks. *Pattern Recognition Letters 29*(8), 1154–1159.

Findler, N. (1977). Studies in machine cognition using the game of poker. *Communications of the ACM 20*, 230–245.

Fischoff, B. and M. Bar-Hillel (1984). Focusing techniques: A shortcut to improving probability judgments? *Organizational Behavior and Human Performance 34*, 339–359.

Flores, J., A. Nicholson, A. Brunskill, S. Mascaro, and K. Korb (2010). Incorporating expert knowledge when learning Bayesian network structure: Heart failure as a case study. Technical Report 2010/3. Bayesian Intelligence. http://www.Bayesian-intelligence.com/publications.php.

Flynn, J. (2002). The effectiveness of a computer games program in teaching decimal notation. Honours thesis, University of Melbourne.

Forbes, J., T. Huang, K. Kanazawa, and S. Russell (1995). The BATmobile: Towards a Bayesian automated taxi. In *Proceedings of the Fourteenth International Joint Conference on Artificial Intelligence (IJCAI'95)*, pp. 1878–1885.

Fowler, M. (2003). *UML distilled: A brief guide to the standard object modeling language* (3rd ed.). Addison-Wesley.

Friedman, N., D. Geiger, and M. Goldszmidt (1997). Bayesian network classifiers. *Machine Learning 29*(2), 131–163.

Friedman, N. and M. Goldszmidt (1999). Learning Bayesian networks with local structure. In M. Jordan (Ed.), *Learning in Graphical Models*, pp. 421–459. MIT.

Friedman, N., I. Nachman, and D. Peér (1999). Learning Bayesian network structure from massive datasets: The "sparse candidate" algorithm. In *Proceedings of the Fifteenth Conference on Uncertainty in Artificial Intelligence*, pp. 206–215.

Fung, R. and K.-C. Chang (1989). Weighting and integrating evidence for stochastic simulation in Bayesian networks. In *Proceedings of the Fifth Conference on Uncertainty in Artificial Intelligence (UAI-89)*, pp. 475–482.

Fung, R. and B. del Favero (1994). Backward simulation in Bayesian networks. In *UAI94 – Proceedings of the Tenth Conference on Uncertainty in Artificial Intelligence*, pp. 227–234.

Galán, S., F. Aguado, F. Diez, and J. Mira (2002). NasoNet, modeling the spread of nasopharyngeal cancer with networks of probabilistic events in discrete time. *Artificial Intelligence in Medicine 25*(3), 247–264.

Garey, M. and D. Johnson (1979). *Computers and Intractability*. New York: W.H. Freeman.

Geiger, D. and D. Heckerman (1991). Advances in probabilistic reasoning. In *Proceedings of the Proceedings of the Seventh Conference Annual Conference on Uncertainty in Artificial Intelligence (UAI-91)*, pp. 118–126.

Geman, S., D. Geman, K. Abend, T. Harley, and L. Kanal (1993). Stochastic relaxation, Gibbs distributions and the Bayesian restoration of images. *Journal of Applied Statistics 20*, 25–62.

Getoor, L., J. Rhee, D. Koller, and P. Small (2004). Understanding tuberculosis epidemiology using structured statistical models. *Artificial Intelligence in Medicine 30*(3), 233–56.

Gigerenzer, G. and U. Hoffrage (1995). How to improve Bayesian reasoning without instruction: Frequency formats. *Psychological Review 102*, 684–704.

Ginac, F. (1997). *Customer Oriented Software Quality Assurance*. New York: Prentice Hall.

Gippel, C. J. and B. L. Finlayson (1993). Downstream environmental impacts of regulation of the Goulburn River, Victoria. In *Hydrology and Water Resources Symposium, Newcastle*.

Glymour, C. (1980). *Theory and Evidence*. Princeton, NJ: Princeton University.

Glymour, C. and G. Cooper (Eds.) (1999). *Computation, causation, and discovery*. MIT Press.

Good, I. (1952). Rational decisions. *Jrn. of the Royal Statistical Society, B 14*, 107–114.

Goodman, N. (1973). *Fact, Fiction and Forecast* (Third ed.). Indianapolis: Bobbs-Merrill.

Gowadia, V., C. Farkas, and M. Valtorta (2005). Paid: A probabilistic agent-based intrusion detection system. *Computers & Security 24*(7), 529–545.

Graham, P. (2003). Better Bayesian filtering. http://www.paulgraham.com/better.html.

Group, B. H. S. (2004, December). The Busselton health study. Website. http://bsn.uwa.edu.au/surveys.htm.

Gupta, S. and H. Kim (2008). Linking structural equation modeling to Bayesian networks: Decision support for customer retention in virtual communities. *European Journal of Operational Research 190*(3), 818–833.

Guyon, I., C. Aliferis, G. Cooper, A. Elisseeff, J.-P. Pellet, P. Spirtes, and A. Statnikov (Eds.) (2008). *JMLR workshop and conference proceedings: Causation and prediction challenge (WCCI 2008)*, Volume 3. Journal of Machine Learning Research.

Haddawy, P., J. Jacobson, and C. E. J. Kahn (1997). BANTER: a Bayesian network tutoring shell. *Artificial Intelligence in Medicine 10*, 177–200.

Hagen, R. and E. Sonenberg (1993). Automated classification of student misconceptions in physics. In C. Rowles, H. Liu, and N. Foo (Eds.), *Proceedings of the 1993 Australian Joint Conference on Artificial Intelligence, AI'93*, pp. 153–159. World Scientific.

Hájek, A. (2008). Dutch book arguments. In P. Anand, P. Pattanaik, and C. Puppe (Eds.), *The Oxford Handbook of Rational and Social Choice*. Oxford University Press.

Halpern, J. and J. Pearl (2005a). Causes and explanations: A structural-model approach. Part I: Causes. *The British Journal for the Philosophy of Science 56*, 843–887.

Halpern, J. and J. Pearl (2005b). Causes and explanations: A structural-model approach. part II: explanations. *The British Journal for the Philosophy of Science 56*, 889–911.

Hand, D. and K. Yu (2001). Idiot's Bayes: Not so stupid after all? *International Statistical Review 69*, 385–398.

Handfield, T., C. Twardy, K. Korb, and G. Oppy (2008). The metaphysics of causal models. *Erkenntnis 68*, 149–168.

Harnad, S. (1989). Minds, machines and Searle. *Journal of Theoretical and Experimental Artificial Intelligence 1*, 5–25.

Hart, B. and C. Pollino (2009). Bayesian modeling for risk-based environmental water allocation. Waterline Report Series No 14, National Water Commission, Canberra, Australia. http://www.nwc.gov.au/resources/documents/Waterlines_14_-_Bayesian_COMPLETE.pdf.

Heckerman, D. (1991). *Probabilistic similarity networks*. Cambridge, MA: MIT Press.

Heckerman, D. (1998). A tutorial on learning with Bayesian networks. In M. Jordan (Ed.), *Learning in Graphical Models*, pp. 301–354. MIT.

Heckerman, D. and D. Geiger (1995). Learning Bayesian networks: A unification for discrete and Gaussian domains. In *UAI95 – Proceedings of the 11th Conference on Uncertainty in Artificial Intelligence*, San Francisco, pp. 274–284.

Heckerman, D. E. (1986). Probabilistic interpretations for MYCIN's certainty factors. In *Uncertainty in Artificial Intelligence Vol 1*, pp. 167–96. Amsterdam: Elsevier.

Helm, S. (2002). The use of computer games to improve understanding about decimal notation. Honours thesis, University of Melbourne.

Helsper, E. M. and L. C. van der Gaag (2007). Ontologies for probabilistic networks: a case study in the oesophageal-cancer domain. *The Knowledge Engineering Review 22*(1), 67–86.

Henriksen, H., P. Rasmusssen, G. Brandt, D. von Bulow, and F. Jensen (2007). Bayesian networks as a participatory modeling tool for groundwater protection. In *Topics on System Analysis and Integrated Water Resource Management*, pp. 49–72. Elsevier Science Ltd.

Henrion, M. (1988). Propagating uncertainty in Bayesian networks by logic sampling. In *Uncertainty in Artificial Intelligence Vol 2*, pp. 149–163. Amsterdam: North-Holland.

Henrion, M., M. Pradhan, B. Del Favero, K. Huang, G. Provan, and P. O'Rorke (1996). Why is diagnosis using belief networks insensitive to imprecision in probabilities? In *UAI96 – Proceedings of the Twelfth Conference on Uncertainty in Artificial Intelligence*, pp. 307–314.

Hesslow, G. (1976). Discussion: Two notes on the probabilistic approcah to causality. *Philosophy of Science 43*, 290–292.

Holmes, D. and L. Jain (Eds.) (2008). *Innovations in Bayesian Networks: Theory and Applications*. Springer Verlag.

Holte, R. (1993). Very simple classification rules perform well on most commonly used datasets. *Machine Learning 11*, 63–90.

Hoover, K. (2003). Nonstationary time series, cointegration, and the principle of the common cause. *The British Journal for the Philosophy of Science 54*, 527–551.

Hope, L. (2008). *Information Measures for Causal Explanation and Prediction*. Ph. D. thesis, Monash University.

Hope, L. and K. Korb (2002). Bayesian information reward. In *15th Australian Joint Conference on Artificial Intelligence (AI'02)*, pp. 272–283. Springer Verlag.

Hope, L. and K. Korb (2004). A Bayesian metric for evaluating machine learning algorithms. In *17th Australian Joint Conference on Artificial Intelligence (AI'04)*, pp. 991–997. Springer Verlag.

Hope, L., A. Nicholson, and K. Korb (2002). Knowledge engineering tools for probability elicitation. Technical report 2002/111, School of Computer Science and Software Engineering, Monash University.

Hope, L., A. Nicholson, and K. Korb (2006). TakeheartII: A tool to support clinical cardiovascular risk assessment. Technical report TR 2006/209, Clayton School of Information Technology, Monash University.

Horvitz, E. and M. Barry (1995). Display of information for time-critical decision making. In *UAI95 – Proceedings of the Eleventh Conference on Uncertainty in Artificial Intelligence*, Montreal, Canada, pp. 296–305.

Horvitz, E., J. Breese, D. Heckerman, D. Hovel, and K. Rommelse (1998). The Lumiere project: Bayesian user modeling for inferring the goals and needs of software users. In *Proceedings of the Fourteenth Conference on Uncertainty in Artificial Intelligence*, pp. 256–265.

Horvitz, E., H. Suermondt, and G. Cooper (1989). Bounded conditioning: Flexible inference for decisions under scarce resources. In *Proceedings of the Fifth Workshop on Uncertainty in Artificial Intelligence*, pp. 182–193.

Howard, C. and M. Stumptner (2005). Situation assessment with Object Oriented Probabilistic Relational Models. In *Proceedings of the Seventh International Conference on Enterprise Information Systems (ICEIS 2005)*, pp. 412–418.

Howard, R. and J. Matheson (1981). Influence diagrams. In R. Howard and J. Matheson (Eds.), *Readings in Decision Analysis*, pp. 763–771. Strategic Decisions Group, Menlo Park, CA.

Howson, C. (2001). The logical basis of uncertainty. In D. Corfield and J. Williamson (Eds.), *Foundations of Bayianism*. Dordrecht: Kluwer Academic.

Howson, C. and P. Urbach (2007). *Scientific Reasoning: The Bayesian Approach* (3rd ed.). La Salle, IL: Open Court.

Hume, D. (1739/1962). *A Treatise of Human Nature*. Cleveland: World Publishing.

Husmeier, D., S. Roberts, and R. Dybowski (2003). *Applications of Probabilistic Modeling in Medical Informatics and Bioinformatics*. London: Springer Verlag.

Innocent (2002). Innocent: Fighting miscarriages of justice. http://www.innocent.org.uk.

Jaakkola, T. S. and M. I. Jordan (1999). Variational probabilistic inference and the QMR-DT network. *Journal of Artificial Intelligence Research 10*, 291–322.

Jaynes, E. (1968). Prior probabilities. *IEEE Transactions on Systems Science and Cybernetics SSC-4*, 227–241.

Jeffrey, R. (1983). *The Logic of Decision* (2nd ed.). New York: McGraw Hill.

Jensen, A. (1995). *A probabilistic model based decision support system for mildew management in winter wheat*. Ph. D. thesis, Aalborg University.

Jensen, F. (1988). Junction trees and decomposable hypergraphs. Research report, Judex Datasystemer A/S, Aalborg, Denmark.

Jensen, F., S. Lauritzen, and K. Olesen (1990). Bayesian updating in causal probabilistic networks by local computations. *Computational Statistics Quarterly 4*, 269–282.

Jensen, F. V. (1996). *An Introduction to Bayesian Networks*. New York: Springer Verlag.

Jensen, F. V. (2001). *Bayesian Networks and Decision Graphs*. New York: Springer Verlag.

Jensen, F. V., B. Chamberlain, T. Nordahl, and F. Jensen (1991). Analysis in HUGIN of data conflict. In P. P. Bonissone, M. Henrion, L. M. Kanal, and J. F. Lemmer (Eds.), *Uncertainty in Artificial Intelligence 6*, pp. 519–528. Elsevier.

Jensen, F. V., U. Kjærulff, K. G. Olesen, and J. Pedersen (1989). Et forprojekt til et ekspertsystem for drift af spildevandsrensning (a prototype expert system for control of waste water treatment). Technical report, Judex Datasystemer A/S, Aalborg, Denmark. In Danish.

Jensen, F. V. and T. D. Nielsen (2007). *Bayesian Networks and Decision Graphs* (2nd ed.). New York: Springer Verlag.

Jensen, K., J. Toftum, and P. Friis-Hansen (2009). A Bayesian network approach to the evaluation of building design and its consequences for employee performance and operational costs. *Building and Environment 44*(3), 456–462.

Jensen, T., A. Kristensen, N. Toft, N. Baadsgaard, S. Østergaard, and H. Houe (2009). An object-oriented Bayesian network modeling the causes of leg disorders in finisher herds. *Preventive Veterinary Medicine 89*(3-4), 237–248.

Jia, H., D. Liu, J. Chen, and X. Liu (2007). A hybrid approach for learning Markov equivalence classes of Bayesian network. *Knowledge Science, Engineering and Management*, 611–616.

Jiang, X., D. Neill, and G. Cooper (2010). A Bayesian network model for spatial event surveillance. *International Journal of Approximate Reasoning 51*(2), 224–239.

Jitnah, N. (1993). Bayesian poker. Honours thesis, Dept. of Computer Science, Monash University.

Jitnah, N. (2000). *Using mutual information for approximate evaluation of Bayesian networks*. Ph. D. thesis, Monash University, School of Computer Science and Software Engineering.

Jordan, M. (1986). An introduction to linear algebra in parallel distributed processing. In J. M. D.E. Rumelhard and the PDP research group (Eds.), *Parallel Distributed Processing, Vol. 1.*, pp. 365–422. Cambridge, MA: MIT Press.

Jordan, M. (Ed.) (1999). *Learning in Graphical Models*. MIT.

K. B. Korb, R. M. I. Z. (1997). A cognitive model of argumentation. In *Proceedings of the Cognitive Science Society Meeting*, pp. 400–405.

Kahn, C. et al. (1997). Construction of a Bayesian network for mammographic diagnosis of breast cancer. *Computers in biology and medicine 27*(1), 19–29.

Kahneman, D. and A. Tversky (1973). On the psychology of prediction. *Psychological Review 80*, 430–454.

Kalman, R. (1960, March). A new approach to linear filtering and prediction problems. *Trans. ASME, J. Basic Engineering 82*, 34–45.

Kanazawa, K. (1992). *Probability, Time, and Action*. Ph. D. thesis, Brown University, Providence, RI.

Kanazawa, K., D. Koller, and S. Russell (1995). Stochastic simulation algorithms for dynamic probabilistic networks. In *UAI95 – Proceedings of the 11th Conference on Uncertainty in Artificial Intelligence*, pp. 346–351.

Kannel, W. B., J. D. Neaton, D. Wentworth, H. E. Thomas, J. Stamler, and S. B. Hulley (1986). Overall and coronary heart disease mortality rates in relation to major risk factors in 325,348 men screend for the MRFIT. *American Heart Journal 112*, 825–836.

Karzanov, A. and L. Khachiyan (1990). On the conductance of order Markov chains. Technical Report DIMACS 90-60, Dept of Computer Science, Rutgers.

Katz, E., V. Heleg-Shabtai, A. Bardea, I. Willner, H. Rau, W. Haehnel, G. Arroyo-Figueroa, L. Sucar, and A. Villavicencio (1998). Probabilistic temporal reasoning and its application to fossil power plant operation. *Expert Systems with Applications 15*(3), 317–324.

Kim, J. and J. Pearl (1983). A computational model for causal and diagnostic reasoning in inference systems. In *Proceedings of the Eighth International Joint Conference on Artificial Intelligence (IJCAI)*, pp. 190–193.

Kim, Y. and M. Valtorta (1995). On the detection of conflicts in diagnostic Bayesian networks. In *UAI95 – Proceedings of the 11th Conference on Uncertainty in Artificial Intelligence*, pp. 362–367.

Kipersztok, O. and H. Wang (2001). Another look at sensitivity of Bayesian networks to imprecise probabilities. In *Proceedings of the Eighth Workshop on Artificial Intelligence and Statistics (AISTAT-2001)*, Florida.

Kjærulff, U. (1992). A computational scheme for reasoning in dynamic probabilistic networks. In *UAI92 – Proceedings of the Eighth Conference on Uncertainty in Artificial Intelligence*, pp. 121–129.

Kjærulff, U. (1994). Reduction of computation complexity in Bayesian networks through removal of weak dependencies. In *UAI94 – Proceedings of the Tenth Conference on Uncertainty in Artificial Intelligence*, pp. 374–382.

Kjærulff, U. (1995). dHugin: A computational system for dynamic time-sliced Bayesian networks. *International Journal of Forecasting, Special Issue on Probability Forecasting 11*, 89–111.

Kjærulff, U. and L. C. van der Gaag (2000). Making sensitivity analysis computationally efficient. In *Proceedings of the Sixteenth Conference on Uncertainty in Artificial Intelligence*, pp. 317–325.

Kjærulff, U. B. and A. L. Madsen (2008). *Bayesian Networks and Influence Diagrams: A Guide to Construction and Analysis*. Springer Verlag.

Kline, J., A. Novobilski, C. Kabrhel, P. Richman, and D. Courtney (2005). Derivation and validation of a Bayesian network to predict pretest probability of venous thromboembolism. *Annals of Emergency Medicine 45*(3), 282–290.

Knuiman, M. (2005). Personal communication.

Knuiman, M. W., H. T. Vu, and H. Bartholomew (1998). Multivariate risk estimation for coronary heart disease: the Busselton Health Study. *Australian and New Zealand Journal of Public Health 22*, 747–753.

Koller, D. and N. Friedman (2009). *Probabilistic Graphical Models: Principles and Techniques*. MIT Press.

Koller, D. and A. Pfeffer (1997a). Object-Oriented Bayesian networks. In *UAI97 – Proceedings of the Thirteenth Conference on Uncertainty in Uncertainty in Artificial Intelligence*, pp. 302–313.

Koller, D. and A. Pfeffer (1997b). Representations and solutions for game-theoretic problems. *Artificial Intelligence 94*, 167–215.

Koller, D. and M. Sahami (1996). Toward optimal feature selection. In *International Conference on Machine Learning*, pp. 284–292.

Kolmogorov, A. (1933). *Grundbegriffe der Warhscheinlichkeitsrechnung*. Berlin: Springer Verlag.

Kolmogorov, A. N. (1965). Three approaches to the quantitative definition of information. *Problems of Information and Transmission 1*, 1–7.

Kononenko, I. and I. Bratko (1991). Information-based evaluation criterion for classifier's performance. *Machine Learning 6*, 67–80.

Korb, K. (1991). Searle's Artificial Intelligence program. *Journal of Theoretical and Experimental Artificial Intelligence 3*, 283–296.

Korb, K., L. Hope, and M. Hughes (2001). The evaluation of predictive learners: some theoretical and empirical results. In *ECML'01 – Twelfth European Conference on Machine Learning*, pp. 276–287. Springer Verlag.

Korb, K., C. Kopp, and L. Allison (1997). A statement on higher education policy in Australia. Technical report, 97/318. School of Computer Science and Software Engineering, Monash University.

Korb, K., R. McConachy, and I. Zukerman (1997). A cognitive model of argumentation. In *Proceedings of the Nineteenth Annual Conference of the Cognitive Science Society*, pp. 400–405.

Korb, K. and E. Nyberg (2006). The power of intervention. *Minds and Machines 16*, 289–302.

Korb, K. B. (1995). Inductive learning and defeasible inference. *Journal for Experimental and Theoretical Artificial Intelligence 7*, 291–324.

Korb, K. B., L. R. Hope, A. E. Nicholson, and K. Axnick (2004). Varieties of causal intervention. In *PRICAI'04—Proceedings of the 8th Pacific Rim International Conference on Artificial Intelligence*, Auckland, New Zealand, pp. 322–331.

Korb, K. B., A. E. Nicholson, and N. Jitnah (1999). Bayesian poker. In *UAI99 – Proceedings of the Fifteenth Conference on Uncertainty in Artificial Intelligence*, pp. 343–350.

Kornfeld, A. (1991, November). Causal diagrams: clarifying uncertainty. *Artificial Intelligence Expert*, 42–49.

Kristensen, K. and I. Rasmussen (2002). The use of a Bayesian network in the design of a decision support system for growing malting barley without use of pesticides. *Computers and Electronics in Agriculture 33*(3), 197–217.

Kuikka, S., M. Hilden, H. Gislason, S. Hanson, H. Sparholt, and O. Varis (1999). Modeling environmentaly driven uncertainties in Baltic cod (Gadus morhua): Management by Bayesian influence diagrams. *Canadian Journal of Fisheries and Aquatic Sciences 54*, 629–641.

Lacave, C., R. Atienza, and F. J. Díez (2000). Graphical explanation in Bayesian networks. *Lecture Notes in Computer Science 1933*, 122–129.

Lacave, C. and F. J. Díez (2002). A review of explanation methods for Bayesian networks. *The Knowledge Engineering Review 17*(2), 107–127.

Lam, W. and F. Bacchus (1993). Learning Bayesian belief networks: An approach based on the MDL principle. *Computational Intelligence 10*, 269–293.

Langseth, H. (2002). *Bayesian networks with applications in reliability analysis*. Ph. D. thesis, Dept. of Mathematical Sciences, NorvegianUniversity of Science and Technology.

Larrañaga, P., C. Kuijpers, R. Murga, and Y. Yurramendi (1996). Learning Bayesian network structures by searching for the bestordering with genetic algorithms. *IEEE Transactions on Systems, Man and Cybernetics, Part A 26*, 487–493.

Laskey, K. and S. Mahoney (1997). Network fragments: Representing knowledge for constructing probabilistic models. In *UAI97 – Proceedings of the Thirteenth Conference on Uncertainty in Uncertainty in Artificial Intelligence*, pp. 334–341.

Laskey, K. and S. Mahoney (2000). Network engineering for agile belief network models. *IEEE: Transactions on Knowledge and Data Engineering 12*(4), 487–498.

Laskey, K. B. (1993). Sensitivity analysis for probability assessments in Bayesian networks. In *UAI93 – Proc of the Ninth Conference on Uncertainty in Artificial Intelligence*, pp. 136–142.

Lauritzen, S. and N. Wermuth (1989). Graphical models for associations between variables, some of which are qualitative and some quantitative. *The Annals of Statistics 17*, 31–57.

Lauritzen, S. L. and D. J. Spiegelhalter (1988). Local computations with probabilities on graphical structures and their application to expert systems. *Journal of the Royal Statistical Society 50*(2), 157–224.

Lee, C. and K. Lee (2006). Application of Bayesian network to the probabilistic risk assessment of nuclear waste disposal. *Reliability Engineering and System Safety 91*(5), 515–532.

Lerner, U., E. Segal, and D. Koller (2001). Exact inference in networks with discrete children of continuous parents. In *UAI01 – Proceedings of the Seventeenth Conference on Uncertainty in Artificial Intelligence*, pp. 319–328.

Lewis, D. (1980). A subjectivist's guide to objective chance. In Jeffrey (Ed.), *Studies in inductive logic and probability, volume II*, pp. 263–293. University of California.

Lichtenstein, S., B. Fischhoff, and L. D. Phillips (1982). Calibration of probabilities: The state of the art to 1980. In *Judgment under Uncertainty: Heuristics and Biases*, pp. 306–334. Cambridge.

Liew, D. and S. Rogers (2004, September).   Personal communication. http://www.csse.monash.edu.au/~ctwardy/bnepi/Epidem/ 20040930.txt.

Lindley, D. (1972). *Bayesian Statistics: A Review*. Philadelphia: Society for Industrial and Applied Mathematics.

Lindley, D. (1985). *Making Decisions* (2nd ed.). John Wiley.

Loehlin, J. C. (1998). *Latent Variable Models: An Introduction to Factor, Path, and Structural Analysis* (Third ed.). Mahwah, N.J.: Lawrence Erlbaum.

Lucas, P., L. van der Gaag, and A. Abu-Hanna (2004).   Bayesian networks in biomedicine and health-care. *Artificial Intelligence in Medicine 30*(3), 201–214.

Luger, G. and W. Stubblefield (1993).   *Artificial intelligence: Structures and Strategies for Complex Problem Solving* (second ed.).   Menlo Park, CA: Benjamin/Cummings.

Mackay, D. (1998).   Introduction to Monte Carlo methods.   In M. Jordan (Ed.), *Learning in Graphical Models*, pp. 175–204. MIT.

Madigan, D., S. A. Andersson, M. D. Perlman, and C. T. Volinsky (1996). Bayesian model averaging and model selection for Markov equivalence classes of acyclic digraphs. *Communications in Statistics: Theory and Methods 25*, 2493–2519.

Madigan, D., K. Mosurski, and R. G. Almond (1997). Graphical explanation in belief networks. *Journal of Computer and Graphical Statistics 6*, 160–181.

Madigan, D. and A. E. Raftery (1994).   Model selection and accounting for model uncertainty in graphical models using Occam's window. *Journal of the American Statistical Association 89*, 1535–1546.

Madigan, D. and I. York (1995). Bayesian graphical models for discrete data. *International Statistical Review 63*, 215–232.

Maglogiannis, I., E. Zafiropoulos, A. Platis, and C. Lambrinoudakis (2006).   Risk analysis of a patient monitoring system using Bayesian network modeling. *Journal of Biomedical Informatics 39*(6), 637–647.

Mahoney, S. and K. Laskey (1999).   Representing and combining partially specified CPTs.   In *UAI99 – Proceedings of the Fifteenth Conference on Uncertainty in Artificial Intelligence*, Sweden, pp. 343–350.

Marcot, B. (1999).   A process for creating Bayian belief network models of species-environment relations. Technical report, USDA Forest Service, Portland, Oregon.

Marcot, B., R. Holthausen, M. Raphael, M. Rowland, and M. Wisdom (2001). Using Bayesian belief networks to evaluate fish and wildlife population viability under land management alternatives from an environmental impact statement. *Forest Ecology and Management 153*(1-3), 29–42.

Marcot, B., J. Steventon, G. Sutherland, and R. McCann (2006). Guidelines for developing and updating Bayesian belief networks applied to ecological modeling and conservation. *Canadian Journal of Forest Research 36*(12), 3063–3074.

Mayo, M. and A. Mitrovic (2001). Optimising ITS behaviour with Bayesian networks and decision theory. *International Journal of Artificial Intelligence in Education 12*, 124–153.

McConachy, R., K. B. Korb, and I. Zukerman (1998). Deciding what not to say: An attentional-probabilistic approach to argument presentation. In *Proceedings of the Cognitive Science Society Meeting*, pp. 669–674.

McDermott, D. (1987). A critique of pure reason. *Computational Intelligence 3*, 151–237.

McGain, K. (2010). Opponent modeling in Bayesian poker. Honours thesis, Clayton School of IT, Monash University (forthcoming).

McGowan, R. D. (1997). Ambulation monitoring and fall detection using dynamic belief networks. honours thesis. dept. of electrical and computer systems engineering, monash university.

McIntosh, J., K. Stacey, C. Tromp, and D. Lightfoot (2000). Designing constructivist computer games for teaching about decimal numbers. In J. Bana and A. Chapman (Eds.), *Proceedings of the Twenty Third Annual Conference of the Mathematical Education Research Group of Australasia*, pp. 409–416. Freemantle: MERGA, Inc.

McNeil, J. J., A. Peeters, D. Liew, S. Lim, and T. Vos (2001). A model for predicting the future incidence of coronary heart disease within percentiles of coronary heart disease risk. *Journal of Cardiovascular Risk 8*, 31–37.

McNeill, J., R. MacEwan, and D. Crawford (2006). Using GIS and a land use impact model to assess risk of soil erosion in West Gippsland. *APPLIED GIS 2*(3).

Mcroy, S. W., S. S. Ali, and S. M. Haller (1997). Uniform knowledge representation for language processing in the b2 system. *Natural Language Engineering 3*(2), 123–145.

Meek, C. and D. M. Chickering (2003). Practically perfect. In *UAI03 – Proceedings of the Nineteenth Conference on Uncertainty in Artificial Intelligence*, San Francisco.

Metropolis, N., A. W. Rosenbluth, M. N. Rosenbluth, A. H. Teller, and E. Teller (1953). Equations of state calculations by fast computing machines. *Journal of Chemical Physics 21*, 1087–1091.

Michie, D., D. Spiegelhalter, C. Taylor, and J. Campbell (1994). *Machine Learning, Neural and Statistical Classification*. Ellis Horwood.

Middleton, B., M. Shwe, D. Heckerman, M. Henrion, E. Horvitz, H. Lehmann, and G. Cooper (1991). Probabilistic diagnosis using a reformulation of the INTERNIST-1/QMR knowledge base II. Evaluation of diagnostic performance. *Methods in Information in Medicine 30*, 256–267.

Miller, A. C., M. M. Merkhofer, R. A. Howard, J. E. Matheson, and T. R. Rice (1976). Development of automated aids for decision analysis. Technical report, SRI International, Menlo Park, CA.

Mitchell, T. (1997). *Machine Learning*. McGraw Hill.

Mittal, A. and A. Kassim (Eds.) (2007). *Bayesian network technologies: Applications and graphical models*. Hershey, PA: IGI Publishing.

Mohan, C., K. Mehrotra, P. Varshney, and J. Yang (2007). Temporal uncertainty reasoning networks for evidence fusion with applications to object detection and tracking. *Information Fusion 8*(3), 281–294.

Montani, S., L. Portinale, A. Bobbio, and D. Codetta-Raiteri (2008). Radyban: A tool for reliability analysis of dynamic fault trees through conversion into dynamic Bayesian networks. *Reliability Engineering & System Safety 93*(7), 922–932.

Montero, J. and L. Sucar (2006). A decision-theoretic video conference system based on gesture recognition. In *Automatic Face and Gesture Recognition, 2006. In FGR 2006, 7th International Conference on Automatic Face and Gesture Recognition*. pp. 387–392.

Monti, S. and G. F. Cooper (1998). A multivariate discretization method for learning hybrid Bayesian networks from data. In *UAI98 – Proceedings of the Fourteenth Conference on Uncertainty in Uncertainty in Artificial Intelligence*, pp. 404–413.

Morgan, M. and M. Henrion (1990). *Uncertainty: A Guide to Dealing with Uncertainty in Quantitative Risk and Policy Analysis*. London: Cambridge University Press.

Morjaria, M. A., F. J. Rink, W. D. Smith, G. Klempner, G. Burns, and J. Stein (1993). Commercialization of EPRI's generator expert monitoring system. In *Proceedings of the EPRI Conference on Advanced Computer Applications in the Electric Utility Industry*, Phoenix.

Mosteller, F. (1973). *Statistics by Example*. Reading, MA: Addison-Wesley.

Murphy, K., Y. Weiss, and M. I. Jordan (1999). Loopy belief propagation for approximate inference: an empirical study. In *UAI99 – Proceedings of the Fifteenth Conference on Uncertainty in Artificial Intelligence*, pp. 467–475.

Murphy, K. P. (2002). *Dynamic Bayesian networks: Representation, inference and learning*. Ph. D. thesis, Department of Computer Science, University of California, Berkeley.

Murphy, P. and D. Aha (1995). UCI repository of machine learning databases. http://www.ics.uci.edu/~mlearn/MLRepository.html.

Nadeau, C. and Y. Bengio (2003). Inference for the generalization error. *Machine Learning 52*(3), 239–281.

Nägele, A., M. Dejori, and M. Stetter (2007). Bayesian substructure learning-approximate learning of very large network structures. In *Proceedings of the 18th European Conference on Machine Learning; Lecture Notes in AI*, Volume 4701, pp. 238–249.

Nagl, S., M. Williams, and J. Williamson (2008). Objective Bayesian nets for systems modeling and prognosis in breast cancer. In D. Holmes and L. Jain (Eds.), *Innovations in Bayesian Networks: Theory and Applications*, pp. 131–167. Springer Verlag.

Naticchia, B., A. Fernandez-Gonzalez, and A. Carbonari (2007). Bayesian network model for the design of roofpond equipped buildings. *Energy and Buildings 39*(3), 258–272.

Neapolitan, R. E. (1990). *Probabilistic Reasoning in Expert Systems*. New York: Wiley and Sons.

Neapolitan, R. E. (2003). *Learning Bayesian Networks*. Prentice Hall.

Neil, J. R. and K. B. Korb (1999). The evolution of causal models. In *Third Pacific-Asia Conf on Knowledge Discovery and Datamining (PAKDD-99)*, pp. 432–437. Springer Verlag.

Neil, J. R., C. S. Wallace, and K. B. Korb (1999). Learning Bayesian networks with restricted causal interactions. In *UAI99 – Proceedings of the Fifteenth Conference on Uncertainty in Artificial Intelligence*, pp. 486–493.

Neil, M., N. Fenton, and L. Nielson (2000). Building large-scale Bayesian networks. *The Knowledge Engineering Review 15*(3), 257–284.

Neil, M., N. Fenton, and M. Tailor (2005). Using Bayesian networks to model expected and unexpected operational losses. *Risk Analysis 25*(4), 963–972.

Neil, M., M. Tailor, D. Marquez, N. Fenton, and P. Hearty (2008). Modeling dependable systems using hybrid Bayesian networks. *Reliability Engineering & System Safety 93*(7), 933–939.

Nicholson, A. (1996). Fall diagnosis using dynamic belief networks. In *Proceedings of the Fourth Pacific Rim International Conference on Artificial Intelligence (PRICAI-96)*, pp. 206–217.

Nicholson, A., T. Boneh, T. Wilkin, K. Stacey, L. Sonenberg, and V. Steinle (2001). A case study in knowledge discovery and elicitation in an intelligent tutoring application. In *UAI01 – Proceedings of the Seventeenth Conference on Uncertainty in Artificial Intelligence*, pp. 386–394.

Nicholson, A. and N. Jitnah (1996). Belief network algorithms: a study of performance using domain characterisation. Technical Report 96/249, Department of Computer Science, Monash University.

Nicholson, A. E. (1992). *Monitoring discrete environments using dynamic belief networks*. Ph. D. thesis, Department of Engineering Sciences, Oxford.

Nicholson, A. E. and J. M. Brady (1992a). The data association problem when monitoring robot vehicles using dynamic belief networks. In *ECAI92 – Proceedings of the Tenth European Conference on Artificial Intelligence*, pp. 689–693.

Nicholson, A. E. and J. M. Brady (1992b). Sensor validation using dynamic belief networks. In *UAI92 – Proceedings of the Eighth Conference on Uncertainty in Artificial Intelligence*, San Mateo, CA, pp. 207–214.

Nicholson, A. E., C. R. Twardy, K. B. Korb, and L. R. Hope (2008). Decision support for clinical cardiovascular risk assessment. In O. Pourret, P. Naim, and B. Marcot (Eds.), *Bayesian Networks: A Practical Guide to Applications*, Statistics in Practice, pp. 33–52. Wiley.

Nikovski, D. (2000). Constructing Bayesian networks for medical diagnosis from incomplete and partially correct statistics. *IEEE Transactions on Knowledge and Data Engineering 12*(4), 509–516.

Norsys (1994-2010). Netica. http://www.norsys.com.

Novak, J. D. and A. J. Cañas (2008). The theory underlying concept maps and how to construct and use them. Technical report IHMC CmapTools 2006-01 Rev 01-2008, Florida Institute for Human and Machine Cognition.

Nyberg, E. P. and K. B. Korb (2006). Informative interventions. Technical Report 2006/204, School of Computer Science, Monash University.

O'Donnell, R. (2010). *Flexible Causal Discovery with MML*. Ph. D. thesis, Monash University.

O'Donnell, R., K. Korb, and L. Allison (2007). Causal KL: Evaluating causal discovery. Technical Report 2007/207, Clayton School of IT, Monash University.

Olesen, K. (1993). Causal probabilistic networks with both discrete and continuous variables. *IEEE Transactions on Pattern Analysis and Machine Intelligence (PAMI) 15*(3), 275–279.

Oliver, J. (1993). Decision graphs – an extension of decision trees. In *Proc. fourth int. conf. on artificial intelligence*, pp. 343–350.

Oliver, J. and D. Hand (1994). Averaging over decision stumps. In *European Conference on Machine Learning*, pp. 231–231. Springer Verlag.

Onisko, A., M. Druzdzel, and H. Wasyluk (1998). A probabilistic model for diagnosis of liver disorders. In *Proceedings of the Seventh Symposium on Intelligent Information Systems (IIS-98)*, pp. 379–387.

Otto, L. and C. Kristensen (2004). A biological network describing infection with Mycoplasma hyopneumoniae in swine herds. *Preventive Veterinary Medicine 66*(1-4), 141–161.

Oukhellou, L., E. Côme, L. Bouillaut, and P. Aknin (2008). Combined use of sensor data and structural knowledge processed by Bayesian network: Application to a railway diagnosis aid scheme. *Transportation Research Part C: Emerging Technologies 16*(6), 755–767.

Ozbay, K. and N. Noyan (2006). Estimation of incident clearance times using Bayesian networks approach. *Accident Analysis & Prevention 38*(3), 542–555.

O'Donnell, R., L. Allison, and K. Korb (2006). Learning hybrid Bayesian networks by MML. In *Proceedings of the 19th Australian Joint Conference on AI*, pp. 192–203. Springer Verlag.

O'Donnell, R., A. Nicholson, B. Han, K. Korb, M. Alam, and L. Hope (2006). Causal discovery with prior information. In *Proceedings of the 19th Australian Joint Conference on AI*, pp. 1162–1167. Springer Verlag.

Park, J. D. (2002). Map complexity results and approximation methods. In *UAI02 – Proceedings of the Eighteenth Conference on Uncertainty in Artificial Intelligence*, pp. 388–396.

Pauker, S. and J. Wong (2005). The influence of influence diagrams in medicine. *Decision Analysis 2*(4), 238–244.

Pearl, J. (1978). An economic basis for certain methods of evaluating probabilistic forecasts. *International Journal of Man-Machine Studies 10*, 175–183.

Pearl, J. (1982). Reverend Bayes on inference engines: a distributed hierarchical approach. In *AAAI82 – Proceedings of the Second National Conference on Artificial Intelligence*, pp. 133–136.

Pearl, J. (1986). Fusion, propagation, and structuring in belief networks. *Artificial Intelligence 29*, 241–288.

Pearl, J. (1988). *Probabilistic Reasoning in Intelligent Systems*. San Mateo, CA: Morgan Kaufmann.

Pearl, J. (2000). *Causality: Models, Reasoning and Inference*. Cambridge, MA.

Pollino, C., P. Feehan, M. Grace, and B. Hart (2004). Fish communities and habitat changes in the highly modified goulburn river, australia. *Marine & Freshwater Research 55*, 769–780.

Pollino, C. and B. Hart (2006). Bayesian approaches can help make better sense of ecotoxicological information in risk assessments. *Australasian Journal of Ecotoxicology 11*, 56–57.

Pollino, C., A. White, and B. Hart (2007). Examination of conflicts and improved strategies for the management of an endangered eucalypt species using Bayesian networks. *Ecological Modeling 201*, 37–59.

Pollino, C., O. Woodberry, A. Nicholson, and K. Korb (2005). Parameterising Bayesian networks: A case study in ecological risk assessment. In *SIMMOD'05 – Proceedings of the 2005 Conference on Simulation and Modeling*, Bangkok, Thailand, pp. 289–297.

Pollino, C., O. Woodberry, A. Nicholson, K. Korb, and B. T. Hart (2007). Parameterisation of a Bayesian network for use in an ecological risk management case study. *Environmental Modeling and Software 22*(8), 1140–1152.

Poole, D., A. Mackworth, and R. Goebel (1998). *Computational Intelligence: A Logical Approach*. Oxford University Press.

Popper, K. (1959). The propensity interpretation of probability. *British Journal for the Philosophy of Science 10*, 25–42.

Pourret, O., P. Naim, and B. Marcot (Eds.) (2008). *Bayesian Networks: A Practical Guide to Applications*. Statistics in Practice. Wiley.

Pradhan, M., M. Henrion, G. Provan, B. D. Favero, and K. Huang (1996). The sensitivity of belief networks to imprecise probabilities: An experimental investigation. *Artificial Intelligence 85*(1-2), 363–397.

Pradhan, M., G. Provan, B. Middleton, and M. Henrion (1994). Knowledge engineering for large belief networks. In *UAI94 – Proceedings of the 10th Conference on Uncertainty in Artificial Intelligence*, pp. 484–490.

Promedas. Promedas. http://www.promedas.nl.

Provost, F. and T. Fawcett (2001). Robust classification for imprecise environments. *Machine Learning 42*, 203–231.

Provost, F., T. Fawcett, and R. Kohavi (1998). The case against accuracy estimation for comparing induction algorithms. In *Proceedings of the Fifteenth International Conference on Machine Learning*, pp. 445–453.

Pullar, D. and T. Phan (2007, 10-13 December 2007). Using a Bayesian network in a GIS to model relationships and threats to koala populations close to urban environments. In *Proceedings of the International Congress on Modeling and Simulation MODSIM2007*.

Pynadeth, D. and M. P. Wellman (1995). Accounting for context in plan recognition, with application to traffic monitoring. In *UAI95 – Proceedings of the 11th Conference on Uncertainty in Artificial Intelligence*, pp. 472–481.

Quinlan, J. (1983). Learning efficient classification procedures and their application to chess endgames. In R. Michalski, J. Carbonell, and T. Mitchell (Eds.), *Machine learning: An artificial intelligence approach*. Palo Alto: Tioga Publishing.

Quinlan, J. (1986). Induction of decision trees. *Machine learning 1*, 81–106.

Quinlan, J. (1993). *C4. 5: Programs for machine learning*. Morgan Kaufmann.

Quinlan, J. R. (1996). Learning decision tree classifiers. *ACM Computing Surveys 28*, 71–72.

Raiffa, H. (1968). *Decision Analysis: Introductory Lectures on Choices under Uncertainty*. New York: Random House.

Ramesh, R., M. Mannan, A. Poo, and S. Keerthi (2003). Thermal error measurement and modeling in machine tools. part ii. hybrid Bayesian network–support vector machine model. *International Journal of Machine Tools and Manufacture 43*(4), 405–419.

Ramsey, F. P. (1931). Truth and probability. In R. Braithwaite (Ed.), *The Foundations of Mathematics and Other Essays*. New York: Humanities Press.

Rawls, J. (1971). *A Theory of Justice*. Cambridge: Harvard University.

Reichenbach, H. (1956). *The Direction of Time*. Berkeley: University of California.

Reichert, P., M. Borsuk, M. Hostmann, S. Schweizer, C. Sporri, K. Tockner, and B. Truffer (2007). Concepts of decision support for river rehabilitation. *Environmental Modeling & Software 22*(2), 188–201.

Renninger, H. and H. von Hasseln (2002). Object-oriented dynamic Bayesian network-templates for modeling mechatronic systems. In *International Workshop on Principles of Diagnosis (DX-02)*. http://www.dbai.tuwien.ac.at/user/dx2002/proceedings/dx02final21.pdf.

Resnick, L. B., P. Nesher, F. Leonard, M. Magone, S. Omanson, and I. Peled (1989). Conceptual bases of arithmetic errors: The case of decimal fractions. *Journal for Research in Mathematics Education 20*(1), 8–27.

Rissanen, J. (1978). Modeling by shortest data description. *Automatica 14*, 465–471.

Rivas, T., J. Matias, J. Taboada, and A. Arguelles (2007). Application of Bayesian networks to the evaluation of roofing slate quality. *Engineering Geology 94*(1-2), 27–37.

Robinson, R. (1977). Counting unlabeled acyclic digraphs. *Combinatorial Mathematics V*, 28–43.

Ross, B. and E. Zuviria (2007). Evolving dynamic Bayesian networks with multi-objective genetic algorithms. *Applied Intelligence 26*, 13–23.

Russell, S. and P. Norvig (1995). *Artificial Intelligence: A Modern Approach* (First ed.). Prentice Hall Series in Artificial Intelligence. Englewood Cliffs, NJ: Prentice Hall.

Russell, S. and P. Norvig (2010). *Artificial Intelligence: A Modern Approach* (3rd ed.). Prentice Hall Series in Artificial Intelligence. Englewood Cliffs, NJ: Prentice Hall.

Sackur-Grisvard, C. and F. Leonard (1985). Intermediate cognitive organization in the process of learning a mathematical concept: The order of positive decimal numbers. *Cognition and Instruction 2*, 157–174.

Sahin, F., M. Yavuz, Z. Arnavut, and O. Uluyol (2007). Fault diagnosis for airplane engines using Bayesian networks and distributed particle swarm optimization. *Parallel Computing 33*(2), 124–143.

Sakamoto, Y. et al. (1986). *Akaike information criterion statistics*. KTK scientific publishers.

Salmon, W. (1984). *Scientific Explanation and the Causal Structure of the World*. Princeton, NJ: Princeton University.

Savage, L. (1971). Elicitation of personal probabilities and expectations. *Journal of the American Statistical Association 66*(336), 783–801.

Schach, S. (2008). *Object-Oriented software engineering*. New York: McGraw-Hill.

Scheines, R., P. Spirtes, C. Glymour, and C. Meek (1994). *TETRAD II: Tools for Discovery*. Hillsdale, NJ: Lawrence Erlbaum Associates.

Schwarz, G. (1978). Estimating the dimension of a model. *The Annals of Statistics 6*, 461–464.

Searle, J. (1980). Minds, brains and programs. *Behavioral and Brain Sciences 3*, 417–457.

Sewell, W. and V. Shah (1968). Social class, parental encouragement, and educational aspirations. *American Journal of Sociology 73*, 559–572.

Shachter, R. and M. Peot (1989). Simulation approaches to general probabilistic inference on belief networks. In *Proceedings of the Fifth Workshop on Uncertainty in Artificial Intelligence*, pp. 311–318.

Shachter, R. D. (1986). Evaluating influence diagrams. *Operations Research 34*, 871–882.

Shachter, R. D. and C. R. Kenley (1989). Gaussian influence diagrams. *Management Science 35*(5), 527–550.

Shafer, G. and P. Shenoy (1990). Probability propagation. *Annals of Mathematics and Artificial Intelligence 2*, 327–352.

Shafer, G. R. and J. Pearl (Eds.) (1990). *Readings in Uncertain Reasoning*. San Mateo, CA: Morgan Kaufmann.

Shimony, S. E. (1994). Finding MAPs for belief networks is NPhard. *Artificial Intelligence 68*, 399–410.

Shiratori, N. and N. Okude (2007). Bayesian networks layer model to represent anesthetic practice. In *Proceedings of the IEEE International Conference on Systems, Man and Cybernetics (ISIC*, pp. 674–679.

Shwe, M. and G. Cooper (1991). An empirical analysis of likelihood-weighting simulation on a large, multiply connected belief network. *Computers and Biomedical Research 24*, 453–475.

Shwe, M., B. Middleton, D. Heckerman, M. Henrion, E. Horvitz, H. Lehmann, and G. Cooper (1991). Probabilistic diagnosis using a reformulation of the INTERNIST-1/QMR knowledge base I. The probabilistic model and inference algorithms. *Methods in Information in Medicine 30*, 241–255.

Simon, H. (1954). Spurious correlation: A causal interpretation. *Journal of the American Statistical Association 49*, 467–479.

Sleeman, D. (1984). Mis-generalization: An explanation of observed mal-rules. In *Sixth Annual Conference of the Cognitive Science Society*, pp. 51–56.

Smith, C., A. Howes, B. Price, and C. McAlpine (2007). Using a Bayesian belief network to predict suitable habitat of an endangered mammalThe Julia Creek dunnart (Sminthopsis douglasi). *Biological Conservation 139*, 333–347.

Smith, S. F. (1983). Flexible learning of problem solving heuristics through adaptive search. In *IJCAI-83*, pp. 422–425.

Smith, W., J. Doctor, J. Meyer, I. Kalet, and M. Phillips (2009). A decision aid for intensity-modulated radiation-therapy plan selection in prostate cancer based on a prognostic Bayesian network and a markov model. *Artificial Intelligence in Medicine 46*(2), 119–130.

Sober, E. (1988). The principle of the common cause. In J. Fetzer (Ed.), *Probability and Causality*. Kluwer, Dordrecht.

Solomonoff, R. (1964). A formal theory of inductive inference, I and II. *Information and Control 7*, 1–22 and 224–254.

Sommerville, I. (2010). *Software Engineering* (Ninth ed.). Reading, MA: Addison-Wesley.

Spiegel, M., J. Schiller, and R. Srinivasan (2008). *Schaum's Outline of Probability and Statistics*. McGraw-Hill.

Spiegelhalter, D. (1986). Probabilistic reasoning in predictive expert systems. In J. Lemmer and L. Kanal (Eds.), *Uncertainty in Artificial Intelligence*. Amsterdam: Elsevier.

Spiegelhalter, D. J. and S. L. Lauritzen (1990). Sequential updating of conditional probabilities on directed graphical structures. *Networks 20*, 579–605.

Spirtes, P., C. Glymour, and R. Scheines (1993). *Causation, Prediction and Search*. Number 81 in Lecture Notes in Statistics. Springer Verlag.

Spirtes, P., C. Glymour, and R. Scheines (2000). *Causation, Prediction and Search* (Second ed.). Cambridge, MA: MIT Press.

Srinivas, S. and J. Breese (1989). IDEAL. Technical Report 23, Rockwell International Science Center.

Stacey, K., E. Sonenberg, A. Nicholson, T. Boneh, and V. Steinle (2003). A teaching model exploiting cognitive conflict driven by a Bayesian network. In *Proceedings of the Nineth International Conference on User Modeling (UM'2003)*, Pittsburgh, pp. 352–362.

Stacey, K. and V. Steinle (1998). Refining the classification of students' interpretations of decimal notation. *Hiroshima Journal of Mathematics Education 6*, 49–69.

Stacey, K. and V. Steinle (1999). A longitudinal study of childen's thinking about decimals: A preliminary analysis. In O. Zaslavsky (Ed.), *Proceedings of the Twenty Third Conference of the International Group for the Psychology of Mathematical Education*, Volume 4, Haifa, pp. 233–241. PME.

Stajduhar, I., B. Dalbelo-Basic, and N. Bogunovic (2009). Impact of censoring on learning Bayesian networks in survival modeling. *Artificial Intelligence in Medicine 47*(3), 199–217.

Stassopoulou, A. and M. Dikaiakos (2009). Web robot detection: A probabilistic reasoning approach. *Computer Networks 53*(3), 265–278.

Steel, D. (2006). Homogeneity, selection and the faithfulness condition. *Minds and Machines 16*, 303–317.

Steeneveld, W., L. van der Gaag, H. Barkema, and H. Hogeveen (2010). Simplify the interpretation of alert lists for clinical mastitis in automatic milking systems. *Computers and Electronics in Agriculture 71*(1), 50–56.

Steinle, V. and K. Stacey (1998). The incidence of misconceptions of decimal notation amongst students in Grades 5 to 10. In C. Kanes, M. Goos, and E. Warren (Eds.), *Teaching Mathematics in New Times, MERGA 21*, pp. 548–555. MERGA.

Suermondt, H. J. (1992). *Explanation in Bayesian belief networks*. Ph. D. thesis, Medical Information Sciences, Stanford University.

Sullivan, J. People v. Collins. Problems, cases, and materials on evidence. http://en.wikipedia.org/wiki/People_v._Collins.

Suzuki, J. (1996). Learning Bayesian belief networks based on the Minimum Description Length principle. In *Proceedings of the Thirteenth International Conference on Machine Learning*, pp. 462–470.

Tang, A., A. Nicholson, Y. Jin, and J. Han (2007). Using Bayesian belief networks for change impact analysis in architecture design. *Journal of Systems and Software 80*(1), 127–148.

Tang, Z., M. Taylor, P. Lisboa, and M. Dyas (2005). Quantitative risk modeling for new pharmaceutical compounds. *Drug discovery today 10*(22), 1520–1526.

Taylor, B. (2007). Opponent modeling in Bayesian poker player. Honours thesis, Clayton School of Information Technology, Monash University.

Teyssier, M. and D. Koller (2005). Ordering-based search: A simple and effective algorithm for learning Bayesian networks. In *Proceedings of the Twenty-First Conference on Uncertainty in Artificial Intelligence (UAI-05)*, pp. 584–590. AUAI Press.

Thomson, S. (1995). Bayesian poker. Honours thesis, Dept of Computer Science, Monash University.

Thrun, S., J. Bala, E. Bloedorn, I. Bratko, B. Cestnik, J. Cheng, K. De Jong, S. Dzeroski, S. Fahlman, D. Fisher, et al. (1991). The MONK's problems: A performance comparison of different learning algorithms. *Revised version*, Technical Report CMU-CS-91-197, Carnegie Mellon University, CS Department.

Ticehurst, J., L. Newham, D. Rissik, R. Letcher, and A. Jakeman (2007). A Bayesian network approach for assessing the sustainability of coastal lakes in New South Wales, Australia. *Environmental Modeling & Software 22*(8), 1129–1139.

Tighe, M., C. Pollino, S. Cuddy, and S. Whitfield (2007, 10-13 December 2007). A Bayesian approach to assessing regional climate change pressures on natural resource conditions in the central west of nsw, australia. In *Proceedings of the International Congress on Modeling and Simulation MODSIM2007*, pp. 233–239.

Tsamardinos, I., L. Brown, and C. Aliferis (2006). The max-min hill-climbing Bayesian network structure learning algorithm. *Machine Learning 65*, 31–78.

Tucker, A., V. Vinciotti, X. Liu, and D. Garway-Heath (2005). A spatio-temporal Bayesian network classifier for understanding visual field deterioration. *Artificial Intelligence in Medicine 34*(2), 163–177.

Turing, A. (1950). Computing machinery and intelligence. *Mind 59*, 433–460.

Turney, P. (1995). Cost-sensitive classification: Empirical evaluation of a hybrid genetic decision tree induction algorithm. *Journal of AI Research*, 369–409.

Tversky, A. and D. Kahneman (1974). Judgment under uncertainty: Heuristics and biases. *Science 185*, 1124–1131.

Tversky, A. and D. Kahneman (1982). Judgments of and by representativeness. In *Judgment under Uncertainty: Heuristics and Biases*, pp. 84–98. Cambridge.

Twardy, C., A. Nicholson, and K. Korb (2005). Knowledge engineering cardiovascular Bayesian networks from the literature. Technical report TR 2005/170, Clayton School of IT, Monash University.

Twardy, C. R., A. E. Nicholson, K. B. Korb, and J. McNeil (2006). Epidemiological data mining of cardiovascular Bayesian networks. *Electronic Journal of Health Informatics 1*(1), 1–13. http://www.ejhi.net.

Van Berlo, R., E. Van Someren, and M. Reinders (2003). Studying the conditions for learning dynamic Bayesian networks to discover genetic regulatory networks. *Simulation 79*, 689–702.

van der Gaag, L. C. and S. Renooij (2001). Analysing sensitivity data from probabilistic networks. In *UAI01 – Proceedings of the Seventeenth Conference on Uncertainty in Artificial Intelligence*, pp. 530–537.

van der Gaag, L. C., S. Renooij, C. L. M. Witteman, B. M. P. Aleman, and B. G. Taal (1999). How to elicit many probabilities. In *UAI99 – Proceedings of the Fifteenth Conference on Uncertainty in Artificial Intelligence*, pp. 647–654.

van der Gaag, L. C., S. Renooij, C. L. M. Witteman, B. M. P. Aleman, and B. G. Taal (2002). Probabilities for a probabilistic network: A case-study in oesophageal cancer. *Artificial Intelligence in Medicine 25*(2), 123–148.

van Fraassen, B. (1989). *Laws and Symmetry*. Oxford: Clarendon Press.

VanLehn, V. and Z. Niu (2001). Bayesian student modeling, user interfaces and feedback: a sensitivity analysis. *International Journal of Artificial Intelligence in Education 12*, 154–184.

Varis, O. (1997). Bayesian decision analysis for environmental and resource management. *Environmental Modeling and Software 12*, 177–185.

Varis, O. and M. Keskinen (2006). Policy analysis for the Tonle Sap lake, Cambodia: A Bayesian network model approach. *International Journal of Water Resources Development 22*(3), 417–431.

Varis, O. and V. Lahtela (2002). Integrated water resources management along the Senegal River: Introducing an analytical framework. *International Journal of Water Resources Development 18*, 501–521.

Venn, J. (1866). *Logic of Chance.*

Verma, T. S. and J. Pearl (1990). Equivalence and synthesis of causal models. In *Proceedings of the Sixth Conference on Uncertainty in Artificial Intelligence*, pp. 220–227.

Vicard, P., A. Dawid, J. Mortera, and S. Lauritzen (2008). Estimating mutation rates from paternity casework. *Forensic Science International: Genetics 2*(1), 9–18.

Visscher, S., P. Lucas, C. Schurink, and M. Bonten (2009). Modeling treatment effects in a clinical Bayesian network using Boolean threshold functions. *Artificial Intelligence in Medicine 46*(3), 251–266.

Voie, A., A. Johnsen, A. Strømseng, and K. Longva (2010). Environmental risk assessment of white phosphorus from the use of munitions – a probabilistic approach. *Science of The Total Environment 408*(8), 1833–1841.

von Mises, R. (1919). Grundlage der Warhscheinlichkeitsrechnung. *Mathematische Zeitschrift.*

von Mises, R. (1928/1957). *Probability, Statistics and Truth.* London: Allen and Unwin.

von Neumann, J. and O. Morgenstern (1947). *Theory of Games and Economic Behavior* (second ed.). Princeton, NJ: Princeton University Press.

Von Winterfeldt, D. and W. Edwards (1986). *Decision analysis and behavioral research.* Cambridge University Press.

Wallace, C. and D. Dowe (2000). MML clustering of multi-state, Poisson, von Mises circular and Gaussian distributions. *Statistics and Computing 10*(1), 73–83.

Wallace, C. S. and D. M. Boulton (1968). An information measure for classification. *The Computer Journal 11*, 185–194.

Wallace, C. S. and D. L. Dowe (1994). Intrinsic classification by MML – the Snob program. In *AI94 – Proceedings of the Seventh Australian Joint Conference on Artificial Intelligence*, Armidale, Australia, pp. 37–44.

Wallace, C. S. and P. R. Freeman (1987). Estimation and inference by compact coding. *Journal of the Royal Statistical Society (Series B) 49*, 240–252.

Wallace, C. S., K. Korb, and H. Dai (1996). Causal discovery via MML. In L. Saitta (Ed.), *Proceedings of the Thirteenth International Conference on Machine Learning*, pp. 516–524. Morgan Kaufman.

Wallace, C. S. and K. B. Korb (1999). Learning linear causal models by MML sampling. In A. Gammerman (Ed.), *Causal Models and Intelligent Data Management*. Springer Verlag.

Wang, X., B. Zheng, W. Good, J. King, and Y. Chang (1999). Computer-assisted diagnosis of breast cancer using a data-driven Bayesian belief network. *International Journal of Medical Informatics 54*(2), 115–126.

Waterman, D. A. (1970). A generalization learning technique for automating the learning of heuristics. *Artificial Intelligence 1*, 121–170.

Webb, G., J. Boughton, and Z. Wang (2005). Not so naive Bayes: Aggregating one-dependence estimators. *Machine Learning 58*, 5–24.

Webb, G., F. Zheng, J. Boughton, and K. Ting (2010). Decreasingly naive Bayes: Aggregating n-dependence estimators. *Machine Learning*. Under submission.

Weber, P. and L. Jouffe (2006). Complex system reliability modeling with dynamic Object Oriented Bayesian networks (DOOBN). *Reliability Engineering & System Safety 91*(2), 149–162.

Weber, P., D. Theilliol, C. Aubrun, and A. Evsukoff (2007). Increasing effectiveness of model-based fault diagnosis: A dynamic Bayesian network design for decision making. In *In Proceedings of the Conference on Fault Detection, Supervision and Safety of Technical Processes*, pp. 90–95.

Weidl, G., A. Madsen, and S. Israelson (2005). Applications of object-oriented Bayesian networks for condition monitoring, root cause analysis and decision support on operation of complex continuous processes. *Computers & chemical engineering 29*(9), 1996–2009.

Wellman, M. (1990). Fundamental concepts of Qualitative Probabilistic Networks. *Artificial Intelligence 44*(3), 257–303.

Wellman, M. P. and C.-L. Liu (1994). State-space abstraction for anytime evaluation of probabilistic networks. In *UAI94 – Proceedings of the Tenth Conference on Uncertainty in Artificial Intelligence*, San Francisco, CA, pp. 567–574.

Wen, Y., K. Korb, and A. Nicholson (2009). Datazapper: Generating incomplete datasets. In *First International Conference on Agents and Artificial Intelligence*, pp. 69–76.

Wermuth, N. (1987). Parametric collapsibility and the lack of moderating effects. *Journal of the Royal Statistical Society B 49*, 353–364.

Weymouth, G., T. Boneh, P. Newham, J. Bally, R. Potts, A. Nicholson, and K. Korb (2007, July). Dealing with uncertainty in fog forecasting for major airports in australia. In *Proceedings of the Fourth International Conference on Fog, Fog Collection and Dew*, Pontificia Universidad Catolica de Chile, Chile, pp. 73–76.

White, A. (2009). *Modeling the Impact of Climate Change on Peatlands in the Bogong High Plains*. Ph. D. thesis, University of Melbourne.

Widarsson, B. and E. Dotzauer (2008). Bayesian network-based early-warning for leakage in recovery boilers. *Applied Thermal Engineering 28*(7), 754–760.

Wikipedia (2010). Conceptual model. http://en.wikipedia.org/wiki/Conceptual_model_(computer_science).

Williamson, J. (2001). Foundations for Bayesian networks. In D. Corfield and J. Williamson (Eds.), *Foundations of Bayesianism*. Dordrecht: Kluwer.

Williamson, J. (2005). *Bayesian Nets and Causality: Philosophical and Computational Foundations*. Oxford: Oxford University.

Willis, D. (2000). Ambulation monitoring and fall detection system using dynamic belief networks. BCSE Thesis report. Department of Electrical and Computer Systems Engineering, Monash University.

Wilson, D., M. Stoddard, and K. Puettmann (2008). Monitoring amphibian populations with incomplete survey information using a Bayesian probabilistic model. *Ecological Modeling 214*(2-4), 210–218.

Winston, P. H. (1977). *Artificial Intelligence*. Reading, MA: Addison-Wesley.

Witten, I. and E. Frank (2005). *Data mining: Practical machine learning tools and techniques* (2nd ed.). Morgan Kaufmann. http://www.cs.waikato.ac.nz/~ml/index.html.

Wong, M., W. Lam, and K. Leung (1999). Using evolutionary programming and Minimum Description Length principle for data mining of Bayesian networks. *IEEE Transactions on Pattern Analysis and Machine Intelligence 21*, 174–178.

Woodberry, O., A. Nicholson, K. Korb, and C. Pollino (2004a). A methodology for parameterising Bayesian networks. Technical report TR 2004/159, School of Computer Science and Software Engineering, Monash University.

Woodberry, O., A. Nicholson, K. Korb, and C. Pollino (2004b). Parameterising Bayesian networks. In G. Webb and X. Yu (Eds.), *Lecture Notes in Artificial Intelligence, (Proceedings of the 17th Australian Joint Conference on Advances in Artificial Intelligence [AI'04], Cairns, Australia, 4-6 December 2004)*, Volume 3339 of *LNCS/LNAI Series*, pp. 1101–1107. Berlin, Germany: Springer Verlag.

Wright, S. (1921). Correlation and causation. *Journal of Agricultural Research 20*, 557–585.

Wright, S. (1934). The method of path coefficients. *Annals of Mathematical Statistics 5*, 161–215.

Xenos, M. (2004). Prediction and assessment of student behaviour in open and distance education in computers using Bayesian networks. *Computers & Education 43*(4), 345–359.

Zadora, G. (2009). Evaluation of evidence value of glass fragments by likelihood ratio and Bayesian network approaches. *Analytica Chimica Acta 642*(1-2), 279–290.

Zhou, H. and S. Sakane (2007). Mobile robot localization using active sensing based on Bayesian network inference. *Robotics and Autonomous Systems 55*(4), 292–305.

Zonneveldt, S., K. Korb, and A. Nicholson (2010). Bayesian network classifiers for the German credit data. Technical report, 2010/1, Bayesian Intelligence. http://www.Bayesian-intelligence.com/publications.php.

Zukerman, I., R. McConachy, and K. B. Korb (1996). Perambulations on the way to an architecture for a nice argument generator. In *ECAI'96 Workshop. Gaps and Bridges: New Directions in Planning and Natural Language Generation*, Budapest, Hungary, pp. 31–36.

Zukerman, I., R. McConachy, and K. B. Korb (1998). Bayesian reasoning in an abductive mechanism for argument generation and analysis. In *Proceedings of the Fifteenth National Conference on Artificial Intelligence (AAAI-98)*, Madison, Wisconsin, pp. 833–838.

Zweig, G. (2003). Bayesian network structures and inference techniques for automatic speech recognition. *Computer Speech & Language 17*(2-3), 173–193.

Index

absorption, 147
abstraction, 90, 122, 123, 322, 356, 367
action, 84, 97, 100–102, 141, 152, 158,
    163, 297, 342
    intervening, 84, 105
    node, 107, 115, 159, 164, 342, 363
    nodes, 100, 123
    non-deterministic, 98
    non-intervening, 104
activation, 41, 174
adaptation, 298, 356
    parameters, 357
    structural, 359
additive, 343
additivity, 6
AgenaRisk, 130, 411, 413
ALARM network, 286
algorithm
    CaMML Metropolis, 270
    DBN updating process, 116
    decision network evaluation, 101
    decision table, 102
    decision tree evaluation, 111
    expectation maximization, 194
    Gibbs sampling, 193
    Gibbs sampling for expected value,
        192
    IC, 242
    junction tree, 71
    K2, 256, 263, 286
    Lauritzen-Spiegelhalter parameteri-
        zation, 383
    likelihood weighting, 76
    logic sampling, 75
    maximum aposteriori probability
        EM, 196
    maximum likelihood EM, 195
    MDL, 263, 286

    message passing, 60
    multinomial parameterization, 189
    PC, 247, 248, 255
    Pearl's network construction, 38
    predictive accuracy, 211
    using a test-action decision net-
        work, 107
ancestor node, 32
approximate inference, 94
    stochastic simulation, 74
arc, 29
    directed, 29
    inter-slice, 113
    intra-slice, 113
    temporal, 113
arc density, 266
arc reversal, 117, 246
    rule, 246
Area Under the Curve (AUC), 216, 399
argument graph, 170
Aristotle, 17, 255
artificial intelligence, 3
    descriptive, 21
    normative, 21
attention, 173

base-rate neglect, 18
Bayes, 10
    Idiot, 4
    naive, 4, 135, 206, 284
Bayes' theorem, 12, 55–59, 61, 65
    odds-likelihood, 14, 67
Bayesia, 65
BayesiaLab, 46, 92, 411
Bayesian metrics, 255
Bayesian model averaging, 222
Bayesian network, 29
    definition, 29